Mathematical Foundations
of Reinforcement Learning

Shiyu Zhao

Mathematical Foundations of Reinforcement Learning

 Springer

 清华大学出版社
Tsinghua University Press

Shiyu Zhao
School of Engineering
Westlake University
Hangzhou, China

Jointly published with Tsinghua University Press
The print edition is not for sale in China (Mainland). Customers from China (Mainland) please order the print book from: Tsinghua University Press.

ISBN 978-981-97-3943-1   ISBN 978-981-97-3944-8   (eBook)
https://doi.org/10.1007/978-981-97-3944-8

Licensed from the English language edition "Mathematical Foundations of Reinforcement Learning" by Shiyu Zhao © Tsinghua University Press 2024. Published by Tsinghua University Press. All Rights Reserved.

This Springer imprint is published by the registered company Springer Nature Singapore Pte Ltd.
The registered company address is: 152 Beach Road, #21-01/04 Gateway East, Singapore 189721, Singapore

If disposing of this product, please recycle the paper.

Contents

Preface

This book aims to provide a *mathematical* but *friendly* introduction to the fundamental concepts, basic problems, and classic algorithms in reinforcement learning. Some essential features of this book are highlighted as follows.

⋄ The book introduces reinforcement learning from a mathematical point of view. Hopefully, readers will not only know the procedure of an algorithm but also understand why the algorithm was designed in the first place and why it works effectively.

⋄ The depth of the mathematics is carefully controlled to an adequate level. The mathematics is also presented in a carefully designed manner to ensure that the book is friendly to read. Readers can read the materials presented in gray boxes selectively according to their interests.

⋄ Many illustrative examples are given to help readers better understand the topics. All the examples in this book are based on a grid world task, which is easy to understand and helpful for illustrating concepts and algorithms.

⋄ When introducing an algorithm, the book aims to separate its core idea from complications that may be distracting. In this way, readers can better grasp the core idea of an algorithm.

⋄ The contents of the book are coherently organized. Each chapter is built based on the preceding chapter and lays a necessary foundation for the subsequent one.

This book is designed for senior undergraduate students, graduate students, researchers, and practitioners who are interested in reinforcement learning. It does not require readers to have any background in reinforcement learning because it starts by introducing the most basic concepts. If the reader already has some background in reinforcement learning, I believe the book can help them understand some topics more deeply or provide different perspectives. This book, however, requires the reader to have some knowledge of probability theory and linear algebra. Some basics of the required mathematics are also included in the appendix of this book.

I have been teaching a graduate-level course on reinforcement learning since 2019. I want to thank the students in my class for their feedback on my teaching. I put the draft of this book online in August 2022. Up to now, I have received valuable feedback from many readers. I want to express my gratitude to these readers. Moreover, I would like to thank my research assistant, Jialing Lv, for her excellent support in editing the book and my lecture videos; my teaching assistants, Jianan Li and Yize Mi, for their help in my teaching; my Ph.D. student Canlun Zheng for his help in the design of a picture in the book; and my family for their wonderful support. Finally, I would like to thank the editors of this book, Dr. Lanlan Chang and Mr. Sai Guo from Springer Nature Press and Tsinghua University Press, for their great support.

Please note that I have created an open course based on this textbook. Both the slides of the open course and the PDF of this textbook are available online for free download. For more information, you can visit the homepage of the textbook: `https://github.com/ MathFoundationRL/Book-Mathematical-Foundation-of-Reinforcement-Learning`

I sincerely hope this book can help readers smoothly enter the exciting field of reinforcement learning.

Shiyu Zhao

Overview of this Book

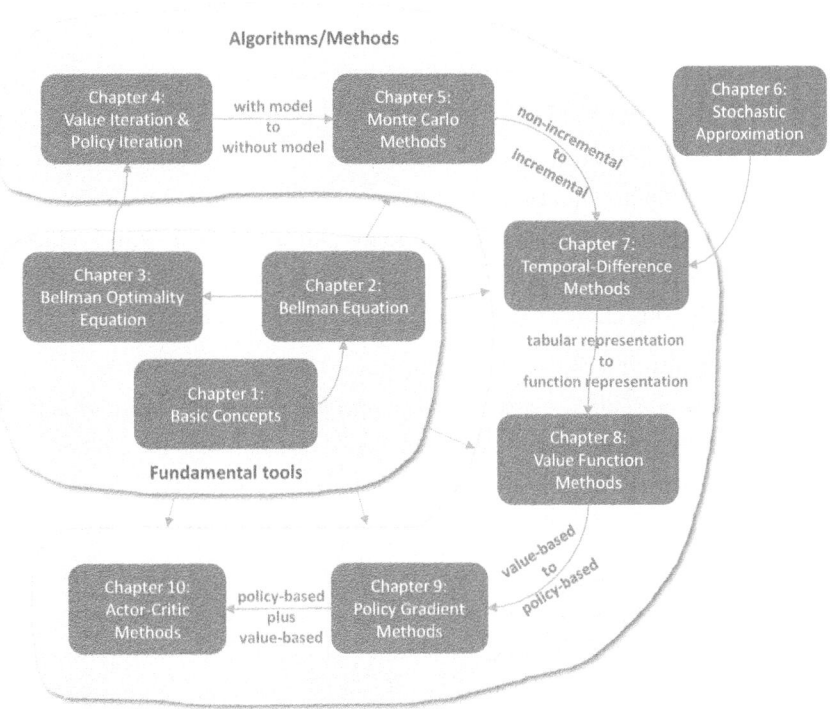

Figure 1: The map of this book.

Before we start the journey, it is important to look at the "map" of the book shown in Figure 1. This book contains ten chapters, which can be classified into two parts: the first part is about basic tools, and the second part is about algorithms. The ten chapters are highly correlated. In general, it is necessary to study the earlier chapters first before the later ones.

Next, please follow me on a quick tour through the ten chapters. Two aspects of each chapter will be covered. The first aspect is the contents introduced in each chapter, and the second aspect is its relationships with the previous and subsequent chapters. A heads up for you to read this overview is as follows. The purpose of this overview is to give you an impression of the contents and structure of this book. It is all right if you encounter many concepts you do not understand. Hopefully, you can make a proper study plan that is suitable for you after reading this overview.

◇ Chapter 1 introduces the basic concepts such as states, actions, rewards, returns, and policies, which are widely used in the subsequent chapters. These concepts are first introduced based on a grid world example, where a robot aims to reach a prespecified target. Then, the concepts are introduced in a more formal manner based on the framework of Markov decision processes.

◇ Chapter 2 introduces two key elements. The first is a key concept, and the second is a key tool. The *key concept* is the *state value*, which is defined as the expected return that an agent can obtain when starting from a state if it follows a given policy. The greater the state value is, the better the corresponding policy is. Thus, state values can be used to evaluate whether a policy is good or not.

The *key tool* is the *Bellman equation*, which can be used to analyze state values. In a nutshell, the Bellman equation describes the relationship between the values of all states. By solving the Bellman equation, we can obtain the state values. Such a process is called *policy evaluation*, which is a fundamental concept in reinforcement learning. Finally, this chapter introduces the concept of action values.

◇ Chapter 3 also introduces two key elements. The first is a key concept, and the second is a key tool. The *key concept* is the *optimal policy*. An optimal policy has the greatest state values compared to other policies. The *key tool* is the *Bellman optimality equation*. As its name suggests, the Bellman optimality equation is a special Bellman equation.

Here is a fundamental question: what is the ultimate goal of reinforcement learning? The answer is to obtain optimal policies. The Bellman optimality equation is important because it can be used to obtain optimal policies. We will see that the Bellman optimality equation is elegant and can help us thoroughly understand many fundamental problems.

The first three chapters constitute the first part of this book. This part lays the necessary foundations for the subsequent chapters. Starting in Chapter 4, the book introduces algorithms for learning optimal policies.

◇ Chapter 4 introduces three algorithms: value iteration, policy iteration, and truncated policy iteration. The three algorithms have close relationships with each other. First, the value iteration algorithm is exactly the algorithm introduced in Chapter 3 for solving the Bellman optimality equation. Second, the policy iteration algorithm is an extension of the value iteration algorithm. It is also the foundation for Monte Carlo (MC) algorithms introduced in Chapter 5. Third, the truncated policy iteration algorithm is a unified version that includes the value iteration and policy iteration algorithms as special cases.

The three algorithms share the same structure. That is, every iteration has two steps. One step is to update the value, and the other step is to update the policy. The idea of the interaction between value and policy updates widely exists in reinforcement learning algorithms. This idea is also known as *generalized policy iteration*. In addition, the algorithms introduced in this chapter are actually *dynamic programming* algorithms, which require system models. By contrast, all the algorithms introduced in the subsequent chapters do not require models. It is important to well understand the contents of this chapter before proceeding to the subsequent ones.

◇ Starting in Chapter 5, we introduce *model-free* reinforcement learning algorithms that do not require system models. While this is the first time we introduce model-free algorithms in this book, we must fill a knowledge gap: how to find optimal policies without models? The philosophy is simple. If we do not have a model, we must have some data. If we do not have data, we must have a model. If we have neither, then we can do nothing. The "data" in reinforcement learning refer to the experience samples generated when the agent interacts with the environment.

This chapter introduces three algorithms based on MC estimation that can learn optimal policies from experience samples. The first and simplest algorithm is MC Basic, which can be readily obtained by extending the policy iteration algorithm introduced in Chapter 4. Understanding the MC Basic algorithm is important for grasping the fundamental idea of MC-based reinforcement learning. By extending this algorithm, we further introduce two more complicated but more efficient MC-based algorithms. The fundamental trade-off between *exploration* and *exploitation* is also elaborated in this chapter.

Up to this point, the reader may have noticed that the contents of these chapters are highly correlated. For example, if we want to study the MC algorithms (Chapter 5), we must first understand the policy iteration algorithm (Chapter 4). To study the policy iteration algorithm, we must first know the value iteration algorithm (Chapter 4). To comprehend the value iteration algorithm, we first need to understand the Bellman optimality equation (Chapter 3). To understand the Bellman optimality equation, we need to study the Bellman equation (Chapter 2) first. Therefore, it is highly recommended to study the chapters one by one. Otherwise, it may be difficult to understand the contents in the later chapters.

◇ There is a knowledge gap when we move from Chapter 5 to Chapter 7: the algorithms in Chapter 7 are *incremental*, but the algorithms in Chapter 5 are *non-incremental*. Chapter 6 is designed to fill this knowledge gap by introducing the stochastic approximation theory. Stochastic approximation refers to a broad class of stochastic iterative algorithms for solving root-finding or optimization problems. The classic Robbins-Monro and stochastic gradient descent algorithms are special stochastic approximation algorithms. Although this chapter does not introduce any reinforcement learning algorithms, it is important because it lays the necessary foundations for studying Chapter 7.

◇ Chapter 7 introduces the classic temporal-difference (TD) algorithms. With the preparation in Chapter 6, I believe the reader will not be surprised when seeing the TD algorithms. From a mathematical point of view, TD algorithms can be viewed as stochastic approximation algorithms for solving the Bellman or Bellman optimality equations. Like Monte Carlo learning, TD learning is also model-free, but it has some advantages due to its incremental form. For example, it can learn in an online manner: it can update the value estimate every time an experience sample is received. This chapter introduces quite a few TD algorithms such as Sarsa and Q-learning. The important concepts of on-policy and off-policy are also introduced.

◇ Chapter 8 introduces the value function approximation method. In fact, this chapter continues to introduce TD algorithms, but it uses a different way to represent state/action values. In the preceding chapters, state/action values are represented by *tables*. The tabular method is straightforward to understand, but it is inefficient for handling large state or action spaces. To solve this problem, we can employ the value function approximation method. The key to understanding this method is to understand the three steps in its optimization formulation. The first step is to select an objective function for defining optimal policies. The second step is to derive the gradient of the objective function. The third step is to apply a gradient-based algorithm to solve the optimization problem. This method is important because it has become the standard technique to represent values. It is also the location in which *artificial neural networks* are incorporated into reinforcement learning as function approximators. The famous deep Q-learning algorithm is also introduced in this chapter.

◇ Chapter 9 introduces the policy gradient method, which is the foundation of many modern reinforcement learning algorithms. The policy gradient method is *policy-based*. It is a large step forward in this book because all the methods in the previous chapters are *value-based*. The basic idea of the policy gradient method is simple: it selects an appropriate scalar metric and then optimizes it via a gradient-ascent algorithm. Chapter 9 has an intimate relationship with Chapter 8 because they both rely on the idea of function approximation. The advantages of the policy gradient method are numerous. For example, it is more efficient for handling large state/action spaces. It has stronger generalization abilities and is more efficient in sample usage.

◇ Chapter 10 introduces actor-critic methods. From one point of view, actor-critic refers to a structure that incorporates both policy-based and value-based methods. From another point of view, actor-critic methods are not new since they still fall into the scope of the policy gradient method. Specifically, they can be obtained by extending the policy gradient algorithm introduced in Chapter 9. It is necessary for the reader to properly understand the contents in Chapters 8 and 9 before studying Chapter 10.

Chapter 1

Basic Concepts

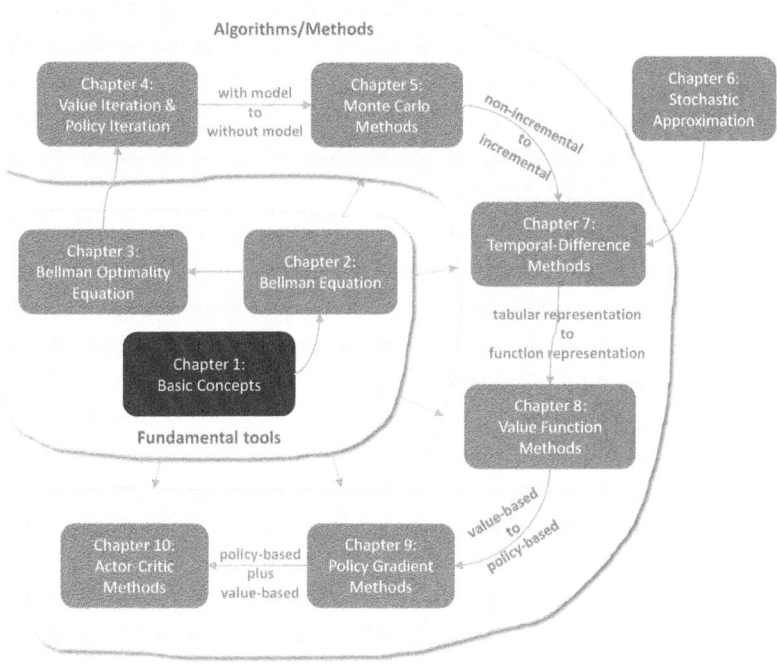

Figure 1.1: Where we are in this book.

This chapter introduces the basic concepts of reinforcement learning. These concepts are important because they will be widely used in this book. We first introduce these concepts using examples and then formalize them in the framework of Markov decision processes.

1.1 A grid world example

Consider an example as shown in Figure 1.2, where a robot moves in a grid world. The robot, called *agent*, can move across adjacent cells in the grid. At each time step, it can

S. Zhao, *Mathematical Foundations of Reinforcement Learning*, https://doi.org/10.1007/978-981-97-3944-8_1

only occupy a single cell. The white cells are *accessible* for entry, and the orange cells are *forbidden*. There is a *target* cell that the robot would like to reach. We will use such grid world examples throughout this book since they are intuitive for illustrating new concepts and algorithms.

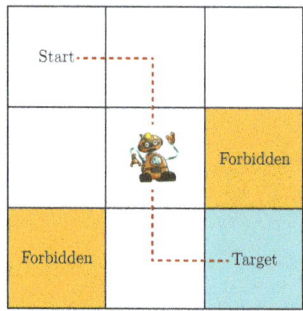

Figure 1.2: The grid world example is used throughout the book.

The ultimate goal of the agent is to find a "good" policy that enables it to reach the target cell when starting from any initial cell. How can the "goodness" of a policy be defined? The idea is that the agent should reach the target without entering any forbidden cells, taking unnecessary detours, or colliding with the boundary of the grid.

It would be trivial to plan a path to reach the target cell if the agent knew the map of the grid world. The task becomes nontrivial if the agent does not know any information about the environment in advance. Then, the agent must interact with the environment to find a good policy by trial and error. To do that, the concepts presented in the rest of the chapter are necessary.

1.2 State and action

The first concept to be introduced is the *state*, which describes the agent's status with respect to the environment. In the grid world example, the state corresponds to the agent's location. Since there are nine cells, there are nine states as well. They are indexed as $s_1, s_2, \ldots, s_9$, as shown in Figure 1.3(a). The set of all the states is called the *state space*, denoted as $\mathcal{S} = \{s_1, \ldots, s_9\}$.

For each state, the agent can take five possible *actions*: moving upward, moving rightward, moving downward, moving leftward, and staying still. These five actions are denoted as $a_1, a_2, \ldots, a_5$, respectively (see Figure 1.3(b)). The set of all actions is called the *action space*, denoted as $\mathcal{A} = \{a_1, \ldots, a_5\}$. Different states can have different action spaces. For instance, considering that taking a_1 or a_4 in state s_1 would lead to a collision with the boundary, we can set the action space for state s_1 as $\mathcal{A}(s_1) = \{a_2, a_3, a_5\}$. In this book, we consider the most general case: $\mathcal{A}(s_i) = \mathcal{A} = \{a_1, \ldots, a_5\}$ for all i.

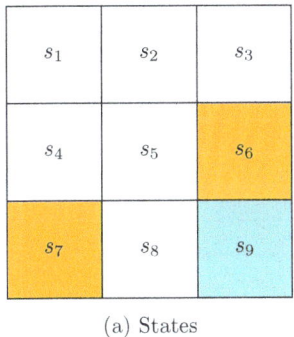
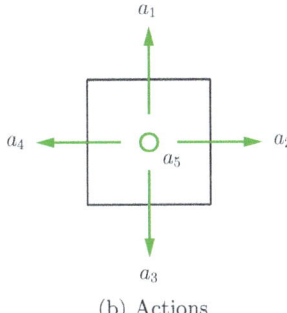

(a) States   (b) Actions

Figure 1.3: Illustrations of the state and action concepts. (a) There are nine states $\{s_1, \ldots, s_9\}$. (b) Each state has five possible actions $\{a_1, a_2, a_3, a_4, a_5\}$.

1.3 State transition

When taking an action, the agent may move from one state to another. Such a process is called *state transition*. For example, if the agent is in state s_1 and selects action a_2 (that is, moving rightward), then the agent moves to state s_2. Such a process can be expressed as

$$s_1 \xrightarrow{a_2} s_2.$$

We next examine two important examples.

◇ What is the next state when the agent attempts to go beyond the boundary, for example, taking action a_1 in state s_1? The answer is that the agent will be bounced back because it is impossible for the agent to exit the state space. Hence, we have $s_1 \xrightarrow{a_1} s_1$.

◇ What is the next state when the agent attempts to enter a forbidden cell, for example, taking action a_2 in state s_5? Two different scenarios may be encountered. In the first scenario, although s_6 is forbidden, it is still *accessible*. In this case, the next state is s_6; hence, the state transition process is $s_5 \xrightarrow{a_2} s_6$. In the second scenario, s_6 is *not accessible* because, for example, it is surrounded by walls. In this case, the agent is bounced back to s_5 if it attempts to move rightward; hence, the state transition process is $s_5 \xrightarrow{a_2} s_5$.

Which scenario should we consider? The answer depends on the physical environment. In this book, we consider the first scenario where the forbidden cells are accessible, although stepping into them may get punished. This scenario is more general and interesting. Moreover, since we are considering a simulation task, we can define the state transition process however we prefer. In real-world applications, the state transition process is determined by real-world dynamics.

The state transition process is defined for each state and its associated actions. This process can be described by a table as shown in Table 1.1. In this table, each row corresponds to a state, and each column corresponds to an action. Each cell indicates the next state to transition to after the agent takes an action at the corresponding state.

	a_1 (upward)	a_2 (rightward)	a_3 (downward)	a_4 (leftward)	a_5 (still)
s_1	s_1	s_2	s_4	s_1	s_1
s_2	s_2	s_3	s_5	s_1	s_2
s_3	s_3	s_3	s_6	s_2	s_3
s_4	s_1	s_5	s_7	s_4	s_4
s_5	s_2	s_6	s_8	s_4	s_5
s_6	s_3	s_6	s_9	s_5	s_6
s_7	s_4	s_8	s_7	s_7	s_7
s_8	s_5	s_9	s_8	s_7	s_8
s_9	s_6	s_9	s_9	s_8	s_9

Table 1.1: A tabular representation of the state transition process. Each cell indicates the next state to transition to after the agent takes an action at a state.

Mathematically, the state transition process can be described by conditional probabilities. For example, for s_1 and a_2, the conditional probability distribution is

$$p(s_1|s_1, a_2) = 0,$$
$$p(s_2|s_1, a_2) = 1,$$
$$p(s_3|s_1, a_2) = 0,$$
$$p(s_4|s_1, a_2) = 0,$$
$$p(s_5|s_1, a_2) = 0,$$

which indicates that, when taking a_2 at s_1, the probability of the agent moving to s_2 is one, and the probabilities of the agent moving to other states are zero. As a result, taking action a_2 at s_1 will certainly cause the agent to transition to s_2. The preliminaries of conditional probability are given in Appendix A. Readers are strongly advised to be familiar with probability theory since it is necessary for studying reinforcement learning.

Although it is intuitive, the tabular representation is only able to describe *deterministic* state transitions. In general, state transitions can be *stochastic* and must be described by conditional probability distributions. For instance, when random wind gusts are applied across the grid, if taking action a_2 at s_1, the agent may be blown to s_5 instead of s_2. We have $p(s_5|s_1, a_2) > 0$ in this case. Nevertheless, we merely consider deterministic state transitions in the grid world examples for simplicity in this book.

1.4 Policy

A *policy* tells the agent which actions to take at every state. Intuitively, policies can be depicted as arrows (see Figure 1.4(a)). Following a policy, the agent can generate a trajectory starting from an initial state (see Figure 1.4(b)).

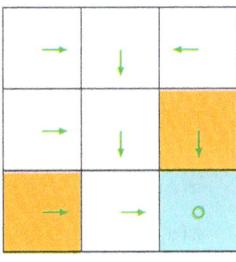

(a) A deterministic policy

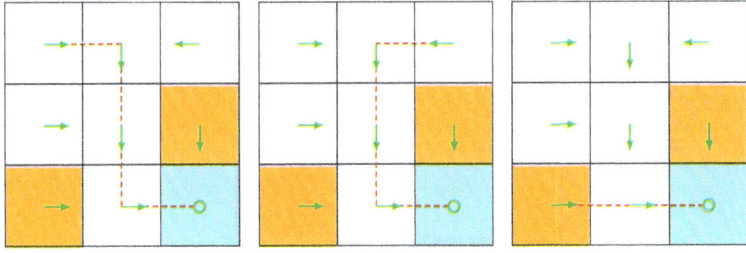

(b) Trajectories obtained from the policy

Figure 1.4: A policy represented by arrows and some trajectories obtained by starting from different initial states.

Mathematically, policies can be described by conditional probabilities. Denote the policy in Figure 1.4 as $\pi(a|s)$, which is a conditional probability distribution function defined for *every* state. For example, the policy for s_1 is

$$\pi(a_1|s_1) = 0,$$
$$\pi(a_2|s_1) = 1,$$
$$\pi(a_3|s_1) = 0,$$
$$\pi(a_4|s_1) = 0,$$
$$\pi(a_5|s_1) = 0,$$

which indicates that the probability of taking action a_2 in state s_1 is one, and the probabilities of taking other actions are zero.

The above policy is *deterministic*. Policies may be *stochastic* in general. For example, the policy shown in Figure 1.5 is stochastic: in state s_1, the agent may take actions to go either rightward or downward. The probabilities of taking these two actions are the same (both are 0.5). In this case, the policy for s_1 is

$$\pi(a_1|s_1) = 0,$$
$$\pi(a_2|s_1) = 0.5,$$
$$\pi(a_3|s_1) = 0.5,$$
$$\pi(a_4|s_1) = 0,$$
$$\pi(a_5|s_1) = 0.$$

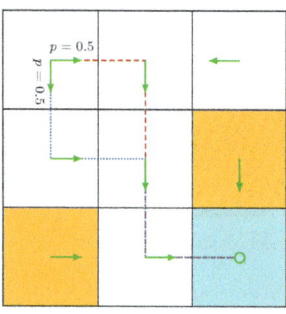

Figure 1.5: A stochastic policy. In state s_1, the agent may move rightward or downward with equal probabilities of 0.5.

Policies represented by conditional probabilities can be stored as tables. For example, Table 1.2 represents the stochastic policy depicted in Figure 1.5. The entry in the ith row and jth column is the probability of taking the jth action at the ith state. Such a representation is called a *tabular representation*. We will introduce another way to represent policies as parameterized functions in Chapter 8.

	a_1 (upward)	a_2 (rightward)	a_3 (downward)	a_4 (leftward)	a_5 (still)
s_1	0	0.5	0.5	0	0
s_2	0	0	1	0	0
s_3	0	0	0	1	0
s_4	0	1	0	0	0
s_5	0	0	1	0	0
s_6	0	0	1	0	0
s_7	0	1	0	0	0
s_8	0	1	0	0	0
s_9	0	0	0	0	1

Table 1.2: A tabular representation of a policy. Each entry indicates the probability of taking an action at a state.

1.5 Reward

Reward is one of the most unique concepts in reinforcement learning.

After executing an action at a state, the agent obtains a reward, denoted as r, as feedback from the environment. The reward is a function of the state s and action a. Hence, it is also denoted as $r(s, a)$. Its value can be a positive or negative real number or zero. Different rewards have different impacts on the policy that the agent would eventually learn. Generally speaking, with a positive reward, we encourage the agent to take the corresponding action. With a negative reward, we discourage the agent from taking that action.

In the grid world example, the rewards are designed as follows:

⋄ If the agent attempts to exit the boundary, let $r_{\text{boundary}} = -1$.

⋄ If the agent attempts to enter a forbidden cell, let $r_{\text{forbidden}} = -1$.

⋄ If the agent reaches the target state, let $r_{\text{target}} = +1$.

⋄ Otherwise, the agent obtains a reward of $r_{\text{other}} = 0$.

Special attention should be given to the target state s_9. The reward process does not have to terminate after the agent reaches s_9. If the agent takes action a_5 at s_9, the next state is again s_9, and the reward is $r_{\text{target}} = +1$. If the agent takes action a_2, the next state is also s_9, but the reward is $r_{\text{boundary}} = -1$.

A reward can be interpreted as a human-machine interface, with which we can guide the agent to behave as we expect. For example, with the rewards designed above, we can expect that the agent tends to avoid exiting the boundary or stepping into the forbidden cells. Designing appropriate rewards is an important step in reinforcement learning. This step is, however, nontrivial for complex tasks since it may require the user to understand the given problem well. Nevertheless, it may still be much easier than solving the problem with other approaches that require a professional background or a deep understanding of the given problem.

The process of getting a reward after executing an action can be intuitively represented as a table, as shown in Table 1.3. Each row of the table corresponds to a state, and each column corresponds to an action. The value in each cell of the table indicates the reward that can be obtained by taking an action at a state.

One question that beginners may ask is as follows: if given the table of rewards, can we find good policies by simply selecting the actions with the greatest rewards? The answer is no. That is because these rewards are *immediate rewards* that can be obtained after taking an action. To determine a good policy, we must consider the *total reward* obtained in the long run (see Section 1.6 for more information). An action with the greatest immediate reward may not lead to the greatest total reward.

Although intuitive, the tabular representation is only able to describe *deterministic* reward processes. A more general approach is to use conditional probabilities $p(r|s, a)$ to describe reward processes. For example, for state s_1, we have

$$p(r = -1|s_1, a_1) = 1, \quad p(r \neq -1|s_1, a_1) = 0.$$

	a_1 (upward)	a_2 (rightward)	a_3 (downward)	a_4 (leftward)	a_5 (still)
s_1	r_{boundary}	0	0	r_{boundary}	0
s_2	r_{boundary}	0	0	0	0
s_3	r_{boundary}	r_{boundary}	$r_{\text{forbidden}}$	0	0
s_4	0	0	$r_{\text{forbidden}}$	r_{boundary}	0
s_5	0	$r_{\text{forbidden}}$	0	0	0
s_6	0	r_{boundary}	r_{target}	0	$r_{\text{forbidden}}$
s_7	0	0	r_{boundary}	r_{boundary}	$r_{\text{forbidden}}$
s_8	0	r_{target}	r_{boundary}	$r_{\text{forbidden}}$	0
s_9	$r_{\text{forbidden}}$	r_{boundary}	r_{boundary}	0	r_{target}

Table 1.3: A tabular representation of the process of obtaining rewards. Here, the process is deterministic. Each cell indicates how much reward can be obtained after the agent takes an action at a given state.

This indicates that, when taking a_1 at s_1, the agent obtains $r = -1$ with certainty. In this example, the reward process is deterministic. In general, it can be stochastic. For example, if a student studies hard, he or she would receive a positive reward (e.g., higher grades on exams), but the specific value of the reward may be uncertain.

1.6 Trajectories, returns, and episodes

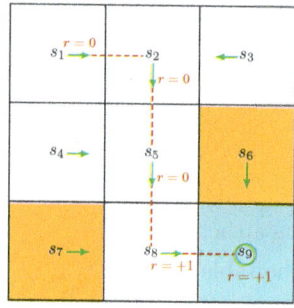
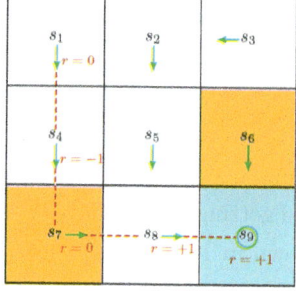

(a) Policy 1 and the trajectory   (b) Policy 2 and the trajectory

Figure 1.6: Trajectories obtained by following two policies. The trajectories are indicated by red dashed lines.

A *trajectory* is a state-action-reward chain. For example, given the policy shown in Figure 1.6(a), the agent can move along a trajectory as follows:

$$s_1 \xrightarrow[r=0]{a_2} s_2 \xrightarrow[r=0]{a_3} s_5 \xrightarrow[r=0]{a_3} s_8 \xrightarrow[r=1]{a_2} s_9.$$

The *return* of this trajectory is defined as the sum of all the rewards collected along the trajectory:

$$\text{return} = 0 + 0 + 0 + 1 = 1. \tag{1.1}$$

Returns are also called *total rewards* or *cumulative rewards*.

Returns can be used to *evaluate* policies. For example, we can evaluate the two policies in Figure 1.6 by comparing their returns. In particular, starting from s_1, the return obtained by the left policy is 1 as calculated above. For the right policy, starting from s_1, the following trajectory is generated:

$$s_1 \xrightarrow[r=0]{a_3} s_4 \xrightarrow[r=-1]{a_3} s_7 \xrightarrow[r=0]{a_2} s_8 \xrightarrow[r=+1]{a_2} s_9.$$

The corresponding return is

$$\text{return} = 0 - 1 + 0 + 1 = 0. \tag{1.2}$$

The returns in (1.1) and (1.2) indicate that the left policy is better than the right one since its return is greater. This mathematical conclusion is consistent with the intuition that the right policy is worse since it passes through a forbidden cell.

A return consists of an *immediate reward* and *future rewards*. Here, the immediate reward is the reward obtained after taking an action at the initial state; the future rewards refer to the rewards obtained after leaving the initial state. It is possible that the immediate reward is negative while the future reward is positive. Thus, which actions to take should be determined by the return (i.e., the total reward) rather than the immediate reward to avoid short-sighted decisions.

The return in (1.1) is defined for a finite-length trajectory. Return can also be defined for infinitely long trajectories. For example, the trajectory in Figure 1.6 stops after reaching s_9. Since the policy is well defined for s_9, the process does not have to stop after the agent reaches s_9. We can design a policy so that the agent stays still after reaching s_9. Then, the policy would generate the following infinitely long trajectory:

$$s_1 \xrightarrow[r=0]{a_2} s_2 \xrightarrow[r=0]{a_3} s_5 \xrightarrow[r=0]{a_3} s_8 \xrightarrow[r=1]{a_2} s_9 \xrightarrow[r=1]{a_5} s_9 \xrightarrow[r=1]{a_5} s_9 \dots$$

The direct sum of the rewards along this trajectory is

$$\text{return} = 0 + 0 + 0 + 1 + 1 + 1 + \dots = \infty,$$

which unfortunately diverges. Therefore, we must introduce the *discounted return* concept for infinitely long trajectories. In particular, the discounted return is the sum of the discounted rewards:

$$\text{discounted return} = 0 + \gamma 0 + \gamma^2 0 + \gamma^3 1 + \gamma^4 1 + \gamma^5 1 + \dots, \tag{1.3}$$

where $\gamma \in (0, 1)$ is called the *discount rate*. When $\gamma \in (0, 1)$, the value of (1.3) can be calculated as

$$\text{discounted return} = \gamma^3(1 + \gamma + \gamma^2 + \dots) = \gamma^3 \frac{1}{1 - \gamma}.$$

The introduction of the discount rate is useful for the following reasons. First, it removes the stop criterion and allows for infinitely long trajectories. Second, the discount rate can be used to adjust the emphasis placed on near- or far-future rewards. In particular, if γ is close to 0, then the agent places more emphasis on rewards obtained in the near future. The resulting policy would be short-sighted. If γ is close to 1, then the agent places more emphasis on the far future rewards. The resulting policy is far-sighted and dares to take risks of obtaining negative rewards in the near future. These points will be demonstrated in Section 3.5.

One important notion that was not explicitly mentioned in the above discussion is the *episode*. When interacting with the environment by following a policy, the agent may stop at some *terminal states*. The resulting trajectory is called an *episode* (or a *trial*). If the environment or policy is stochastic, we obtain different episodes when starting from the same state. However, if everything is deterministic, we always obtain the same episode when starting from the same state.

An episode is usually assumed to be a finite trajectory. Tasks with episodes are called *episodic tasks*. However, some tasks may have no terminal states, meaning that the process of interacting with the environment will never end. Such tasks are called *continuing tasks*. In fact, we can treat episodic and continuing tasks in a unified mathematical manner by converting episodic tasks to continuing ones. To do that, we need well define the process after the agent reaches the terminal state. Specifically, after reaching the terminal state in an episodic task, the agent can continue taking actions in the following two ways.

⋄ First, if we treat the terminal state as a special state, we can specifically design its action space or state transition so that the agent stays in this state forever. Such states are called *absorbing states*, meaning that the agent never leaves a state once reached. For example, for the target state s_9, we can specify $\mathcal{A}(s_9) = \{a_5\}$ or set $\mathcal{A}(s_9) = \{a_1, \dots, a_5\}$ with $p(s_9|s_9, a_i) = 1$ for all $i = 1, \dots, 5$.

⋄ Second, if we treat the terminal state as a normal state, we can simply set its action space to the same as the other states, and the agent may leave the state and come back again. Since a positive reward of $r = 1$ can be obtained every time s_9 is reached, the agent will eventually learn to stay at s_9 forever to collect more rewards. Notably, when an episode is infinitely long and the reward received for staying at s_9 is positive, a discount rate must be used to calculate the discounted return to avoid divergence.

In this book, we consider the second scenario where the target state is treated as a normal state whose action space is $\mathcal{A}(s_9) = \{a_1, \dots, a_5\}$.

1.7 Markov decision processes

The previous sections of this chapter illustrated some fundamental concepts in reinforcement learning through examples. This section presents these concepts in a more formal way under the framework of Markov decision processes (MDPs).

An MDP is a general framework for describing stochastic dynamical systems. The key ingredients of an MDP are listed below.

◇ Sets:

- State space: the set of all states, denoted as $\mathcal{S}$.
- Action space: a set of actions, denoted as $\mathcal{A}(s)$, associated with each state $s \in \mathcal{S}$.
- Reward set: a set of rewards, denoted as $\mathcal{R}(s, a)$, associated with each state-action pair (s, a).

◇ Model:

- State transition probability: In state s, when taking action a, the probability of transitioning to state s' is $p(s'|s, a)$. It holds that $\sum_{s' \in \mathcal{S}} p(s'|s, a) = 1$ for any (s, a).
- Reward probability: In state s, when taking action a, the probability of obtaining reward r is $p(r|s, a)$. It holds that $\sum_{r \in \mathcal{R}(s,a)} p(r|s, a) = 1$ for any (s, a).

◇ Policy: In state s, the probability of choosing action a is $\pi(a|s)$. It holds that $\sum_{a \in \mathcal{A}(s)} \pi(a|s) = 1$ for any $s \in \mathcal{S}$.

◇ Markov property: The *Markov property* refers to the memoryless property of a stochastic process. Mathematically, it means that

$$p(s_{t+1}|s_t, a_t, s_{t-1}, a_{t-1}, \ldots, s_0, a_0) = p(s_{t+1}|s_t, a_t),$$
$$p(r_{t+1}|s_t, a_t, s_{t-1}, a_{t-1}, \ldots, s_0, a_0) = p(r_{t+1}|s_t, a_t), \tag{1.4}$$

where t represents the current time step and $t + 1$ represents the next time step. Equation (1.4) indicates that the next state or reward depends merely on the current state and action and is independent of the previous ones. The Markov property is important for deriving the fundamental Bellman equation of MDPs, as shown in the next chapter.

Here, $p(s'|s, a)$ and $p(r|s, a)$ for all (s, a) are called the *model* or *dynamics*. The model can be either *stationary* or *nonstationary* (or in other words, time-invariant or time-variant). A stationary model does not change over time; a nonstationary model may vary over time. For instance, in the grid world example, if a forbidden area may pop up or disappear sometimes, the model is nonstationary. In this book, we only consider stationary models.

One may have heard about the Markov processes (MPs). What is the difference between an MDP and an MP? The answer is that, once the policy in an MDP is fixed, the MDP degenerates into an MP. For example, the grid world example in Figure 1.7 can be abstracted as a Markov process. In the literature on stochastic processes, a Markov process is also called a Markov chain if it is a discrete-time process and the number of states is finite or countable [1]. In this book, the terms "Markov process" and "Markov chain" are used interchangeably when the context is clear. Moreover, this book mainly considers *finite* MDPs where the numbers of states and actions are finite. This is the simplest case that should be fully understood.

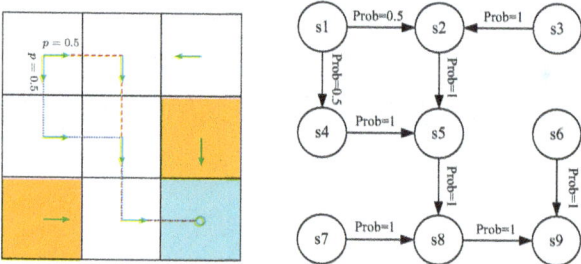

Figure 1.7: Abstraction of the grid world example as a Markov process. Here, the circles represent states and the links with arrows represent state transitions.

Finally, reinforcement learning can be described as an agent-environment interaction process. The *agent* is a decision-maker that can sense its state, maintain policies, and execute actions. Everything outside of the agent is regarded as the *environment*. In the grid world examples, the agent and environment correspond to the robot and grid world, respectively. After the agent decides to take an action, the actuator executes such a decision. Then, the state of the agent would be changed and a reward can be obtained. By using interpreters, the agent can interpret the new state and the reward. Thus, a closed loop can be formed.

1.8 Summary

This chapter introduced the basic concepts that will be widely used in the remainder of the book. We used intuitive grid world examples to demonstrate these concepts and then formalized them in the framework of MDPs. For more information about MDPs, readers can see [1,2].

1.9 Q&A

◇ Q: Can we set all the rewards as negative or positive?

A: In this chapter, we mentioned that a positive reward would encourage the agent to take an action and that a negative reward would discourage the agent from taking the action. In fact, it is the *relative* reward values instead of the *absolute* values that determine encouragement or discouragement.

More specifically, we set $r_{\text{boundary}} = -1$, $r_{\text{forbidden}} = -1$, $r_{\text{target}} = +1$, and $r_{\text{other}} = 0$ in this chapter. We can also add a common value to all these values without changing the resulting optimal policy. For example, we can add -2 to all the rewards to obtain $r_{\text{boundary}} = -3$, $r_{\text{forbidden}} = -3$, $r_{\text{target}} = -1$, and $r_{\text{other}} = -2$. Although the rewards are all negative, the resulting optimal policy is unchanged. That is because optimal policies are invariant to affine transformations of the rewards. Details will be given in Chapter 3.5.

◇ Q: Is the reward a function of the next state?

A: We mentioned that the reward r depends only on s and a but not the next state s'. However, this may be counterintuitive since it is the next state that determines the reward in many cases. For example, the reward is positive when the next state is the target state. As a result, a question that naturally follows is whether a reward should depend on the next state. A mathematical rephrasing of this question is whether we should use $p(r|s, a, s')$ where s' is the next state rather than $p(r|s, a)$. The answer is that r depends on s, a, and s'. However, since s' also depends on s and a, we can equivalently write r as a function of s and a: $p(r|s, a) = \sum_{s'} p(r|s, a, s')p(s'|s, a)$. In this way, the Bellman equation can be easily established as shown in Chapter 2.

Chapter 2

State Values and Bellman Equation

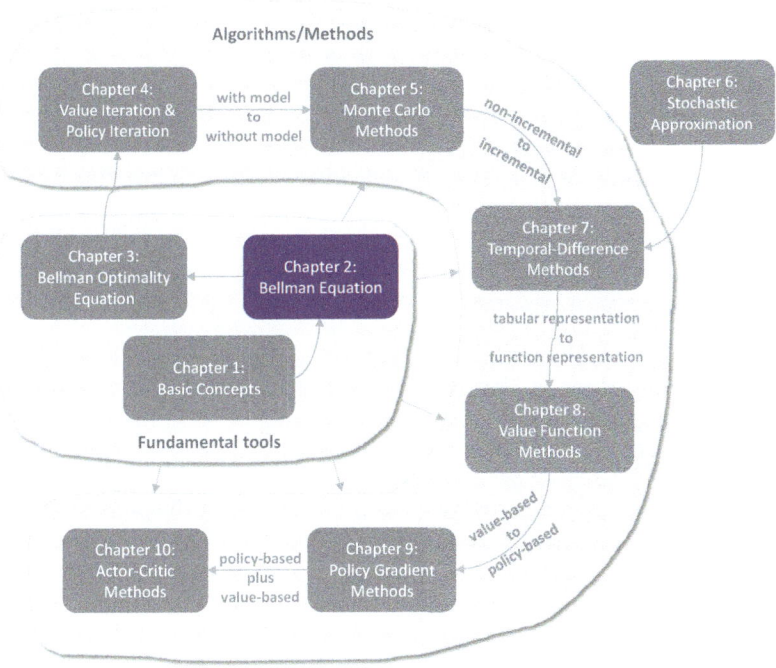

Figure 2.1: Where we are in this book.

This chapter introduces *a core concept* and *an important tool*. The core concept is the *state value*, which is defined as the average reward that an agent can obtain if it follows a given policy. The greater the state value is, the better the corresponding policy is. State values can be used as a metric to evaluate whether a policy is good or not. While state values are important, how can we analyze them? The answer is the *Bellman equation*, which is an important tool for analyzing state values. In a nutshell, the Bellman equation describes the relationships between the values of all states. By solving the Bellman equation, we can obtain the state values. This process is called *policy evaluation*, which is a fundamental concept in reinforcement learning. Finally, this

S. Zhao, *Mathematical Foundations of Reinforcement Learning*, https://doi.org/10.1007/978-981-97-3944-8_2

chapter introduces another important concept called the *action value*.

2.1 Motivating example 1: Why are returns important?

The previous chapter introduced the concept of returns. In fact, returns play a fundamental role in reinforcement learning since they can evaluate whether a policy is good or not. This is demonstrated by the following examples.

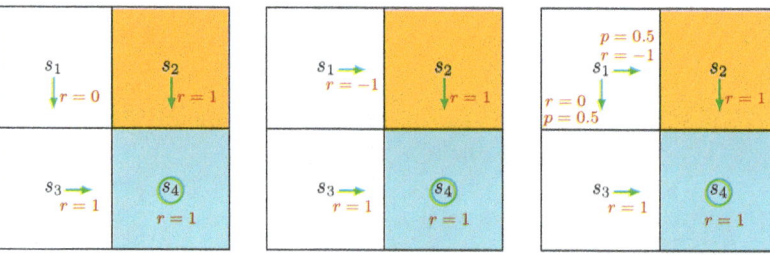

Figure 2.2: Examples for demonstrating the importance of returns. The three examples have different policies for s_1.

Consider the three policies shown in Figure 2.2. It can be seen that the three policies are different at s_1. Which is the best and which is the worst? Intuitively, the leftmost policy is the best because the agent starting from s_1 can avoid the forbidden area. The middle policy is intuitively worse because the agent starting from s_1 moves to the forbidden area. The rightmost policy is in between the others because it has a probability of 0.5 to go to the forbidden area.

While the above analysis is based on intuition, a question that immediately follows is whether we can use mathematics to describe such intuition. The answer is yes and relies on the return concept. In particular, suppose that the agent starts from s_1.

⋄ Following the first policy, the trajectory is $s_1 \to s_3 \to s_4 \to s_4 \cdots$. The corresponding discounted return is

$$\begin{aligned} \text{return}_1 &= 0 + \gamma 1 + \gamma^2 1 + \dots \\ &= \gamma(1 + \gamma + \gamma^2 + \dots) \\ &= \frac{\gamma}{1-\gamma}, \end{aligned}$$

where $\gamma \in (0,1)$ is the discount rate.

⋄ Following the second policy, the trajectory is $s_1 \to s_2 \to s_4 \to s_4 \cdots$. The discounted return is

$$\begin{aligned}
\text{return}_2 &= -1 + \gamma 1 + \gamma^2 1 + \dots \\
&= -1 + \gamma(1 + \gamma + \gamma^2 + \dots) \\
&= -1 + \frac{\gamma}{1 - \gamma}.
\end{aligned}$$

◇ Following the third policy, two trajectories can possibly be obtained. One is $s_1 \rightarrow s_3 \rightarrow s_4 \rightarrow s_4 \cdots$, and the other is $s_1 \rightarrow s_2 \rightarrow s_4 \rightarrow s_4 \cdots$. The probability of either of the two trajectories is 0.5. Then, the average return that can be obtained starting from s_1 is

$$\begin{aligned}
\text{return}_3 &= 0.5 \left(-1 + \frac{\gamma}{1 - \gamma}\right) + 0.5 \left(\frac{\gamma}{1 - \gamma}\right) \\
&= -0.5 + \frac{\gamma}{1 - \gamma}.
\end{aligned}$$

By comparing the returns of the three policies, we notice that

$$\text{return}_1 > \text{return}_3 > \text{return}_2 \tag{2.1}$$

for any value of γ. Inequality (2.1) suggests that the first policy is the best because its return is the greatest, and the second policy is the worst because its return is the smallest. This mathematical conclusion is consistent with the aforementioned intuition: the first policy is the best since it can avoid entering the forbidden area, and the second policy is the worst because it leads to the forbidden area.

The above examples demonstrate that returns can be used to evaluate policies: a policy is better if the return obtained by following that policy is greater. Finally, it is notable that return_3 does not strictly comply with the definition of returns because it is more like an expected value. It will become clear later that return_3 is actually a state value.

2.2 Motivating example 2: How to calculate returns?

While we have demonstrated the importance of returns, a question that immediately follows is how to calculate the returns when following a given policy.

There are two ways to calculate returns.

◇ The first is simply by definition: a return equals the discounted sum of all the rewards collected along a trajectory. Consider the example in Figure 2.3. Let v_i denote the return obtained by starting from s_i for $i = 1, 2, 3, 4$. Then, the returns obtained when

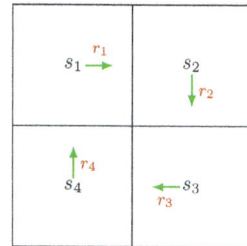

Figure 2.3: An example for demonstrating how to calculate returns. There are no target or forbidden cells in this example.

starting from the four states in Figure 2.3 can be calculated as

$$
\begin{aligned}
v_1 &= r_1 + \gamma r_2 + \gamma^2 r_3 + \ldots, \\
v_2 &= r_2 + \gamma r_3 + \gamma^2 r_4 + \ldots, \\
v_3 &= r_3 + \gamma r_4 + \gamma^2 r_1 + \ldots, \\
v_4 &= r_4 + \gamma r_1 + \gamma^2 r_2 + \ldots.
\end{aligned}
\tag{2.2}
$$

◇ The second way, which is more important, is based on the idea of *bootstrapping*. By observing the expressions of the returns in (2.2), we can rewrite them as

$$
\begin{aligned}
v_1 &= r_1 + \gamma(r_2 + \gamma r_3 + \ldots) = r_1 + \gamma v_2, \\
v_2 &= r_2 + \gamma(r_3 + \gamma r_4 + \ldots) = r_2 + \gamma v_3, \\
v_3 &= r_3 + \gamma(r_4 + \gamma r_1 + \ldots) = r_3 + \gamma v_4, \\
v_4 &= r_4 + \gamma(r_1 + \gamma r_2 + \ldots) = r_4 + \gamma v_1.
\end{aligned}
\tag{2.3}
$$

The above equations indicate an interesting phenomenon that the values of the returns rely on each other. More specifically, v_1 relies on v_2, v_2 relies on v_3, v_3 relies on v_4, and v_4 relies on v_1. This reflects the idea of bootstrapping, which is to obtain the values of some quantities from themselves.

At first glance, bootstrapping is an endless loop because the calculation of an unknown value relies on another unknown value. In fact, bootstrapping is easier to understand if we view it from a mathematical perspective. In particular, the equations in (2.3) can be reformed into a linear matrix-vector equation:

$$
\underbrace{\begin{bmatrix} v_1 \\ v_2 \\ v_3 \\ v_4 \end{bmatrix}}_{v} = \begin{bmatrix} r_1 \\ r_2 \\ r_3 \\ r_4 \end{bmatrix} + \begin{bmatrix} \gamma v_2 \\ \gamma v_3 \\ \gamma v_4 \\ \gamma v_1 \end{bmatrix} = \underbrace{\begin{bmatrix} r_1 \\ r_2 \\ r_3 \\ r_4 \end{bmatrix}}_{r} + \gamma \underbrace{\begin{bmatrix} 0 & 1 & 0 & 0 \\ 0 & 0 & 1 & 0 \\ 0 & 0 & 0 & 1 \\ 1 & 0 & 0 & 0 \end{bmatrix}}_{P} \underbrace{\begin{bmatrix} v_1 \\ v_2 \\ v_3 \\ v_4 \end{bmatrix}}_{v},
$$

which can be written compactly as

$$v = r + \gamma P v.$$

Thus, the value of v can be calculated easily as $v = (I - \gamma P)^{-1} r$, where I is the identity matrix with appropriate dimensions. One may ask whether $I - \gamma P$ is always invertible. The answer is yes and explained in Section 2.7.1.

In fact, (2.3) is the Bellman equation for this simple example. Although it is simple, (2.3) demonstrates the core idea of the Bellman equation: the return obtained by starting from one state depends on those obtained when starting from other states. The idea of bootstrapping and the Bellman equation for general scenarios will be formalized in the following sections.

2.3 State values

We mentioned that returns can be used to evaluate policies. However, they are inapplicable to stochastic systems because starting from one state may lead to different returns. Motivated by this problem, we introduce the concept of state value in this section.

First, we need to introduce some necessary notations. Consider a sequence of time steps $t = 0, 1, 2, \ldots$. At time t, the agent is in state S_t, and the action taken following a policy π is A_t. The next state is S_{t+1}, and the immediate reward obtained is R_{t+1}. This process can be expressed concisely as

$$S_t \xrightarrow{A_t} S_{t+1}, R_{t+1}.$$

Note that $S_t, S_{t+1}, A_t, R_{t+1}$ are all *random variables*. Moreover, $S_t, S_{t+1} \in \mathcal{S}$, $A_t \in \mathcal{A}(S_t)$, and $R_{t+1} \in \mathcal{R}(S_t, A_t)$.

Starting from t, we can obtain a state-action-reward trajectory:

$$S_t \xrightarrow{A_t} S_{t+1}, R_{t+1} \xrightarrow{A_{t+1}} S_{t+2}, R_{t+2} \xrightarrow{A_{t+2}} S_{t+3}, R_{t+3} \ldots.$$

By definition, the discounted return along the trajectory is

$$G_t \doteq R_{t+1} + \gamma R_{t+2} + \gamma^2 R_{t+3} + \cdots,$$

where $\gamma \in (0, 1)$ is the discount rate. Note that G_t is a random variable since $R_{t+1}, R_{t+2}, \ldots$ are all random variables.

Since G_t is a random variable, we can calculate its expected value (also called the expectation or mean):

$$v_\pi(s) \doteq \mathbb{E}[G_t | S_t = s].$$

Here, $v_\pi(s)$ is called the *state-value function* or simply the *state value* of s. Some important remarks are given below.

⋄ $v_\pi(s)$ depends on s. This is because its definition is a conditional expectation with the condition that the agent starts from $S_t = s$.

⋄ $v_\pi(s)$ depends on π. This is because the trajectories are generated by following the policy π. For a different policy, the state value may be different.

⋄ $v_\pi(s)$ does not depend on t. If the agent moves in the state space, t represents the current time step. The value of $v_\pi(s)$ is determined once the policy is given.

The relationship between state values and returns is further clarified as follows. When both the policy and the system model are deterministic, starting from a state always leads to the same trajectory. In this case, the return obtained starting from a state is equal to the value of that state. By contrast, when either the policy or the system model is stochastic, starting from the same state may generate different trajectories. In this case, the returns of different trajectories are different, and the state value is the mean of these returns.

Although returns can be used to evaluate policies as shown in Section 2.1, it is more formal to use state values to evaluate policies: policies that generate greater state values are better. Therefore, state values constitute a core concept in reinforcement learning. While state values are important, a question that immediately follows is how to calculate them. This question is answered in the next section.

2.4   Bellman equation

We now introduce the Bellman equation, a mathematical tool for analyzing state values. In a nutshell, the Bellman equation is a set of linear equations that describe the relationships between the values of all the states.

We next derive the Bellman equation. First, note that G_t can be rewritten as

$$
\begin{aligned}
G_t &= R_{t+1} + \gamma R_{t+2} + \gamma^2 R_{t+3} + \dots \\
&= R_{t+1} + \gamma(R_{t+2} + \gamma R_{t+3} + \dots) \\
&= R_{t+1} + \gamma G_{t+1},
\end{aligned}
$$

where $G_{t+1} = R_{t+2} + \gamma R_{t+3} + \dots$. This equation establishes the relationship between G_t and G_{t+1}. Then, the state value can be written as

$$
\begin{aligned}
v_\pi(s) &= \mathbb{E}[G_t | S_t = s] \\
&= \mathbb{E}[R_{t+1} + \gamma G_{t+1} | S_t = s] \\
&= \mathbb{E}[R_{t+1} | S_t = s] + \gamma \mathbb{E}[G_{t+1} | S_t = s].
\end{aligned} \tag{2.4}
$$

The two terms in (2.4) are analyzed below.

⋄ The first term, $\mathbb{E}[R_{t+1}|S_t = s]$, is the expectation of the immediate rewards. By using the law of total expectation (Appendix A), it can be calculated as

$$\mathbb{E}[R_{t+1}|S_t = s] = \sum_{a \in \mathcal{A}} \pi(a|s) \mathbb{E}[R_{t+1}|S_t = s, A_t = a]$$
$$= \sum_{a \in \mathcal{A}} \pi(a|s) \sum_{r \in \mathcal{R}} p(r|s, a) r. \tag{2.5}$$

Here, $\mathcal{A}$ and $\mathcal{R}$ are the sets of possible actions and rewards, respectively. It should be noted that $\mathcal{A}$ may be different for different states. In this case, $\mathcal{A}$ should be written as $\mathcal{A}(s)$. Similarly, $\mathcal{R}$ may also depend on (s, a). We drop the dependence on s or (s, a) for the sake of simplicity in this book. Nevertheless, the conclusions are still valid in the presence of dependence.

⋄ The second term, $\mathbb{E}[G_{t+1}|S_t = s]$, is the expectation of the future rewards. It can be calculated as

$$\mathbb{E}[G_{t+1}|S_t = s] = \sum_{s' \in \mathcal{S}} \mathbb{E}[G_{t+1}|S_t = s, S_{t+1} = s'] p(s'|s)$$
$$= \sum_{s' \in \mathcal{S}} \mathbb{E}[G_{t+1}|S_{t+1} = s'] p(s'|s) \qquad \text{(due to the Markov property)}$$
$$= \sum_{s' \in \mathcal{S}} v_\pi(s') p(s'|s)$$
$$= \sum_{s' \in \mathcal{S}} v_\pi(s') \sum_{a \in \mathcal{A}} p(s'|s, a) \pi(a|s). \tag{2.6}$$

The above derivation uses the fact that $\mathbb{E}[G_{t+1}|S_t = s, S_{t+1} = s'] = \mathbb{E}[G_{t+1}|S_{t+1} = s']$, which is due to the Markov property that the future rewards depend merely on the present state rather than the previous ones.

Substituting (2.5)-(2.6) into (2.4) yields

$$v_\pi(s) = \mathbb{E}[R_{t+1}|S_t = s] + \gamma \mathbb{E}[G_{t+1}|S_t = s],$$
$$= \underbrace{\sum_{a \in \mathcal{A}} \pi(a|s) \sum_{r \in \mathcal{R}} p(r|s, a) r}_{\text{mean of immediate rewards}} + \gamma \underbrace{\sum_{a \in \mathcal{A}} \pi(a|s) \sum_{s' \in \mathcal{S}} p(s'|s, a) v_\pi(s')}_{\text{mean of future rewards}}$$
$$= \sum_{a \in \mathcal{A}} \pi(a|s) \left[\sum_{r \in \mathcal{R}} p(r|s, a) r + \gamma \sum_{s' \in \mathcal{S}} p(s'|s, a) v_\pi(s') \right], \qquad \text{for all } s \in \mathcal{S}. \tag{2.7}$$

This equation is the *Bellman equation*, which characterizes the relationships of state values. It is a fundamental tool for designing and analyzing reinforcement learning algorithms.

The Bellman equation seems complex at first glance. In fact, it has a clear structure. Some remarks are given below.

◇ $v_\pi(s)$ and $v_\pi(s')$ are unknown state values to be calculated. It may be confusing to beginners how to calculate the unknown $v_\pi(s)$ given that it relies on another unknown $v_\pi(s')$. It must be noted that the Bellman equation refers to a set of linear equations for all states rather than a single equation. If we put these equations together, it becomes clear how to calculate all the state values. Details will be given in Section 2.7.

◇ $\pi(a|s)$ is a given policy. Since state values can be used to evaluate a policy, solving the state values from the Bellman equation is a *policy evaluation* process, which is an important process in many reinforcement learning algorithms, as we will see later in the book.

◇ $p(r|s, a)$ and $p(s'|s, a)$ represent the system model. We will first show how to calculate the state values *with* this model in Section 2.7, and then show how to do that *without* the model by using model-free algorithms later in this book.

In addition to the expression in (2.7), readers may also encounter other expressions of the Bellman equation in the literature. We next introduce two equivalent expressions. First, it follows from the law of total probability that

$$p(s'|s, a) = \sum_{r \in \mathcal{R}} p(s', r|s, a),$$

$$p(r|s, a) = \sum_{s' \in \mathcal{S}} p(s', r|s, a).$$

Then, equation (2.7) can be rewritten as

$$v_\pi(s) = \sum_{a \in \mathcal{A}} \pi(a|s) \sum_{s' \in \mathcal{S}} \sum_{r \in \mathcal{R}} p(s', r|s, a) \left[r + \gamma v_\pi(s') \right].$$

This is the expression used in [3].

Second, the reward r may depend solely on the next state s' in some problems. As a result, we can write the reward as $r(s')$ and hence $p(r(s')|s, a) = p(s'|s, a)$, substituting which into (2.7) gives

$$v_\pi(s) = \sum_{a \in \mathcal{A}} \pi(a|s) \sum_{s' \in \mathcal{S}} p(s'|s, a) \left[r(s') + \gamma v_\pi(s') \right].$$

2.5 Examples for illustrating the Bellman equation

We next use two examples to demonstrate how to write out the Bellman equation and calculate the state values step by step. Readers are advised to carefully go through the examples to gain a better understanding of the Bellman equation.

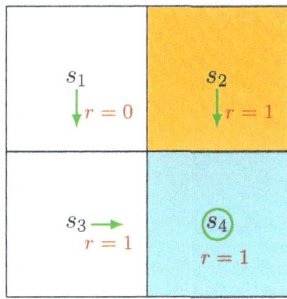

Figure 2.4: An example for demonstrating the Bellman equation. The policy in this example is deterministic.

◇ Consider the first example shown in Figure 2.4, where the policy is deterministic. We next write out the Bellman equation and then solve the state values from it.

First, consider state s_1. Under the policy, the probabilities of taking the actions are $\pi(a = a_3|s_1) = 1$ and $\pi(a \neq a_3|s_1) = 0$. The state transition probabilities are $p(s' = s_3|s_1, a_3) = 1$ and $p(s' \neq s_3|s_1, a_3) = 0$. The reward probabilities are $p(r = 0|s_1, a_3) = 1$ and $p(r \neq 0|s_1, a_3) = 0$. Substituting these values into (2.7) gives

$$v_\pi(s_1) = 0 + \gamma v_\pi(s_3).$$

Interestingly, although the expression of the Bellman equation in (2.7) seems complex, the expression for this specific state is very simple.

Similarly, it can be obtained that

$$v_\pi(s_2) = 1 + \gamma v_\pi(s_4),$$
$$v_\pi(s_3) = 1 + \gamma v_\pi(s_4),$$
$$v_\pi(s_4) = 1 + \gamma v_\pi(s_4).$$

We can solve the state values from these equations. Since the equations are simple, we can manually solve them. More complicated equations can be solved by the algorithms presented in Section 2.7. Here, the state values can be solved as

$$v_\pi(s_4) = \frac{1}{1 - \gamma},$$
$$v_\pi(s_3) = \frac{1}{1 - \gamma},$$
$$v_\pi(s_2) = \frac{1}{1 - \gamma},$$
$$v_\pi(s_1) = \frac{\gamma}{1 - \gamma}.$$

Furthermore, if we set $\gamma = 0.9$, then

$$v_\pi(s_4) = \frac{1}{1 - 0.9} = 10,$$
$$v_\pi(s_3) = \frac{1}{1 - 0.9} = 10,$$
$$v_\pi(s_2) = \frac{1}{1 - 0.9} = 10,$$
$$v_\pi(s_1) = \frac{0.9}{1 - 0.9} = 9.$$

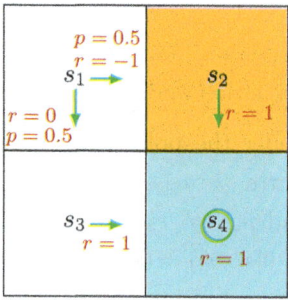

Figure 2.5: An example for demonstrating the Bellman equation. The policy in this example is stochastic.

◇ Consider the second example shown in Figure 2.5, where the policy is stochastic. We next write out the Bellman equation and then solve the state values from it.

In state s_1, the probabilities of going right and down equal 0.5. Mathematically, we have $\pi(a = a_2|s_1) = 0.5$ and $\pi(a = a_3|s_1) = 0.5$. The state transition probability is deterministic since $p(s' = s_3|s_1, a_3) = 1$ and $p(s' = s_2|s_1, a_2) = 1$. The reward probability is also deterministic since $p(r = 0|s_1, a_3) = 1$ and $p(r = -1|s_1, a_2) = 1$. Substituting these values into (2.7) gives

$$v_\pi(s_1) = 0.5[0 + \gamma v_\pi(s_3)] + 0.5[-1 + \gamma v_\pi(s_2)].$$

Similarly, it can be obtained that

$$v_\pi(s_2) = 1 + \gamma v_\pi(s_4),$$
$$v_\pi(s_3) = 1 + \gamma v_\pi(s_4),$$
$$v_\pi(s_4) = 1 + \gamma v_\pi(s_4).$$

The state values can be solved from the above equations. Since the equations are simple, we can solve the state values manually and obtain

$$v_\pi(s_4) = \frac{1}{1-\gamma},$$
$$v_\pi(s_3) = \frac{1}{1-\gamma},$$
$$v_\pi(s_2) = \frac{1}{1-\gamma},$$
$$v_\pi(s_1) = 0.5[0 + \gamma v_\pi(s_3)] + 0.5[-1 + \gamma v_\pi(s_2)],$$
$$= -0.5 + \frac{\gamma}{1-\gamma}.$$

Furthermore, if we set $\gamma = 0.9$, then

$$v_\pi(s_4) = 10,$$
$$v_\pi(s_3) = 10,$$
$$v_\pi(s_2) = 10,$$
$$v_\pi(s_1) = -0.5 + 9 = 8.5.$$

If we compare the state values of the two policies in the above examples, it can be seen that

$$v_{\pi_1}(s_i) \geq v_{\pi_2}(s_i), \quad i = 1, 2, 3, 4,$$

which indicates that the policy in Figure 2.4 is better because it has greater state values. This mathematical conclusion is consistent with the intuition that the first policy is better because it can avoid entering the forbidden area when the agent starts from s_1. As a result, the above two examples demonstrate that state values can be used to evaluate policies.

2.6 Matrix-vector form of the Bellman equation

The Bellman equation in (2.7) is in an *elementwise form*. Since it is valid for every state, we can combine all these equations and write them concisely in a *matrix-vector form*, which will be frequently used to analyze the Bellman equation.

To derive the matrix-vector form, we first rewrite the Bellman equation in (2.7) as

$$v_\pi(s) = r_\pi(s) + \gamma \sum_{s' \in \mathcal{S}} p_\pi(s'|s) v_\pi(s'), \tag{2.8}$$

where

$$r_\pi(s) \doteq \sum_{a \in \mathcal{A}} \pi(a|s) \sum_{r \in \mathcal{R}} p(r|s, a)r,$$

$$p_\pi(s'|s) \doteq \sum_{a \in \mathcal{A}} \pi(a|s)p(s'|s, a).$$

Here, $r_\pi(s)$ denotes the mean of the immediate rewards, and $p_\pi(s'|s)$ is the probability of transitioning from s to s' under policy π.

Suppose that the states are indexed as s_i with $i = 1, \ldots, n$, where $n = |\mathcal{S}|$. For state s_i, (2.8) can be written as

$$v_\pi(s_i) = r_\pi(s_i) + \gamma \sum_{s_j \in \mathcal{S}} p_\pi(s_j|s_i)v_\pi(s_j). \tag{2.9}$$

Let $v_\pi = [v_\pi(s_1), \ldots, v_\pi(s_n)]^T \in \mathbb{R}^n$, $r_\pi = [r_\pi(s_1), \ldots, r_\pi(s_n)]^T \in \mathbb{R}^n$, and $P_\pi \in \mathbb{R}^{n \times n}$ with $[P_\pi]_{ij} = p_\pi(s_j|s_i)$. Then, (2.9) can be written in the following matrix-vector form:

$$v_\pi = r_\pi + \gamma P_\pi v_\pi, \tag{2.10}$$

where v_π is the unknown to be solved, and r_π, P_π are known.

The matrix P_π has some interesting properties. First, it is a nonnegative matrix, meaning that all its elements are equal to or greater than zero. This property is denoted as $P_\pi \geq 0$, where 0 denotes a zero matrix with appropriate dimensions. In this book, $\geq$ or $\leq$ represents an elementwise comparison operation. Second, P_π is a stochastic matrix, meaning that the sum of the values in every row is equal to one. This property is denoted as $P_\pi \mathbf{1} = \mathbf{1}$, where $\mathbf{1} = [1, \ldots, 1]^T$ has appropriate dimensions.

Consider the example shown in Figure 2.6. The matrix-vector form of the Bellman equation is

$$\underbrace{\begin{bmatrix} v_\pi(s_1) \\ v_\pi(s_2) \\ v_\pi(s_3) \\ v_\pi(s_4) \end{bmatrix}}_{v_\pi} = \underbrace{\begin{bmatrix} r_\pi(s_1) \\ r_\pi(s_2) \\ r_\pi(s_3) \\ r_\pi(s_4) \end{bmatrix}}_{r_\pi} + \gamma \underbrace{\begin{bmatrix} p_\pi(s_1|s_1) & p_\pi(s_2|s_1) & p_\pi(s_3|s_1) & p_\pi(s_4|s_1) \\ p_\pi(s_1|s_2) & p_\pi(s_2|s_2) & p_\pi(s_3|s_2) & p_\pi(s_4|s_2) \\ p_\pi(s_1|s_3) & p_\pi(s_2|s_3) & p_\pi(s_3|s_3) & p_\pi(s_4|s_3) \\ p_\pi(s_1|s_4) & p_\pi(s_2|s_4) & p_\pi(s_3|s_4) & p_\pi(s_4|s_4) \end{bmatrix}}_{P_\pi} \underbrace{\begin{bmatrix} v_\pi(s_1) \\ v_\pi(s_2) \\ v_\pi(s_3) \\ v_\pi(s_4) \end{bmatrix}}_{v_\pi}.$$

Substituting the specific values into the above equation gives

$$\begin{bmatrix} v_\pi(s_1) \\ v_\pi(s_2) \\ v_\pi(s_3) \\ v_\pi(s_4) \end{bmatrix} = \begin{bmatrix} 0.5(0) + 0.5(-1) \\ 1 \\ 1 \\ 1 \end{bmatrix} + \gamma \begin{bmatrix} 0 & 0.5 & 0.5 & 0 \\ 0 & 0 & 0 & 1 \\ 0 & 0 & 0 & 1 \\ 0 & 0 & 0 & 1 \end{bmatrix} \begin{bmatrix} v_\pi(s_1) \\ v_\pi(s_2) \\ v_\pi(s_3) \\ v_\pi(s_4) \end{bmatrix}.$$

It can be seen that P_π satisfies $P_\pi \mathbf{1} = \mathbf{1}$.

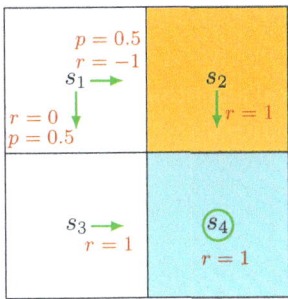

Figure 2.6: An example for demonstrating the matrix-vector form of the Bellman equation.

2.7 Solving state values from the Bellman equation

Calculating the state values of a given policy is a fundamental problem in reinforcement learning. This problem is often referred to as *policy evaluation*. In this section, we present two methods for calculating state values from the Bellman equation.

2.7.1 Closed-form solution

Since $v_\pi = r_\pi + \gamma P_\pi v_\pi$ is a simple linear equation, its *closed-form solution* can be easily obtained as

$$v_\pi = (I - \gamma P_\pi)^{-1} r_\pi.$$

Some properties of $(I - \gamma P_\pi)^{-1}$ are given below.

◇ $I - \gamma P_\pi$ is invertible. The proof is as follows. According to the Gershgorin circle theorem [4], every eigenvalue of $I - \gamma P_\pi$ lies within at least one of the Gershgorin circles. The ith Gershgorin circle has a center at $[I - \gamma P_\pi]_{ii} = 1 - \gamma p_\pi(s_i|s_i)$ and a radius equal to $\sum_{j \neq i}[I - \gamma P_\pi]_{ij} = -\sum_{j \neq i} \gamma p_\pi(s_j|s_i)$. Since $\gamma < 1$, we know that the radius is less than the magnitude of the center: $\sum_{j \neq i} \gamma p_\pi(s_j|s_i) < 1 - \gamma p_\pi(s_i|s_i)$. Therefore, all Gershgorin circles do not encircle the origin, and hence no eigenvalue of $I - \gamma P_\pi$ is zero.

◇ $(I - \gamma P_\pi)^{-1} \geq I$, meaning that every element of $(I - \gamma P_\pi)^{-1}$ is nonnegative and, more specifically, no less than that of the identity matrix. This is because P_π has nonnegative entries, and hence $(I - \gamma P_\pi)^{-1} = I + \gamma P_\pi + \gamma^2 P_\pi^2 + \cdots \geq I \geq 0$.

◇ For any vector $r \geq 0$, it holds that $(I - \gamma P_\pi)^{-1} r \geq r \geq 0$. This property follows from the second property because $[(I - \gamma P_\pi)^{-1} - I]r \geq 0$. As a consequence, if $r_1 \geq r_2$, we have $(I - \gamma P_\pi)^{-1} r_1 \geq (I - \gamma P_\pi)^{-1} r_2$.

2.7.2   Iterative solution

Although the closed-form solution is useful for theoretical analysis purposes, it is not applicable in practice because it involves a matrix inversion operation, which still needs to be calculated by other numerical algorithms. In fact, we can directly solve the Bellman equation using the following iterative algorithm:

$$v_{k+1} = r_\pi + \gamma P_\pi v_k, \quad k = 0, 1, 2, \ldots \tag{2.11}$$

This algorithm generates a sequence of values $\{v_0, v_1, v_2, \ldots\}$, where $v_0 \in \mathbb{R}^n$ is an initial guess of v_π. It holds that

$$v_k \to v_\pi = (I - \gamma P_\pi)^{-1} r_\pi, \quad \text{as } k \to \infty. \tag{2.12}$$

Interested readers may see the proof in Box 2.1.

Box 2.1: Convergence proof of (2.12)

Define the error as $\delta_k = v_k - v_\pi$. We only need to show that $\delta_k \to 0$. Substituting $v_{k+1} = \delta_{k+1} + v_\pi$ and $v_k = \delta_k + v_\pi$ into $v_{k+1} = r_\pi + \gamma P_\pi v_k$ gives

$$\delta_{k+1} + v_\pi = r_\pi + \gamma P_\pi (\delta_k + v_\pi),$$

which can be rewritten as

$$
\begin{aligned}
\delta_{k+1} &= -v_\pi + r_\pi + \gamma P_\pi \delta_k + \gamma P_\pi v_\pi, \\
&= \gamma P_\pi \delta_k - v_\pi + (r_\pi + \gamma P_\pi v_\pi), \\
&= \gamma P_\pi \delta_k.
\end{aligned}
$$

As a result,

$$\delta_{k+1} = \gamma P_\pi \delta_k = \gamma^2 P_\pi^2 \delta_{k-1} = \cdots = \gamma^{k+1} P_\pi^{k+1} \delta_0.$$

Since every entry of P_π is nonnegative and no greater than one, we have that $0 \leq P_\pi^k \leq 1$ for any k. That is, every entry of P_π^k is no greater than 1. On the other hand, since $\gamma < 1$, we know that $\gamma^k \to 0$, and hence $\delta_{k+1} = \gamma^{k+1} P_\pi^{k+1} \delta_0 \to 0$ as $k \to \infty$.

2.7.3   Illustrative examples

We next apply the algorithm in (2.11) to solve the state values of some examples.

The examples are shown in Figure 2.7. The orange cells represent forbidden areas. The blue cell represents the target area. The reward settings are $r_{\text{boundary}} = r_{\text{forbidden}} = -1$

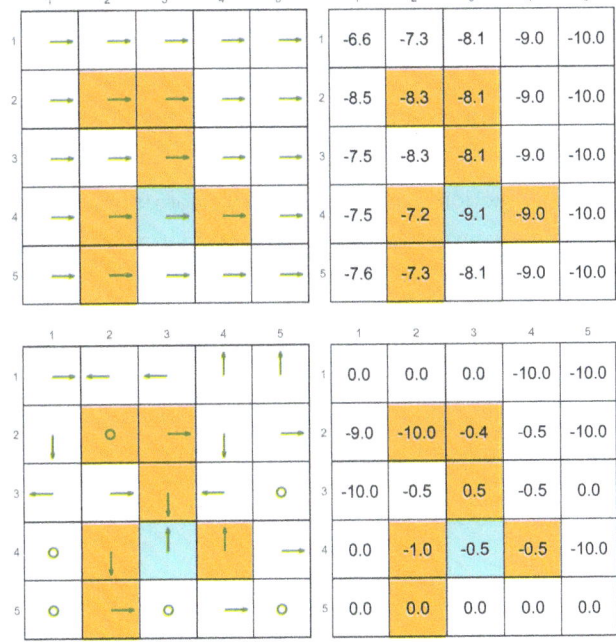

(a) Two "good" policies and their state values. The state values of the two policies are the same, but the two policies are different at the top two states in the fourth column.

(b) Two "bad" policies and their state values. The state values are smaller than those of the "good" policies.

Figure 2.7: Examples of policies and their corresponding state values.

and $r_{\text{target}} = 1$. Here, the discount rate is $\gamma = 0.9$.

Figure 2.7(a) shows two "good" policies and their corresponding state values obtained by (2.11). The two policies have the same state values but differ at the top two states in the fourth column. Therefore, we know that different policies may have the same state values.

Figure 2.7(b) shows two "bad" policies and their corresponding state values. These two policies are bad because the actions of many states are intuitively unreasonable. Such intuition is supported by the obtained state values. As can be seen, the state values of these two policies are negative and much smaller than those of the good policies in Figure 2.7(a).

2.8   From state value to action value

While we have been discussing state values thus far in this chapter, we now turn to the *action value*, which indicates the "value" of taking an action at a state. While the concept of action value is important, the reason why it is introduced in the last section of this chapter is that it heavily relies on the concept of state values. It is important to understand state values well first before studying action values.

The action value of a state-action pair (s, a) is defined as

$$q_\pi(s, a) \doteq \mathbb{E}[G_t | S_t = s, A_t = a].$$

As can be seen, the action value is defined as the expected return that can be obtained after taking an action at a state. It must be noted that $q_\pi(s, a)$ depends on a state-action pair (s, a) rather than an action alone. It may be more rigorous to call this value a state-action value, but it is conventionally called an action value for simplicity.

What is the relationship between action values and state values?

⋄ First, it follows from the properties of conditional expectation that

$$\underbrace{\mathbb{E}[G_t | S_t = s]}_{v_\pi(s)} = \sum_{a \in \mathcal{A}} \underbrace{\mathbb{E}[G_t | S_t = s, A_t = a]}_{q_\pi(s,a)} \pi(a|s).$$

It then follows that

$$v_\pi(s) = \sum_{a \in \mathcal{A}} \pi(a|s) q_\pi(s, a). \tag{2.13}$$

As a result, a state value is the expectation of the action values associated with that state.

⋄ Second, since the state value is given by

$$v_\pi(s) = \sum_{a \in \mathcal{A}} \pi(a|s) \left[\sum_{r \in \mathcal{R}} p(r|s, a) r + \gamma \sum_{s' \in \mathcal{S}} p(s'|s, a) v_\pi(s') \right],$$

comparing it with (2.13) leads to

$$q_\pi(s, a) = \sum_{r \in \mathcal{R}} p(r|s, a)r + \gamma \sum_{s' \in \mathcal{S}} p(s'|s, a)v_\pi(s'). \tag{2.14}$$

It can be seen that the action value consists of two terms. The first term is the mean of the immediate rewards, and the second term is the mean of the future rewards.

Both (2.13) and (2.14) describe the relationship between state values and action values. They are the two sides of the same coin: (2.13) shows how to obtain state values from action values, whereas (2.14) shows how to obtain action values from state values.

2.8.1 Illustrative examples

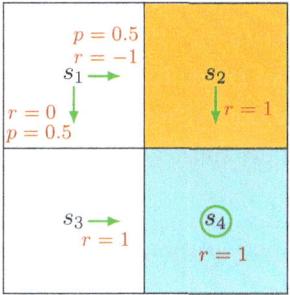

Figure 2.8: An example for demonstrating the process of calculating action values.

We next present an example to illustrate the process of calculating action values and discuss a common mistake that beginners may make.

Consider the stochastic policy shown in Figure 2.8. We next only examine the actions of s_1. The other states can be examined similarly. The action value of (s_1, a_2) is

$$q_\pi(s_1, a_2) = -1 + \gamma v_\pi(s_2),$$

where s_2 is the next state. Similarly, it can be obtained that

$$q_\pi(s_1, a_3) = 0 + \gamma v_\pi(s_3).$$

A common mistake that beginners may make is about the values of the actions that the given policy does not select. For example, the policy in Figure 2.8 can only select a_2 or a_3 and cannot select a_1, a_4, a_5. One may argue that since the policy does not select a_1, a_4, a_5, we do not need to calculate their action values, or we can simply set $q_\pi(s_1, a_1) = q_\pi(s_1, a_4) = q_\pi(s_1, a_5) = 0$. This is wrong.

⋄ First, even if an action would not be selected by a policy, it still has an action value. In this example, although policy π does not take a_1 at s_1, we can still calculate its action value by observing what we would obtain after taking this action. Specifically, after taking a_1, the agent is bounced back to s_1 (hence, the immediate reward is -1) and then continues moving in the state space starting from s_1 by following π (hence, the future reward is $\gamma v_\pi(s_1)$). As a result, the action value of (s_1, a_1) is

$$q_\pi(s_1, a_1) = -1 + \gamma v_\pi(s_1).$$

Similarly, for a_4 and a_5, which cannot be possibly selected by the given policy either, we have

$$q_\pi(s_1, a_4) = -1 + \gamma v_\pi(s_1),$$
$$q_\pi(s_1, a_5) = 0 + \gamma v_\pi(s_1).$$

⋄ Second, why do we care about the actions that the given policy would not select? Although some actions cannot be possibly selected by a given policy, this does not mean that these actions are not good. It is possible that the given policy is not good, so it cannot select the best action. The purpose of reinforcement learning is to find optimal policies. To that end, we must keep exploring all actions to determine better actions for each state.

Finally, after computing the action values, we can also calculate the state value according to (2.13):

$$v_\pi(s_1) = 0.5q_\pi(s_1, a_2) + 0.5q_\pi(s_1, a_3),$$
$$= 0.5[0 + \gamma v_\pi(s_3)] + 0.5[-1 + \gamma v_\pi(s_2)].$$

2.8.2 The Bellman equation in terms of action values

The Bellman equation that we previously introduced was defined based on state values. In fact, it can also be expressed in terms of action values.

In particular, substituting (2.13) into (2.14) yields

$$q_\pi(s, a) = \sum_{r \in \mathcal{R}} p(r|s, a)r + \gamma \sum_{s' \in \mathcal{S}} p(s'|s, a) \sum_{a' \in \mathcal{A}(s')} \pi(a'|s')q_\pi(s', a'),$$

which is an equation of action values. The above equation is valid for every state-action pair. If we put all these equations together, their matrix-vector form is

$$q_\pi = \tilde{r} + \gamma P \Pi q_\pi, \tag{2.15}$$

where q_π is the action value vector indexed by the state-action pairs: its (s, a)th element is $[q_\pi]_{(s,a)} = q_\pi(s, a)$. $\tilde{r}$ is the immediate reward vector indexed by the state-action pairs: $[\tilde{r}]_{(s,a)} = \sum_{r \in \mathcal{R}} p(r|s, a)r$. The matrix P is the probability transition matrix, whose

32

row is indexed by the state-action pairs and whose column is indexed by the states: $[P]_{(s,a),s'} = p(s'|s,a)$. Moreover, Π is a block diagonal matrix in which each block is a $1 \times |\mathcal{A}|$ vector: $\Pi_{s',(s',a')} = \pi(a'|s')$ and the other entries of Π are zero.

Compared to the Bellman equation defined in terms of state values, the equation defined in terms of action values has some unique features. For example, $\tilde{r}$ and P are independent of the policy and are merely determined by the system model. The policy is embedded in Π. It can be verified that (2.15) is also a contraction mapping and has a unique solution that can be iteratively solved. More details can be found in [5].

2.9 Summary

The most important concept introduced in this chapter is the state value. Mathematically, a state value is the expected return that the agent can obtain by starting from a state. The values of different states are related to each other. That is, the value of state s relies on the values of some other states, which may further rely on the value of state s itself. This phenomenon might be the most confusing part of this chapter for beginners. It is related to an important concept called bootstrapping, which involves calculating something from itself. Although bootstrapping may be intuitively confusing, it is clear if we examine the matrix-vector form of the Bellman equation. In particular, the Bellman equation is a set of linear equations that describe the relationships between the values of all states.

Since state values can be used to evaluate whether a policy is good or not, the process of solving the state values of a policy from the Bellman equation is called policy evaluation. As we will see later in this book, policy evaluation is an important step in many reinforcement learning algorithms.

Another important concept, action value, was introduced to describe the value of taking one action at a state. As we will see later in this book, action values play a more direct role than state values when we attempt to find optimal policies. Finally, the Bellman equation is not restricted to the reinforcement learning field. Instead, it widely exists in many fields such as control theories and operation research. In different fields, the Bellman equation may have different expressions. In this book, the Bellman equation is studied under discrete Markov decision processes. More information about this topic can be found in [2].

2.10 Q&A

⋄ Q: What is the relationship between state values and returns?

A: The value of a state is the mean of the returns that can be obtained if the agent starts from that state.

⋄ Q: Why do we care about state values?

A: State values can be used to evaluate policies. In fact, optimal policies are defined based on state values. This point will become clearer in the next chapter.

⋄ Q: Why do we care about the Bellman equation?

A: The Bellman equation describes the relationships among the values of all states. It is the tool for analyzing state values.

⋄ Q: Why is the process of solving the Bellman equation called policy evaluation?

A: Solving the Bellman equation yields state values. Since state values can be used to evaluate a policy, solving the Bellman equation can be interpreted as evaluating the corresponding policy.

⋄ Q: Why do we need to study the matrix-vector form of the Bellman equation?

A: The Bellman equation refers to a set of linear equations established for all the states. To solve state values, we must put all the linear equations together. The matrix-vector form is a concise expression of these linear equations.

⋄ Q: What is the relationship between state values and action values?

A: On the one hand, a state value is the mean of the action values for that state. On the other hand, an action value relies on the values of the next states that the agent may transition to after taking the action.

⋄ Q: Why do we care about the values of the actions that a given policy cannot select?

A: Although a given policy cannot select some actions, this does not mean that these actions are not good. On the contrary, it is possible that the given policy is not good and misses the best action. To find better policies, we must keep exploring different actions even though some of them may not be selected by the given policy.

Chapter 3

Optimal State Values and Bellman Optimality Equation

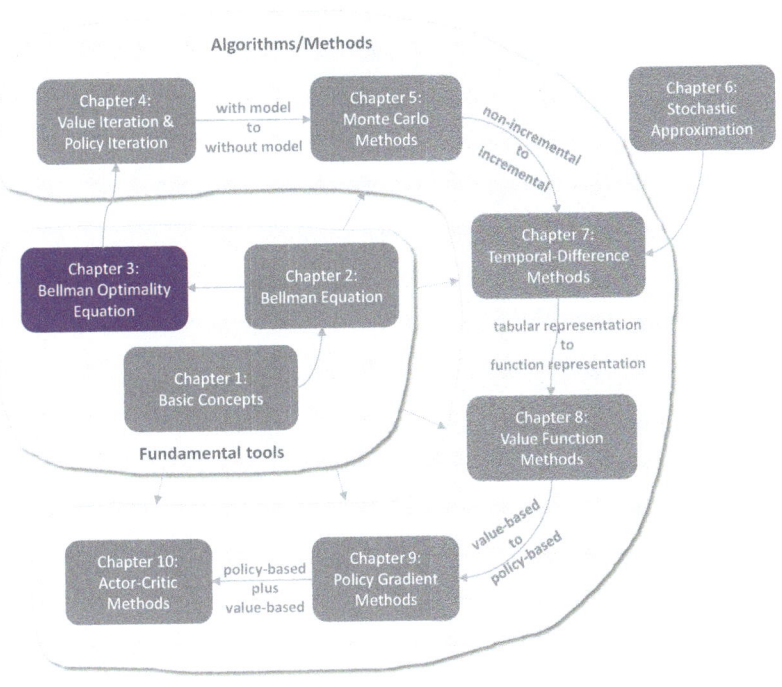

Figure 3.1: Where we are in this book.

The ultimate goal of reinforcement learning is to seek *optimal policies*. It is, therefore, necessary to define what optimal policies are. In this chapter, we introduce *a core concept* and *an important tool*. The core concept is the *optimal state value*, based on which we can define *optimal policies*. The important tool is the *Bellman optimality equation*, from which we can solve the optimal state values and policies.

The relationship between the previous, present, and subsequent chapters is as follows. The previous chapter (Chapter 2) introduced the Bellman equation of any given policy.

© The Author(s), under exclusive license to Springer Nature Singapore Pte Ltd. 2025
S. Zhao, *Mathematical Foundations of Reinforcement Learning*, https://doi.org/10.1007/978-981-97-3944-8_3

The present chapter introduces the Bellman optimality equation, which is a special Bellman equation whose corresponding policy is optimal. The next chapter (Chapter 4) will introduce an important algorithm called value iteration, which is exactly the algorithm for solving the Bellman optimality equation as introduced in the present chapter.

Be prepared that this chapter is slightly mathematically intensive. However, it is worth it because many fundamental questions can be clearly answered.

3.1   Motivating example: How to improve policies?

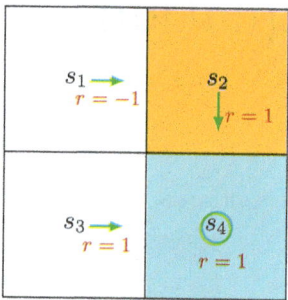

Figure 3.2: An example for demonstrating policy improvement.

Consider the policy shown in Figure 3.2. Here, the orange and blue cells represent the forbidden and target areas, respectively. The policy here is *not good* because it selects a_2 (rightward) in state s_1. How can we improve the given policy to obtain a better policy? The answer lies in state values and action values.

◇ *Intuition:* It is intuitively clear that the policy can improve if it selects a_3 (downward) instead of a_2 (rightward) at s_1. This is because moving downward enables the agent to avoid entering the forbidden area.

◇ *Mathematics:* The above intuition can be realized based on the calculation of state values and action values.

First, we calculate the state values of the given policy. In particular, the Bellman equation of this policy is

$$v_\pi(s_1) = -1 + \gamma v_\pi(s_2),$$
$$v_\pi(s_2) = +1 + \gamma v_\pi(s_4),$$
$$v_\pi(s_3) = +1 + \gamma v_\pi(s_4),$$
$$v_\pi(s_4) = +1 + \gamma v_\pi(s_4).$$

Let $\gamma = 0.9$. It can be easily solved that

$$v_\pi(s_4) = v_\pi(s_3) = v_\pi(s_2) = 10,$$
$$v_\pi(s_1) = 8.$$

Second, we calculate the action values for state s_1:

$$q_\pi(s_1, a_1) = -1 + \gamma v_\pi(s_1) = 6.2,$$
$$q_\pi(s_1, a_2) = -1 + \gamma v_\pi(s_2) = 8,$$
$$q_\pi(s_1, a_3) = 0 + \gamma v_\pi(s_3) = 9,$$
$$q_\pi(s_1, a_4) = -1 + \gamma v_\pi(s_1) = 6.2,$$
$$q_\pi(s_1, a_5) = 0 + \gamma v_\pi(s_1) = 7.2.$$

It is notable that action a_3 has the greatest action value:

$$q_\pi(s_1, a_3) \geq q_\pi(s_1, a_i), \quad \text{for all } i \neq 3.$$

Therefore, we can update the policy to select a_3 at s_1.

This example illustrates that we can obtain a better policy if we update the policy to select the action with the *greatest action value*. This is the basic idea of many reinforcement learning algorithms.

This example is very simple in the sense that the given policy is only not good for state s_1. If the policy is also not good for the other states, will selecting the action with the greatest action value still generate a better policy? Moreover, whether there always exist optimal policies? What does an optimal policy look like? We will answer all of these questions in this chapter.

3.2   Optimal state values and optimal policies

While the ultimate goal of reinforcement learning is to obtain optimal policies, it is necessary to first define what an optimal policy is. The definition is based on state values. In particular, consider two given policies π_1 and π_2. If the state value of π_1 is greater than or equal to that of π_2 for any state:

$$v_{\pi_1}(s) \geq v_{\pi_2}(s), \quad \text{for all } s \in \mathcal{S},$$

then π_1 is said to be better than π_2. Furthermore, if a policy is better than all the other possible policies, then this policy is optimal. This is formally stated below.

Definition 3.1 (Optimal policy and optimal state value). *A policy π^* is optimal if $v_{\pi^*}(s) \geq v_\pi(s)$ for all $s \in \mathcal{S}$ and for any other policy π. The state values of π^* are the optimal state values.*

The above definition indicates that an optimal policy has the greatest state value for every state compared to all the other policies. This definition also leads to many questions:

⬦ Existence: Does the optimal policy exist?

⬦ Uniqueness: Is the optimal policy unique?

⬦ Stochasticity: Is the optimal policy stochastic or deterministic?

⬦ Algorithm: How to obtain the optimal policy and the optimal state values?

These fundamental questions must be clearly answered to thoroughly understand optimal policies. For example, regarding the existence of optimal policies, if optimal policies do not exist, then we do not need to bother to design algorithms to find them.

We will answer all these questions in the remainder of this chapter.

3.3 Bellman optimality equation

The tool for analyzing optimal policies and optimal state values is the *Bellman optimality equation* (BOE). By solving this equation, we can obtain optimal policies and optimal state values. We next present the expression of the BOE and then analyze it in detail.

For every $s \in \mathcal{S}$, the elementwise expression of the BOE is

$$
\begin{aligned}
v(s) &= \max_{\pi(s) \in \Pi(s)} \sum_{a \in \mathcal{A}} \pi(a|s) \left(\sum_{r \in \mathcal{R}} p(r|s,a)r + \gamma \sum_{s' \in \mathcal{S}} p(s'|s,a)v(s') \right) \\
&= \max_{\pi(s) \in \Pi(s)} \sum_{a \in \mathcal{A}} \pi(a|s)q(s,a),
\end{aligned}
\tag{3.1}
$$

where $v(s), v(s')$ are unknown variables to be solved and

$$
q(s,a) \doteq \sum_{r \in \mathcal{R}} p(r|s,a)r + \gamma \sum_{s' \in \mathcal{S}} p(s'|s,a)v(s').
$$

Here, $\pi(s)$ denotes a policy for state s, and $\Pi(s)$ is the set of all possible policies for s.

The BOE is an elegant and powerful tool for analyzing optimal policies. However, it may be nontrivial to understand this equation. For example, this equation has two unknown variables $v(s)$ and $\pi(a|s)$. It may be confusing to beginners how to solve two unknown variables from one equation. Moreover, the BOE is actually a special Bellman equation. However, it is nontrivial to see that since its expression is quite different from that of the Bellman equation. We also need to answer the following fundamental questions about the BOE.

◇ Existence: Does this equation have a solution?

◇ Uniqueness: Is the solution unique?

◇ Algorithm: How to solve this equation?

◇ Optimality: How is the solution related to optimal policies?

Once we can answer these questions, we will clearly understand optimal state values and optimal policies.

3.3.1 Maximization of the right-hand side of the BOE

We next clarify how to solve the maximization problem on the right-hand side of the BOE in (3.1). At first glance, it may be confusing to beginners how to solve *two* unknown variables $v(s)$ and $\pi(a|s)$ from *one* equation. In fact, these two unknown variables can be solved one by one. This idea is illustrated by the following example.

Example 3.1. *Consider two unknown variables* $x, y \in \mathbb{R}$ *that satisfy*

$$x = \max_{y \in \mathbb{R}}(2x - 1 - y^2).$$

The first step is to solve y *on the right-hand side of the equation. Regardless of the value of* x, *we always have* $\max_y(2x - 1 - y^2) = 2x - 1$, *where the maximum is achieved when* $y = 0$. *The second step is to solve* x. *When* $y = 0$, *the equation becomes* $x = 2x - 1$, *which leads to* $x = 1$. *Therefore,* $y = 0$ *and* $x = 1$ *are the solutions of the equation.* □

We now turn to the maximization problem on the right-hand side of the BOE. The BOE in (3.1) can be written concisely as

$$v(s) = \max_{\pi(s) \in \Pi(s)} \sum_{a \in \mathcal{A}} \pi(a|s)q(s, a), \quad s \in \mathcal{S}.$$

Inspired by Example 3.1, we can first solve the optimal π on the right-hand side. How to do that? The following example demonstrates its basic idea.

Example 3.2. *Given* $q_1, q_2, q_3 \in \mathbb{R}$, *we would like to find the optimal values of* c_1, c_2, c_3 *to maximize*

$$\sum_{i=1}^{3} c_i q_i = c_1 q_1 + c_2 q_2 + c_3 q_3,$$

where $c_1 + c_2 + c_3 = 1$ *and* $c_1, c_2, c_3 \geq 0$.

Without loss of generality, suppose that $q_3 \geq q_1, q_2$. *Then, the optimal solution is* $c_3^* = 1$ *and* $c_1^* = c_2^* = 0$. *This is because*

$$q_3 = (c_1 + c_2 + c_3)q_3 = c_1 q_3 + c_2 q_3 + c_3 q_3 \geq c_1 q_1 + c_2 q_2 + c_3 q_3$$

for any c_1, c_2, c_3. □

Inspired by the above example, since $\sum_a \pi(a|s) = 1$, we have

$$\sum_{a \in \mathcal{A}} \pi(a|s) q(s,a) \leq \sum_{a \in \mathcal{A}} \pi(a|s) \max_{a \in \mathcal{A}} q(s,a) = \max_{a \in \mathcal{A}} q(s,a),$$

where equality is achieved when

$$\pi(a|s) = \begin{cases} 1, & a = a^*, \\ 0, & a \neq a^*. \end{cases}$$

Here, $a^* = \arg\max_a q(s,a)$. In summary, the optimal policy $\pi(s)$ is the one that selects the action that has the greatest value of $q(s,a)$.

3.3.2 Matrix-vector form of the BOE

The BOE refers to a set of equations defined for all states. If we combine these equations, we can obtain a concise matrix-vector form, which will be extensively used in this chapter.

The matrix-vector form of the BOE is

$$v = \max_{\pi \in \Pi}(r_\pi + \gamma P_\pi v), \tag{3.2}$$

where $v \in \mathbb{R}^{|\mathcal{S}|}$ and $\max_\pi$ is performed in an elementwise manner. The structures of r_π and P_π are the same as those in the matrix-vector form of the normal Bellman equation:

$$[r_\pi]_s \doteq \sum_{a \in \mathcal{A}} \pi(a|s) \sum_{r \in \mathcal{R}} p(r|s,a) r, \qquad [P_\pi]_{s,s'} = p(s'|s) \doteq \sum_{a \in \mathcal{A}} \pi(a|s) p(s'|s,a).$$

Since the optimal value of π is determined by v, the right-hand side of (3.2) is a function of v, denoted as

$$f(v) \doteq \max_{\pi \in \Pi}(r_\pi + \gamma P_\pi v).$$

Then, the BOE can be expressed in a concise form as

$$v = f(v). \tag{3.3}$$

In the remainder of this section, we show how to solve this nonlinear equation.

3.3.3 Contraction mapping theorem

Since the BOE can be expressed as a nonlinear equation $v = f(v)$, we next introduce the contraction mapping theorem [6] to analyze it. The contraction mapping theorem is a powerful tool for analyzing general nonlinear equations. It is also known as the fixed-point theorem. Readers who already know this theorem can skip this part. Otherwise, the reader is advised to be familiar with this theorem since it is the key to analyzing the

BOE.

Consider a function $f(x)$, where $x \in \mathbb{R}^d$ and $f : \mathbb{R}^d \to \mathbb{R}^d$. A point x^* is called a *fixed point* if

$$f(x^*) = x^*.$$

The interpretation of the above equation is that the map of x^* is itself. This is the reason why x^* is called "fixed". The function f is a *contraction mapping* (or contractive function) if there exists $\gamma \in (0, 1)$ such that

$$\|f(x_1) - f(x_2)\| \le \gamma \|x_1 - x_2\|$$

for any $x_1, x_2 \in \mathbb{R}^d$. In this book, $\| \cdot \|$ denotes a vector or matrix norm.

Example 3.3. *We present three examples to demonstrate fixed points and contraction mappings.*

◇ $x = f(x) = 0.5x$, $x \in \mathbb{R}$.

  It is easy to verify that $x = 0$ is a fixed point since $0 = 0.5 \cdot 0$. Moreover, $f(x) = 0.5x$ is a contraction mapping because $\|0.5x_1 - 0.5x_2\| = 0.5\|x_1 - x_2\| \le \gamma\|x_1 - x_2\|$ for any $\gamma \in [0.5, 1)$.

◇ $x = f(x) = Ax$, where $x \in \mathbb{R}^n$, $A \in \mathbb{R}^{n \times n}$ and $\|A\| \le \gamma < 1$.

  It is easy to verify that $x = 0$ is a fixed point since $0 = A0$. To see the contraction property, $\|Ax_1 - Ax_2\| = \|A(x_1 - x_2)\| \le \|A\|\|x_1 - x_2\| \le \gamma\|x_1 - x_2\|$. Therefore, $f(x) = Ax$ is a contraction mapping.

◇ $x = f(x) = 0.5\sin x$, $x \in \mathbb{R}$.

  It is easy to see that $x = 0$ is a fixed point since $0 = 0.5\sin 0$. Moreover, it follows from the mean value theorem [7, 8] that

$$\left| \frac{0.5\sin x_1 - 0.5\sin x_2}{x_1 - x_2} \right| = |0.5\cos x_3| \le 0.5, \quad x_3 \in [x_1, x_2].$$

  As a result, $|0.5\sin x_1 - 0.5\sin x_2| \le 0.5|x_1 - x_2|$ and hence $f(x) = 0.5\sin x$ is a contraction mapping. □

The relationship between a fixed point and the contraction property is characterized by the following classic theorem.

Theorem 3.1 (Contraction mapping theorem). *For any equation that has the form $x = f(x)$ where x and $f(x)$ are real vectors, if f is a contraction mapping, then the following properties hold.*

◇ *Existence: There exists a fixed point x^* satisfying $f(x^*) = x^*$.*

◇ *Uniqueness: The fixed point x^* is unique.*

◇ *Algorithm: Consider the iterative process:*

$$x_{k+1} = f(x_k),$$

where $k = 0, 1, 2, \ldots$. Then, $x_k \to x^*$ as $k \to \infty$ for any initial guess x_0. Moreover, the convergence rate is exponentially fast.

The contraction mapping theorem not only can tell whether the solution of a nonlinear equation exists but also suggests a numerical algorithm for solving the equation. The proof of the theorem is given in Box 3.1.

The following example demonstrates how to calculate the fixed points of some equations using the iterative algorithm suggested by the contraction mapping theorem.

Example 3.4. *Let us revisit the abovementioned examples: $x = 0.5x$, $x = Ax$, and $x = 0.5 \sin x$. While it has been shown that the right-hand sides of these three equations are all contraction mappings, it follows from the contraction mapping theorem that they each have a unique fixed point, which can be easily verified to be $x^* = 0$. Moreover, the fixed points of the three equations can be iteratively solved by the following algorithms:*

$$x_{k+1} = 0.5x_k,$$
$$x_{k+1} = Ax_k,$$
$$x_{k+1} = 0.5 \sin x_k,$$

given any initial guess x_0. □

Box 3.1: Proof of the contraction mapping theorem

Part 1: We prove that the sequence $\{x_k\}_{k=1}^{\infty}$ with $x_k = f(x_{k-1})$ is convergent.

The proof relies on *Cauchy sequences*. A sequence $x_1, x_2, \ldots$ is called *Cauchy* if for any small $\varepsilon > 0$, there exists N such that $\|x_m - x_n\| < \varepsilon$ for all $m, n > N$. The intuitive interpretation is that there exists a finite integer N such that all the elements after N are sufficiently close to each other. Cauchy sequences are important because it is guaranteed that a Cauchy sequence converges to a limit. Its convergence property will be used to prove the contraction mapping theorem. Note that we must have $\|x_m - x_n\| < \varepsilon$ for all $m, n > N$. If we simply have $x_{n+1} - x_n \to 0$, it is insufficient to claim that the sequence is a Cauchy sequence. For example, it holds that $x_{n+1} - x_n \to 0$ for $x_n = \sqrt{n}$, but apparently, $x_n = \sqrt{n}$ diverges.

We next show that $\{x_k = f(x_{k-1})\}_{k=1}^{\infty}$ is a Cauchy sequence and hence converges.

First, since f is a contraction mapping, we have

$$\|x_{k+1} - x_k\| = \|f(x_k) - f(x_{k-1})\| \leq \gamma \|x_k - x_{k-1}\|.$$

Similarly, we have $\|x_k - x_{k-1}\| \leq \gamma \|x_{k-1} - x_{k-2}\|$, ..., $\|x_2 - x_1\| \leq \gamma \|x_1 - x_0\|$. Thus, we have

$$\|x_{k+1} - x_k\| \leq \gamma \|x_k - x_{k-1}\|$$
$$\leq \gamma^2 \|x_{k-1} - x_{k-2}\|$$
$$\vdots$$
$$\leq \gamma^k \|x_1 - x_0\|.$$

Since $\gamma < 1$, we know that $\|x_{k+1} - x_k\|$ converges to zero exponentially fast as $k \to \infty$ given any x_1, x_0. Notably, the convergence of $\{\|x_{k+1} - x_k\|\}$ is not sufficient for implying the convergence of $\{x_k\}$. Therefore, we need to further consider $\|x_m - x_n\|$ for any $m > n$. In particular,

$$\|x_m - x_n\| = \|x_m - x_{m-1} + x_{m-1} - \cdots - x_{n+1} + x_{n+1} - x_n\|$$
$$\leq \|x_m - x_{m-1}\| + \cdots + \|x_{n+1} - x_n\|$$
$$\leq \gamma^{m-1} \|x_1 - x_0\| + \cdots + \gamma^n \|x_1 - x_0\|$$
$$= \gamma^n (\gamma^{m-1-n} + \cdots + 1) \|x_1 - x_0\|$$
$$\leq \gamma^n (1 + \cdots + \gamma^{m-1-n} + \gamma^{m-n} + \gamma^{m-n+1} + \ldots) \|x_1 - x_0\|$$
$$= \frac{\gamma^n}{1 - \gamma} \|x_1 - x_0\|. \tag{3.4}$$

As a result, for any ε, we can always find N such that $\|x_m - x_n\| < \varepsilon$ for all $m, n > N$. Therefore, this sequence is Cauchy and hence converges to a limit point denoted as $x^* = \lim_{k \to \infty} x_k$.

Part 2: We show that the limit $x^* = \lim_{k \to \infty} x_k$ is a fixed point. To do that, since

$$\|f(x_k) - x_k\| = \|x_{k+1} - x_k\| \leq \gamma^k \|x_1 - x_0\|,$$

we know that $\|f(x_k) - x_k\|$ converges to zero exponentially fast. Hence, we have $f(x^*) = x^*$ at the limit.

Part 3: We show that the fixed point is unique. Suppose that there is another fixed point x' satisfying $f(x') = x'$. Then,

$$\|x' - x^*\| = \|f(x') - f(x^*)\| \leq \gamma \|x' - x^*\|.$$

Since $\gamma < 1$, this inequality holds if and only if $\|x' - x^*\| = 0$. Therefore, $x' = x^*$.

Part 4: We show that x_k converges to x^* exponentially fast. Recall that $\|x_m - x_n\| \le \frac{\gamma^n}{1-\gamma}\|x_1 - x_0\|$, as proven in (3.4). Since m can be arbitrarily large, we have

$$\|x^* - x_n\| = \lim_{m \to \infty} \|x_m - x_n\| \le \frac{\gamma^n}{1-\gamma}\|x_1 - x_0\|.$$

Since $\gamma < 1$, the error converges to zero exponentially fast as $n \to \infty$.

3.3.4 Contraction property of the right-hand side of the BOE

We next show that $f(v)$ in the BOE in (3.3) is a contraction mapping. Thus, the contraction mapping theorem introduced in the previous subsection can be applied.

Theorem 3.2 (Contraction property of $f(v)$). *The function $f(v)$ on the right-hand side of the BOE in (3.3) is a contraction mapping. In particular, for any $v_1, v_2 \in \mathbb{R}^{|\mathcal{S}|}$, it holds that*

$$\|f(v_1) - f(v_2)\|_\infty \le \gamma\|v_1 - v_2\|_\infty,$$

where $\gamma \in (0,1)$ is the discount rate, and $\|\cdot\|_\infty$ is the maximum norm, which is the maximum absolute value of the elements of a vector.

The proof of the theorem is given in Box 3.2. This theorem is important because we can use the powerful contraction mapping theorem to analyze the BOE.

Box 3.2: Proof of Theorem 3.2

Consider any two vectors $v_1, v_2 \in \mathbb{R}^{|\mathcal{S}|}$, and suppose that $\pi_1^* \doteq \arg\max_\pi (r_\pi + \gamma P_\pi v_1)$ and $\pi_2^* \doteq \arg\max_\pi (r_\pi + \gamma P_\pi v_2)$. Then,

$$f(v_1) = \max_\pi (r_\pi + \gamma P_\pi v_1) = r_{\pi_1^*} + \gamma P_{\pi_1^*} v_1 \ge r_{\pi_2^*} + \gamma P_{\pi_2^*} v_1,$$

$$f(v_2) = \max_\pi (r_\pi + \gamma P_\pi v_2) = r_{\pi_2^*} + \gamma P_{\pi_2^*} v_2 \ge r_{\pi_1^*} + \gamma P_{\pi_1^*} v_2,$$

where $\ge$ is an elementwise comparison. As a result,

$$\begin{aligned}
f(v_1) - f(v_2) &= r_{\pi_1^*} + \gamma P_{\pi_1^*} v_1 - (r_{\pi_2^*} + \gamma P_{\pi_2^*} v_2)\\
&\le r_{\pi_1^*} + \gamma P_{\pi_1^*} v_1 - (r_{\pi_1^*} + \gamma P_{\pi_1^*} v_2)\\
&= \gamma P_{\pi_1^*}(v_1 - v_2).
\end{aligned}$$

Similarly, it can be shown that $f(v_2) - f(v_1) \leq \gamma P_{\pi_2^*}(v_2 - v_1)$. Therefore,

$$\gamma P_{\pi_2^*}(v_1 - v_2) \leq f(v_1) - f(v_2) \leq \gamma P_{\pi_1^*}(v_1 - v_2).$$

Define

$$z \doteq \max\left\{|\gamma P_{\pi_2^*}(v_1 - v_2)|, |\gamma P_{\pi_1^*}(v_1 - v_2)|\right\} \in \mathbb{R}^{|\mathcal{S}|},$$

where $\max(\cdot)$, $|\cdot|$, and $\geq$ are all elementwise operators. By definition, $z \geq 0$. On the one hand, it is easy to see that

$$-z \leq \gamma P_{\pi_2^*}(v_1 - v_2) \leq f(v_1) - f(v_2) \leq \gamma P_{\pi_1^*}(v_1 - v_2) \leq z,$$

which implies

$$|f(v_1) - f(v_2)| \leq z.$$

It then follows that

$$\|f(v_1) - f(v_2)\|_\infty \leq \|z\|_\infty, \tag{3.5}$$

where $\|\cdot\|_\infty$ is the maximum norm.

On the other hand, suppose that z_i is the ith entry of z, and p_i^T and q_i^T are the ith row of $P_{\pi_1^*}$ and $P_{\pi_2^*}$, respectively. Then,

$$z_i = \max\{\gamma|p_i^T(v_1 - v_2)|, \gamma|q_i^T(v_1 - v_2)|\}.$$

Since p_i is a vector with all nonnegative elements and the sum of the elements is equal to one, it follows that

$$|p_i^T(v_1 - v_2)| \leq p_i^T|v_1 - v_2| \leq \|v_1 - v_2\|_\infty.$$

Similarly, we have $|q_i^T(v_1 - v_2)| \leq \|v_1 - v_2\|_\infty$. Therefore, $z_i \leq \gamma\|v_1 - v_2\|_\infty$ and hence

$$\|z\|_\infty = \max_i |z_i| \leq \gamma\|v_1 - v_2\|_\infty.$$

Substituting this inequality to (3.5) gives

$$\|f(v_1) - f(v_2)\|_\infty \leq \gamma\|v_1 - v_2\|_\infty,$$

which concludes the proof of the contraction property of $f(v)$.

45

3.4 Solving an optimal policy from the BOE

With the preparation in the last section, we are ready to solve the BOE to obtain the optimal state value v^* and an optimal policy π^*.

$\diamond$ Solving v^*: If v^* is a solution of the BOE, then it satisfies

$$v^* = \max_{\pi \in \Pi}(r_\pi + \gamma P_\pi v^*).$$

Clearly, v^* is a fixed point because $v^* = f(v^*)$. Then, the contraction mapping theorem suggests the following results.

Theorem 3.3 (Existence, uniqueness, and algorithm). *For the BOE $v = f(v) = \max_{\pi \in \Pi}(r_\pi + \gamma P_\pi v)$, there always exists a unique solution v^*, which can be solved iteratively by*

$$v_{k+1} = f(v_k) = \max_{\pi \in \Pi}(r_\pi + \gamma P_\pi v_k), \quad k = 0, 1, 2, \ldots.$$

The value of v_k converges to v^* exponentially fast as $k \to \infty$ given any initial guess v_0.

The proof of this theorem directly follows from the contraction mapping theorem since $f(v)$ is a contraction mapping. This theorem is important because it answers some fundamental questions.

- Existence of v^*: The solution of the BOE always exists.

- Uniqueness of v^*: The solution v^* is always unique.

- Algorithm for solving v^*: The value of v^* can be solved by the iterative algorithm suggested by Theorem 3.3. This iterative algorithm has a specific name called *value iteration*. Its implementation will be introduced in detail in Chapter 4. We mainly focus on the fundamental properties of the BOE in the present chapter.

$\diamond$ Solving π^*: Once the value of v^* has been obtained, we can easily obtain π^* by solving

$$\pi^* = \arg \max_{\pi \in \Pi}(r_\pi + \gamma P_\pi v^*). \tag{3.6}$$

The value of π^* will be given in Theorem 3.5. Substituting (3.6) into the BOE yields

$$v^* = r_{\pi^*} + \gamma P_{\pi^*} v^*.$$

Therefore, $v^* = v_{\pi^*}$ is the state value of π^*, and the BOE is a special Bellman equation whose corresponding policy is π^*.

At this point, although we can solve v^* and π^*, it is still unclear whether the solution is optimal. The following theorem reveals the optimality of the solution.

Theorem 3.4 (Optimality of v^* and π^*). *The solution v^* is the optimal state value, and π^* is an optimal policy. That is, for any policy π, it holds that*

$$v^* = v_{\pi^*} \geq v_\pi,$$

where v_π is the state value of π, and $\geq$ is an elementwise comparison.

Now, it is clear why we must study the BOE: its solution corresponds to optimal state values and optimal policies. The proof of the above theorem is given in the following box.

Box 3.3: Proof of Theorem 3.4

For any policy π, it holds that

$$v_\pi = r_\pi + \gamma P_\pi v_\pi.$$

Since

$$v^* = \max_\pi (r_\pi + \gamma P_\pi v^*) = r_{\pi^*} + \gamma P_{\pi^*} v^* \geq r_\pi + \gamma P_\pi v^*,$$

we have

$$v^* - v_\pi \geq (r_\pi + \gamma P_\pi v^*) - (r_\pi + \gamma P_\pi v_\pi) = \gamma P_\pi (v^* - v_\pi).$$

Repeatedly applying the above inequality gives $v^* - v_\pi \geq \gamma P_\pi (v^* - v_\pi) \geq \gamma^2 P_\pi^2 (v^* - v_\pi) \geq \cdots \geq \gamma^n P_\pi^n (v^* - v_\pi)$. It follows that

$$v^* - v_\pi \geq \lim_{n \to \infty} \gamma^n P_\pi^n (v^* - v_\pi) = 0,$$

where the last equality is true because $\gamma < 1$ and P_π^n is a nonnegative matrix with all its elements less than or equal to 1 (because $P_\pi^n \mathbf{1} = \mathbf{1}$). Therefore, $v^* \geq v_\pi$ for any π.

We next examine π^* in (3.6) more closely. In particular, the following theorem shows that there always exists a deterministic greedy policy that is optimal.

Theorem 3.5 (Greedy optimal policy). *For any $s \in \mathcal{S}$, the deterministic greedy policy*

$$\pi^*(a|s) = \begin{cases} 1, & a = a^*(s), \\ 0, & a \neq a^*(s), \end{cases} \tag{3.7}$$

is an optimal policy for solving the BOE. Here,

$$a^*(s) = \arg\max_a q^*(a, s),$$

where

$$q^*(s, a) \doteq \sum_{r \in \mathcal{R}} p(r|s, a)r + \gamma \sum_{s' \in \mathcal{S}} p(s'|s, a)v^*(s').$$

Box 3.4: Proof of Theorem 3.5

While the matrix-vector form of the optimal policy is $\pi^* = \arg\max_\pi (r_\pi + \gamma P_\pi v^*)$, its elementwise form is

$$\pi^*(s) = \arg\max_{\pi \in \Pi} \sum_{a \in \mathcal{A}} \pi(a|s) \underbrace{\left(\sum_{r \in \mathcal{R}} p(r|s, a)r + \gamma \sum_{s' \in \mathcal{S}} p(s'|s, a)v^*(s') \right)}_{q^*(s,a)}, \quad s \in \mathcal{S}.$$

It is clear that $\sum_{a \in \mathcal{A}} \pi(a|s)q^*(s, a)$ is maximized if $\pi(s)$ selects the action with the greatest $q^*(s, a)$.

The policy in (3.7) is called *greedy* because it seeks the actions with the greatest $q^*(s, a)$. Finally, we discuss two important properties of π^*.

⋄ Uniqueness of optimal policies: Although the value of v^* is unique, the optimal policy that corresponds to v^* may not be unique. This can be easily verified by counterexamples. For example, the two policies shown in Figure 3.3 are both optimal.

⋄ Stochasticity of optimal policies: An optimal policy can be either stochastic or deterministic, as demonstrated in Figure 3.3. However, it is certain that there always exists a deterministic optimal policy according to Theorem 3.5.

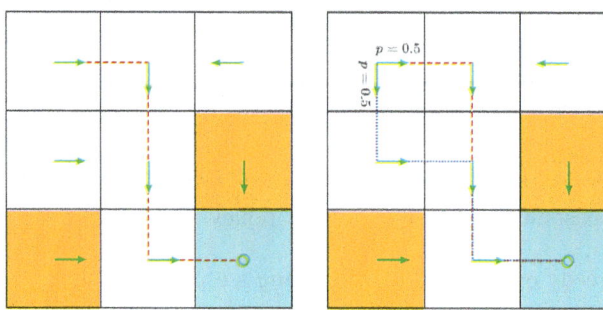

Figure 3.3: Examples for demonstrating that optimal policies may not be unique. The two policies are different but are both optimal.

3.5 Factors that influence optimal policies

The BOE is a powerful tool for analyzing optimal policies. We next apply the BOE to study what factors can influence optimal policies. This question can be easily answered by observing the elementwise expression of the BOE:

$$v(s) = \max_{\pi(s) \in \Pi(s)} \sum_{a \in \mathcal{A}} \pi(a|s) \left(\sum_{r \in \mathcal{R}} p(r|s,a)r + \gamma \sum_{s' \in \mathcal{S}} p(s'|s,a)v(s') \right), \quad s \in \mathcal{S}.$$

The optimal state value and optimal policy are determined by the following parameters: 1) the immediate reward r, 2) the discount rate γ, and 3) the system model $p(s'|s,a), p(r|s,a)$. While the system model is fixed, we next discuss how the optimal policy varies when we change the values of r and γ. All the optimal policies presented in this section can be obtained via the algorithm in Theorem 3.3. The implementation details of the algorithm will be given in Chapter 4. The present chapter mainly focuses on the fundamental properties of optimal policies.

A baseline example

Consider the example in Figure 3.4. The reward settings are $r_{\text{boundary}} = r_{\text{forbidden}} = -1$ and $r_{\text{target}} = 1$. In addition, the agent receives a reward of $r_{\text{other}} = 0$ for every movement step. The discount rate is selected as $\gamma = 0.9$.

With the above parameters, the optimal policy and optimal state values are given in Figure 3.4(a). It is interesting that the agent is not afraid of passing through forbidden areas to reach the target area. More specifically, starting from the state at (row=4, column=1), the agent has two options for reaching the target area. The first option is to avoid all the forbidden areas and travel a long distance to the target area. The second option is to pass through forbidden areas. Although the agent obtains negative rewards when entering forbidden areas, the cumulative reward of the second trajectory is greater than that of the first trajectory. Therefore, the optimal policy is *far-sighted* due to the relatively large value of γ.

Impact of the discount rate

If we change the discount rate from $\gamma = 0.9$ to $\gamma = 0.5$ and keep other parameters unchanged, the optimal policy becomes the one shown in Figure 3.4(b). It is interesting that the agent does not dare to take risks anymore. Instead, it would travel a long distance to reach the target while avoiding all the forbidden areas. This is because the optimal policy becomes *short-sighted* due to the relatively small value of γ.

In the extreme case where $\gamma = 0$, the corresponding optimal policy is shown in Figure 3.4(c). In this case, the agent is not able to reach the target area. This is

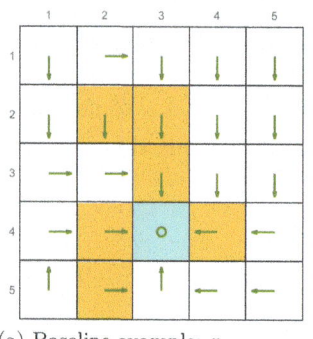

(a) Baseline example: $r_{\text{boundary}} = r_{\text{forbidden}} = -1$, $r_{\text{target}} = 1$, $\gamma = 0.9$.

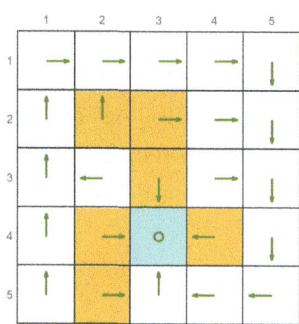

(b) The discount rate is changed to $\gamma = 0.5$. The other parameters are the same as those in (a).

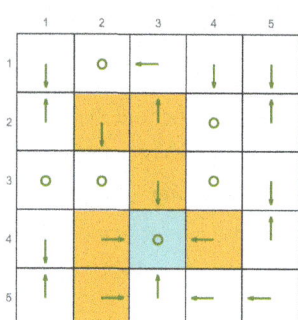
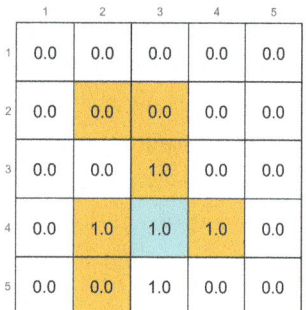

(c) The discount rate is changed to $\gamma = 0$. The other parameters are the same as those in (a).

(d) $r_{\text{forbidden}}$ is changed from -1 to -10. The other parameters are the same as those in (a).

Figure 3.4: The optimal policies and optimal state values given different parameter values.

because the optimal policy for each state is *extremely short-sighted* and merely selects the action with the greatest *immediate* reward instead of the greatest *total* reward.

In addition, the spatial distribution of the state values exhibits an interesting pattern: the states close to the target have greater state values, whereas those far away have lower values. This pattern can be observed from all the examples shown in Figure 3.4. It can be explained by using the discount rate: if a state must travel along a longer trajectory to reach the target, its state value is smaller due to the discount rate.

Impact of the reward values

If we want to strictly prohibit the agent from entering any forbidden area, we can increase the punishment received for doing so. For instance, if $r_{\text{forbidden}}$ is changed from -1 to -10, the resulting optimal policy can avoid all the forbidden areas (see Figure 3.4(d)).

However, changing the rewards does not always lead to different optimal policies. One important fact is that optimal policies are *invariant* to affine transformations of the rewards. In other words, if we scale all the rewards or add the same value to all the rewards, the optimal policy remains the same.

Theorem 3.6 (Optimal policy invariance). *Consider a Markov decision process with $v^* \in \mathbb{R}^{|S|}$ as the optimal state value satisfying $v^* = \max_{\pi \in \Pi}(r_\pi + \gamma P_\pi v^*)$. If every reward $r \in \mathcal{R}$ is changed by an affine transformation to $\alpha r + \beta$, where $\alpha, \beta \in \mathbb{R}$ and $\alpha > 0$, then the corresponding optimal state value v' is also an affine transformation of v^*:*

$$v' = \alpha v^* + \frac{\beta}{1 - \gamma}\mathbf{1}, \tag{3.8}$$

where $\gamma \in (0, 1)$ is the discount rate and $\mathbf{1} = [1, \ldots, 1]^T$. Consequently, the optimal policy derived from v' is invariant to the affine transformation of the reward values.

Box 3.5: Proof of Theorem 3.6

For any policy π, define $r_\pi = [\ldots, r_\pi(s), \ldots]^T$ where

$$r_\pi(s) = \sum_{a \in \mathcal{A}} \pi(a|s) \sum_{r \in \mathcal{R}} p(r|s, a)r, \quad s \in \mathcal{S}.$$

If $r \to \alpha r + \beta$, then $r_\pi(s) \to \alpha r_\pi(s) + \beta$ and hence $r_\pi \to \alpha r_\pi + \beta \mathbf{1}$, where $\mathbf{1} = [1, \ldots, 1]^T$. In this case, the BOE becomes

$$v' = \max_{\pi \in \Pi}(\alpha r_\pi + \beta \mathbf{1} + \gamma P_\pi v'). \tag{3.9}$$

We next solve the new BOE in (3.9) by showing that $v' = \alpha v^* + c\mathbf{1}$ with $c = \beta/(1-\gamma)$ is a solution of (3.9). In particular, substituting $v' = \alpha v^* + c\mathbf{1}$ into (3.9) gives

$$\alpha v^* + c\mathbf{1} = \max_{\pi \in \Pi}(\alpha r_\pi + \beta\mathbf{1} + \gamma P_\pi(\alpha v^* + c\mathbf{1})) = \max_{\pi \in \Pi}(\alpha r_\pi + \beta\mathbf{1} + \alpha\gamma P_\pi v^* + c\gamma\mathbf{1}),$$

where the last equality is due to the fact that $P_\pi\mathbf{1} = \mathbf{1}$. The above equation can be reorganized as

$$\alpha v^* = \max_{\pi \in \Pi}(\alpha r_\pi + \alpha\gamma P_\pi v^*) + \beta\mathbf{1} + c\gamma\mathbf{1} - c\mathbf{1},$$

which is equivalent to

$$\beta\mathbf{1} + c\gamma\mathbf{1} - c\mathbf{1} = 0.$$

Since $c = \beta/(1-\gamma)$, the above equation is valid and hence $v' = \alpha v^* + c\mathbf{1}$ is the solution of (3.9). Since (3.9) is the BOE, v' is also the unique solution. Finally, since v' is an affine transformation of v^*, the relative relationships between the action values remain the same. Hence, the greedy optimal policy derived from v' is the same as that from v^*: $\arg\max_{\pi \in \Pi}(r_\pi + \gamma P_\pi v')$ is the same as $\arg\max_\pi(r_\pi + \gamma P_\pi v^*)$.

Readers may refer to [9] for a further discussion on the conditions under which modifications to the reward values preserve the optimal policy.

Avoiding meaningless detours

In the reward setting, the agent receives a reward of $r_{\text{other}} = 0$ for every movement step (unless it enters a forbidden area or the target area or attempts to go beyond the boundary). Since a zero reward is not a punishment, would the optimal policy take meaningless detours before reaching the target? Should we set r_{other} to be negative to encourage the agent to reach the target as quickly as possible?

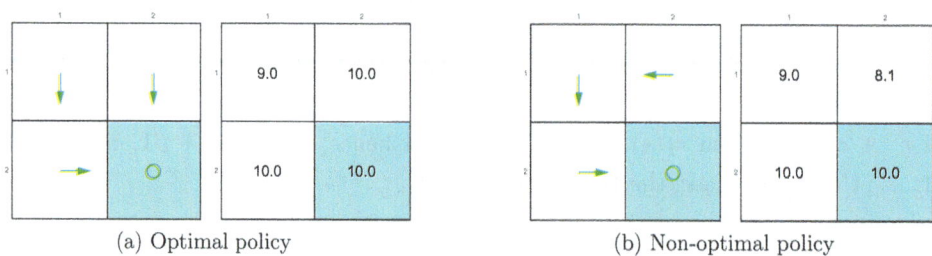

(a) Optimal policy   (b) Non-optimal policy

Figure 3.5: Examples illustrating that optimal policies do not take meaningless detours due to the discount rate.

Consider the examples in Figure 3.5, where the bottom-right cell is the target area to reach. The two policies here are the same except for state s_2. By the policy in Figure 3.5(a), the agent moves downward at s_2 and the resulting trajectory is $s_2 \to s_4$. By the policy in Figure 3.5(b), the agent moves leftward and the resulting trajectory is $s_2 \to s_1 \to s_3 \to s_4$.

It is notable that the second policy takes a detour before reaching the target area. If we merely consider the immediate rewards, taking this detour does not matter because no negative immediate rewards will be obtained. However, if we consider the discounted return, then this detour matters. In particular, for the first policy, the discounted return is

$$\text{return} = 1 + \gamma 1 + \gamma^2 1 + \cdots = 1/(1 - \gamma) = 10.$$

As a comparison, the discounted return for the second policy is

$$\text{return} = 0 + \gamma 0 + \gamma^2 1 + \gamma^3 1 + \cdots = \gamma^2/(1 - \gamma) = 8.1.$$

It is clear that the shorter the trajectory is, the greater the return is. Therefore, although the immediate reward of every step does not encourage the agent to approach the target as quickly as possible, the discount rate does encourage it to do so.

A misunderstanding that beginners may have is that adding a negative reward (e.g., -1) on top of the rewards obtained for every movement is necessary to encourage the agent to reach the target as quickly as possible. This is a misunderstanding because adding the same reward on top of all rewards is an affine transformation, which preserves the optimal policy. Moreover, optimal policies do not take meaningless detours due to the discount rate, even though a detour may not receive any immediate negative rewards.

3.6 Summary

The core concepts in this chapter include optimal policies and optimal state values. In particular, a policy is optimal if its state values are greater than or equal to those of any other policy. The state values of an optimal policy are the optimal state values. The BOE is the core tool for analyzing optimal policies and optimal state values. This equation is a nonlinear equation with a nice contraction property. We can apply the contraction mapping theorem to analyze this equation. It was shown that the solutions of the BOE correspond to the optimal state value and optimal policy. This is the reason why we need to study the BOE.

The contents of this chapter are important for thoroughly understanding many fundamental ideas of reinforcement learning. For example, Theorem 3.3 suggests an iterative algorithm for solving the BOE. This algorithm is exactly the value iteration algorithm that will be introduced in Chapter 4. A further discussion about the BOE can be found in [2].

3.7 Q&A

⋄ Q: What is the definition of optimal policies?

A: A policy is optimal if its corresponding state values are greater than or equal to any other policy.

It should be noted that this specific definition of optimality is valid only for tabular reinforcement learning algorithms. When the values or policies are approximated by functions, different metrics must be used to define optimal policies. This will become clearer in Chapters 8 and 9.

⋄ Q: Why is the Bellman optimality equation important?

A: It is important because it characterizes both optimal policies and optimal state values. Solving this equation yields an optimal policy and the corresponding optimal state value.

⋄ Q: Is the Bellman optimality equation a Bellman equation?

A: Yes. The Bellman optimality equation is a special Bellman equation whose corresponding policy is optimal.

⋄ Q: Is the solution of the Bellman optimality equation unique?

A: The Bellman optimality equation has two unknown variables. The first unknown variable is a value, and the second is a policy. The value solution, which is the optimal state value, is unique. The policy solution, which is an optimal policy, may not be unique.

⋄ Q: What is the key property of the Bellman optimality equation for analyzing its solution?

A: The key property is that the right-hand side of the Bellman optimality equation is a contraction mapping. As a result, we can apply the contraction mapping theorem to analyze its solution.

⋄ Q: Do optimal policies exist?

A: Yes. Optimal policies always exist according to the analysis of the BOE.

⋄ Q: Are optimal policies unique?

A: No. There may exist multiple or infinite optimal policies that have the same optimal state values.

⋄ Q: Are optimal policies stochastic or deterministic?

A: An optimal policy can be either deterministic or stochastic. A nice fact is that there always exist deterministic greedy optimal policies.

◇ Q: How to obtain an optimal policy?

A: Solving the BOE using the iterative algorithm suggested by Theorem 3.3 yields an optimal policy. The detailed implementation of this iterative algorithm will be given in Chapter 4. Notably, all the reinforcement learning algorithms introduced in this book aim to obtain optimal policies under different settings.

◇ Q: What is the general impact on the optimal policies if we reduce the value of the discount rate?

A: The optimal policy becomes more short-sighted when we reduce the discount rate. That is, the agent does not dare to take risks even though it may obtain greater cumulative rewards afterward.

◇ Q: What happens if we set the discount rate to zero?

A: The resulting optimal policy would become extremely short-sighted. The agent would take the action with the greatest immediate reward, even though that action is not good in the long run.

◇ Q: If we increase all the rewards by the same amount, will the optimal state value change? Will the optimal policy change?

A: Increasing all the rewards by the same amount is an affine transformation of the rewards, which would not affect the optimal policies. However, the optimal state value would increase, as shown in (3.8).

◇ Q: If we hope that the optimal policy can avoid meaningless detours before reaching the target, should we add a negative reward to every step so that the agent reaches the target as quickly as possible?

A: First, introducing an additional negative reward to every step is an affine transformation of the rewards, which does not change the optimal policy. Second, the discount rate can automatically encourage the agent to reach the target as quickly as possible. This is because meaningless detours would increase the trajectory length and reduce the discounted return.

Chapter 4

Value Iteration and Policy Iteration

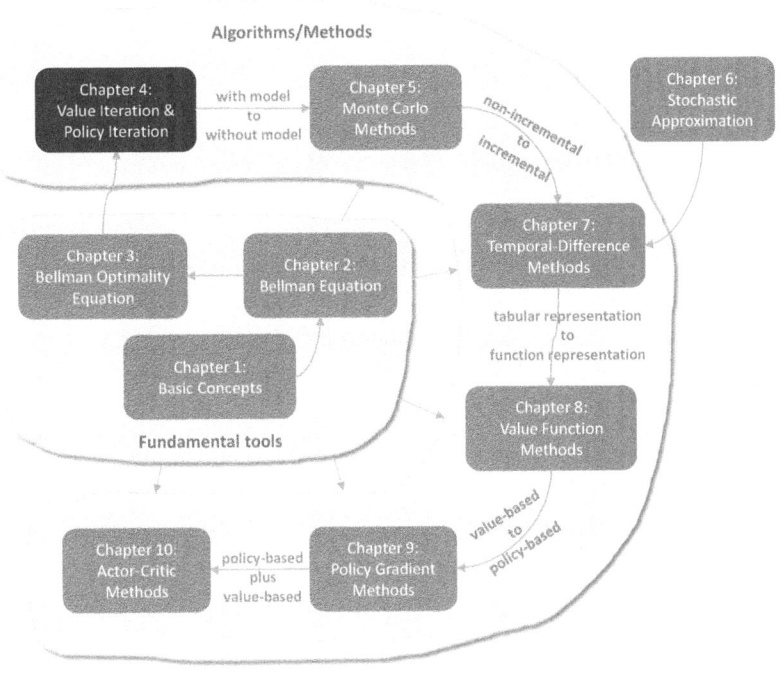

Figure 4.1: Where we are in this book.

With the preparation in the previous chapters, we are now ready to present the first algorithms that can find optimal policies. This chapter introduces three algorithms that are closely related to each other. The first is the value iteration algorithm, which is exactly the algorithm suggested by the contraction mapping theorem for solving the Bellman optimality equation as discussed in the last chapter. We focus more on the implementation details of this algorithm in the present chapter. The second is the policy iteration algorithm, whose idea is widely used in reinforcement learning algorithms. The third is the truncated policy iteration algorithm, which is a unified algorithm that includes the value iteration and policy iteration algorithms as special cases.

S. Zhao, *Mathematical Foundations of Reinforcement Learning*, https://doi.org/10.1007/978-981-97-3944-8_4

The algorithms introduced in this chapter are called *dynamic programming* algorithms [10, 11], which require the system model. These algorithms are important foundations of the model-free reinforcement learning algorithms introduced in the subsequent chapters. For example, the Monte Carlo algorithms introduced in Chapter 5 can be immediately obtained by extending the policy iteration algorithm introduced in this chapter.

4.1   Value iteration

This section introduces the *value iteration* algorithm. It is exactly the algorithm suggested by the contraction mapping theorem for solving the Bellman optimality equation, as introduced in the last chapter (Theorem 3.3). In particular, the algorithm is

$$v_{k+1} = \max_{\pi \in \Pi}(r_\pi + \gamma P_\pi v_k), \quad k = 0, 1, 2, \ldots$$

It is guaranteed by Theorem 3.3 that v_k and π_k converge to the optimal state value and an optimal policy as $k \to \infty$, respectively.

This algorithm is iterative and has two steps in every iteration.

◇ The first step in every iteration is a *policy update* step. Mathematically, it aims to find a policy that can solve the following optimization problem:

$$\pi_{k+1} = \arg\max_{\pi}(r_\pi + \gamma P_\pi v_k),$$

where v_k is obtained in the previous iteration.

◇ The second step is called a *value update* step. Mathematically, it calculates a new value v_{k+1} by

$$v_{k+1} = r_{\pi_{k+1}} + \gamma P_{\pi_{k+1}} v_k, \tag{4.1}$$

where v_{k+1} will be used in the next iteration.

The value iteration algorithm introduced above is in a matrix-vector form. To implement this algorithm, we need to further examine its elementwise form. While the matrix-vector form is useful for understanding the core idea of the algorithm, the elementwise form is necessary for explaining the implementation details.

4.1.1   Elementwise form and implementation

Consider the time step k and a state s.

◇ First, the elementwise form of the *policy update step* $\pi_{k+1} = \arg\max_{\pi}(r_{\pi} + \gamma P_{\pi} v_k)$ is

$$\pi_{k+1}(s) = \arg\max_{\pi} \sum_a \pi(a|s) \underbrace{\left(\sum_r p(r|s,a)r + \gamma \sum_{s'} p(s'|s,a)v_k(s') \right)}_{q_k(s,a)}, \quad s \in \mathcal{S}.$$

We showed in Section 3.3.1 that the optimal policy that can solve the above optimization problem is

$$\pi_{k+1}(a|s) = \begin{cases} 1, & a = a_k^*(s), \\ 0, & a \neq a_k^*(s), \end{cases} \quad (4.2)$$

where $a_k^*(s) = \arg\max_a q_k(s,a)$. If $a_k^*(s) = \arg\max_a q_k(s,a)$ has multiple solutions, we can select any of them without affecting the convergence of the algorithm. Since the new policy π_{k+1} selects the action with the greatest $q_k(s,a)$, such a policy is called greedy.

◇ Second, the elementwise form of the *value update step* $v_{k+1} = r_{\pi_{k+1}} + \gamma P_{\pi_{k+1}} v_k$ is

$$v_{k+1}(s) = \sum_a \pi_{k+1}(a|s) \underbrace{\left(\sum_r p(r|s,a)r + \gamma \sum_{s'} p(s'|s,a)v_k(s') \right)}_{q_k(s,a)}, \quad s \in \mathcal{S}.$$

Substituting (4.2) into the above equation gives

$$v_{k+1}(s) = \max_a q_k(s,a).$$

In summary, the above steps can be illustrated as

$$v_k(s) \to q_k(s,a) \to \text{new greedy policy } \pi_{k+1}(s) \to \text{new value } v_{k+1}(s) = \max_a q_k(s,a)$$

The implementation details are summarized in Algorithm 4.1.

One problem that may be confusing is whether v_k in (4.1) is a state value. The answer is no. Although v_k eventually converges to the optimal state value, it is not ensured to satisfy the Bellman equation of any policy. For example, it does not satisfy $v_k = r_{\pi_{k+1}} + \gamma P_{\pi_{k+1}} v_k$ or $v_k = r_{\pi_k} + \gamma P_{\pi_k} v_k$ in general. It is merely an intermediate value generated by the algorithm. In addition, since v_k is not a state value, q_k is not an action value.

4.1.2 Illustrative examples

We next present an example to illustrate the step-by-step implementation of the value iteration algorithm. This example is a two-by-two grid with one forbidden area (Fig-

---

Algorithm 4.1: Value iteration algorithm

Initialization: The probability models $p(r|s,a)$ and $p(s'|s,a)$ for all (s,a) are known. Initial guess v_0.

Goal: Search for the optimal state value and an optimal policy for solving the Bellman optimality equation.

While v_k has not converged in the sense that $\|v_k - v_{k-1}\|$ is greater than a predefined small threshold, for the kth iteration, do

    For every state $s \in \mathcal{S}$, do

        For every action $a \in \mathcal{A}(s)$, do

            q-value: $q_k(s,a) = \sum_r p(r|s,a)r + \gamma \sum_{s'} p(s'|s,a)v_k(s')$

        Maximum action value: $a_k^*(s) = \arg\max_a q_k(s,a)$

        Policy update: $\pi_{k+1}(a|s) = 1$ if $a = a_k^*$, and $\pi_{k+1}(a|s) = 0$ otherwise

        Value update: $v_{k+1}(s) = \max_a q_k(s,a)$

q-table	a_1	a_2	a_3	a_4	a_5
s_1	$-1 + \gamma v(s_1)$	$-1 + \gamma v(s_2)$	$0 + \gamma v(s_3)$	$-1 + \gamma v(s_1)$	$0 + \gamma v(s_1)$
s_2	$-1 + \gamma v(s_2)$	$-1 + \gamma v(s_2)$	$1 + \gamma v(s_4)$	$0 + \gamma v(s_1)$	$-1 + \gamma v(s_2)$
s_3	$0 + \gamma v(s_1)$	$1 + \gamma v(s_4)$	$-1 + \gamma v(s_3)$	$-1 + \gamma v(s_3)$	$0 + \gamma v(s_3)$
s_4	$-1 + \gamma v(s_2)$	$-1 + \gamma v(s_4)$	$-1 + \gamma v(s_4)$	$0 + \gamma v(s_3)$	$1 + \gamma v(s_4)$

Table 4.1: The expression of $q(s,a)$ for the example as shown in Figure 4.2.

ure 4.2). The target area is s_4. The reward settings are $r_{\text{boundary}} = r_{\text{forbidden}} = -1$ and $r_{\text{target}} = 1$. The discount rate is $\gamma = 0.9$.

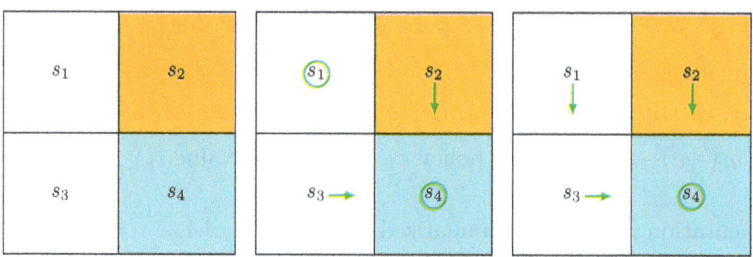

Figure 4.2: An example for demonstrating the implementation of the value iteration algorithm.

The expression of the q-value for each state-action pair is shown in Table 4.1.

$\diamond$ $k = 0$:

Without loss of generality, select the initial values as $v_0(s_1) = v_0(s_2) = v_0(s_3) = v_0(s_4) = 0$.

q-value calculation: Substituting $v_0(s_i)$ into Table 4.1 gives the q-values shown in Table 4.2.

q-table	a_1	a_2	a_3	a_4	a_5
s_1	-1	-1	0	-1	0
s_2	-1	-1	1	0	-1
s_3	0	1	-1	-1	0
s_4	-1	-1	-1	0	1

Table 4.2: The value of $q(s,a)$ at $k=0$.

q-table	a_1	a_2	a_3	a_4	a_5
s_1	$-1+\gamma 0$	$-1+\gamma 1$	$0+\gamma 1$	$-1+\gamma 0$	$0+\gamma 0$
s_2	$-1+\gamma 1$	$-1+\gamma 1$	$1+\gamma 1$	$0+\gamma 0$	$-1+\gamma 1$
s_3	$0+\gamma 0$	$1+\gamma 1$	$-1+\gamma 1$	$-1+\gamma 1$	$0+\gamma 1$
s_4	$-1+\gamma 1$	$-1+\gamma 1$	$-1+\gamma 1$	$0+\gamma 1$	$1+\gamma 1$

Table 4.3: The value of $q(s,a)$ at $k=1$.

Policy update: π_1 is obtained by selecting the actions with the greatest q-values for every state:

$$\pi_1(a_5|s_1) = 1, \quad \pi_1(a_3|s_2) = 1, \quad \pi_1(a_2|s_3) = 1, \quad \pi_1(a_5|s_4) = 1.$$

This policy is visualized in Figure 4.2 (the middle subfigure). It is clear that this policy is not optimal because it selects to stay still at s_1. Notably, the q-values for (s_1, a_5) and (s_1, a_3) are actually the same, and we can randomly select either action.

Value update: v_1 is obtained by updating the v-value to the greatest q-value for each state:

$$v_1(s_1) = 0, \quad v_1(s_2) = 1, \quad v_1(s_3) = 1, \quad v_1(s_4) = 1.$$

⋄ $k=1$:

q-value calculation: Substituting $v_1(s_i)$ into Table 4.1 yields the q-values shown in Table 4.3.

Policy update: π_2 is obtained by selecting the greatest q-values:

$$\pi_2(a_3|s_1) = 1, \quad \pi_2(a_3|s_2) = 1, \quad \pi_2(a_2|s_3) = 1, \quad \pi_2(a_5|s_4) = 1.$$

This policy is visualized in Figure 4.2 (the right subfigure).

Value update: v_2 is obtained by updating the v-value to the greatest q-value for each state:

$$v_2(s_1) = \gamma 1, \quad v_2(s_2) = 1 + \gamma 1, \quad v_2(s_3) = 1 + \gamma 1, \quad v_2(s_4) = 1 + \gamma 1.$$

⋄ $k = 2, 3, 4, \ldots$

It is notable that policy π_2, as illustrated in Figure 4.2, is already optimal. Therefore, we only need to run two iterations to obtain an optimal policy in this simple example. For more complex examples, we need to run more iterations until the value of v_k converges (e.g., until $\|v_{k+1} - v_k\|$ is smaller than a pre-specified threshold).

4.2 Policy iteration

This section presents another important algorithm: *policy iteration*. Unlike value iteration, policy iteration is not for directly solving the Bellman optimality equation. However, it has an intimate relationship with value iteration, as shown later. Moreover, the idea of policy iteration is very important since it is widely utilized in reinforcement learning algorithms.

4.2.1 Algorithm analysis

Policy iteration is an iterative algorithm. Each iteration has two steps.

⋄ The first is a *policy evaluation* step. As its name suggests, this step evaluates a given policy by calculating the corresponding state value. That is to solve the following Bellman equation:

$$v_{\pi_k} = r_{\pi_k} + \gamma P_{\pi_k} v_{\pi_k}, \qquad (4.3)$$

where π_k is the policy obtained in the last iteration and v_{π_k} is the state value to be calculated. The values of r_{π_k} and P_{π_k} can be obtained from the system model.

⋄ The second is a *policy improvement* step. As its name suggests, this step is used to improve the policy. In particular, once v_{π_k} has been calculated in the first step, a new policy π_{k+1} can be obtained as

$$\pi_{k+1} = \arg\max_{\pi}(r_\pi + \gamma P_\pi v_{\pi_k}).$$

Three questions naturally follow the above description of the algorithm.

⋄ In the policy evaluation step, how to solve the state value v_{π_k}?

⋄ In the policy improvement step, why is the new policy π_{k+1} better than π_k?

⋄ Why can this algorithm finally converge to an optimal policy?

We next answer these questions one by one.

In the policy evaluation step, how to calculate v_{π_k}?

We introduced two methods in Chapter 2 for solving the Bellman equation in (4.3). We next revisit the two methods briefly. The first method is a closed-form solution:

$v_{\pi_k} = (I - \gamma P_{\pi_k})^{-1} r_{\pi_k}$. This closed-form solution is useful for theoretical analysis purposes, but it is inefficient to implement since it requires other numerical algorithms to compute the matrix inverse. The second method is an iterative algorithm that can be easily implemented:

$$v_{\pi_k}^{(j+1)} = r_{\pi_k} + \gamma P_{\pi_k} v_{\pi_k}^{(j)}, \quad j = 0, 1, 2, \dots \tag{4.4}$$

where $v_{\pi_k}^{(j)}$ denotes the jth estimate of v_{π_k}. Starting from any initial guess $v_{\pi_k}^{(0)}$, it is ensured that $v_{\pi_k}^{(j)} \to v_{\pi_k}$ as $j \to \infty$. Details can be found in Section 2.7.

Interestingly, policy iteration is an iterative algorithm with another iterative algorithm (4.4) embedded in the policy evaluation step. In theory, this embedded iterative algorithm requires an *infinite* number of steps (that is, $j \to \infty$) to converge to the true state value v_{π_k}. This is, however, impossible to realize. In practice, the iterative process terminates when a certain criterion is satisfied. For example, the termination criterion can be that $\|v_{\pi_k}^{(j+1)} - v_{\pi_k}^{(j)}\|$ is less than a prespecified threshold or that j exceeds a prespecified value. If we do not run an infinite number of iterations, we can only obtain an imprecise value of v_{π_k}, which will be used in the subsequent policy improvement step. Would this cause problems? The answer is no. The reason will become clear when we introduce the truncated policy iteration algorithm later in Section 4.3.

In the policy improvement step, why is π_{k+1} better than π_k?

The policy improvement step can improve the given policy, as shown below.

Lemma 4.1 (Policy improvement). *If $\pi_{k+1} = \arg\max_\pi (r_\pi + \gamma P_\pi v_{\pi_k})$, then $v_{\pi_{k+1}} \geq v_{\pi_k}$.*

Here, $v_{\pi_{k+1}} \geq v_{\pi_k}$ means that $v_{\pi_{k+1}}(s) \geq v_{\pi_k}(s)$ for all s. The proof of this lemma is given in Box 4.1.

Box 4.1: Proof of Lemma 4.1

Since $v_{\pi_{k+1}}$ and v_{π_k} are state values, they satisfy the Bellman equations:

$$v_{\pi_{k+1}} = r_{\pi_{k+1}} + \gamma P_{\pi_{k+1}} v_{\pi_{k+1}},$$
$$v_{\pi_k} = r_{\pi_k} + \gamma P_{\pi_k} v_{\pi_k}.$$

Since $\pi_{k+1} = \arg\max_\pi (r_\pi + \gamma P_\pi v_{\pi_k})$, we know that

$$r_{\pi_{k+1}} + \gamma P_{\pi_{k+1}} v_{\pi_k} \geq r_{\pi_k} + \gamma P_{\pi_k} v_{\pi_k}.$$

It then follows that

$$
\begin{aligned}
v_{\pi_k} - v_{\pi_{k+1}} &= (r_{\pi_k} + \gamma P_{\pi_k} v_{\pi_k}) - (r_{\pi_{k+1}} + \gamma P_{\pi_{k+1}} v_{\pi_{k+1}}) \\
&\leq (r_{\pi_{k+1}} + \gamma P_{\pi_{k+1}} v_{\pi_k}) - (r_{\pi_{k+1}} + \gamma P_{\pi_{k+1}} v_{\pi_{k+1}}) \\
&\leq \gamma P_{\pi_{k+1}} (v_{\pi_k} - v_{\pi_{k+1}}).
\end{aligned}
$$

Therefore,

$$
\begin{aligned}
v_{\pi_k} - v_{\pi_{k+1}} \leq \gamma^2 P_{\pi_{k+1}}^2 (v_{\pi_k} - v_{\pi_{k+1}}) &\leq \ldots \leq \gamma^n P_{\pi_{k+1}}^n (v_{\pi_k} - v_{\pi_{k+1}}) \\
&\leq \lim_{n \to \infty} \gamma^n P_{\pi_{k+1}}^n (v_{\pi_k} - v_{\pi_{k+1}}) = 0.
\end{aligned}
$$

The limit is due to the facts that $\gamma^n \to 0$ as $n \to \infty$ and $P_{\pi_{k+1}}^n$ is a nonnegative stochastic matrix for any n. Here, a stochastic matrix refers to a nonnegative matrix whose row sums are equal to one for all rows.

Why can the policy iteration algorithm eventually find an optimal policy?

The policy iteration algorithm generates two sequences. The first is a sequence of policies: $\{\pi_0, \pi_1, \ldots, \pi_k, \ldots\}$. The second is a sequence of state values: $\{v_{\pi_0}, v_{\pi_1}, \ldots, v_{\pi_k}, \ldots\}$. Suppose that v^* is the optimal state value. Then, $v_{\pi_k} \leq v^*$ for all k. Since the policies are continuously improved according to Lemma 4.1, we know that

$$
v_{\pi_0} \leq v_{\pi_1} \leq v_{\pi_2} \leq \cdots \leq v_{\pi_k} \leq \cdots \leq v^*.
$$

Since v_{π_k} is nondecreasing and always bounded from above by v^*, it follows from the monotone convergence theorem [12] (Appendix C) that v_{π_k} converges to a constant value, denoted as v_∞, when $k \to \infty$. The following analysis shows that $v_\infty = v^*$.

Theorem 4.1 (Convergence of policy iteration). *The state value sequence $\{v_{\pi_k}\}_{k=0}^{\infty}$ generated by the policy iteration algorithm converges to the optimal state value v^*. As a result, the policy sequence $\{\pi_k\}_{k=0}^{\infty}$ converges to an optimal policy.*

The proof of this theorem is given in Box 4.2. The proof not only shows the convergence of the policy iteration algorithm but also reveals the relationship between the policy iteration and value iteration algorithms. Loosely speaking, if both algorithms start from the same initial guess, policy iteration will converge faster than value iteration due to the additional iterations embedded in the policy evaluation step. This point will become clearer when we introduce the truncated policy iteration algorithm in Section 4.3.

Box 4.2: Proof of Theorem 4.1

The idea of the proof is to show that the policy iteration algorithm converges faster than the value iteration algorithm.

In particular, to prove the convergence of $\{v_{\pi_k}\}_{k=0}^{\infty}$, we introduce another sequence $\{v_k\}_{k=0}^{\infty}$ generated by

$$v_{k+1} = f(v_k) = \max_{\pi}(r_{\pi} + \gamma P_{\pi} v_k).$$

This iterative algorithm is exactly the value iteration algorithm. We already know that v_k converges to v^* when given any initial value v_0.

For $k = 0$, we can always find a v_0 such that $v_{\pi_0} \geq v_0$ for any π_0.

We next show that $v_k \leq v_{\pi_k} \leq v^*$ for all k by induction.

For $k \geq 0$, suppose that $v_{\pi_k} \geq v_k$.

For $k + 1$, we have

$$v_{\pi_{k+1}} - v_{k+1} = (r_{\pi_{k+1}} + \gamma P_{\pi_{k+1}} v_{\pi_{k+1}}) - \max_{\pi}(r_{\pi} + \gamma P_{\pi} v_k)$$

$$\geq (r_{\pi_{k+1}} + \gamma P_{\pi_{k+1}} v_{\pi_k}) - \max_{\pi}(r_{\pi} + \gamma P_{\pi} v_k)$$

$$\left(\text{because } v_{\pi_{k+1}} \geq v_{\pi_k} \text{ by Lemma 4.1 and } P_{\pi_{k+1}} \geq 0\right)$$

$$= (r_{\pi_{k+1}} + \gamma P_{\pi_{k+1}} v_{\pi_k}) - (r_{\pi_k'} + \gamma P_{\pi_k'} v_k)$$

$$\left(\text{suppose } \pi_k' = \arg\max_{\pi}(r_{\pi} + \gamma P_{\pi} v_k)\right)$$

$$\geq (r_{\pi_k'} + \gamma P_{\pi_k'} v_{\pi_k}) - (r_{\pi_k'} + \gamma P_{\pi_k'} v_k)$$

$$\left(\text{because } \pi_{k+1} = \arg\max_{\pi}(r_{\pi} + \gamma P_{\pi} v_{\pi_k})\right)$$

$$= \gamma P_{\pi_k'}(v_{\pi_k} - v_k).$$

Since $v_{\pi_k} - v_k \geq 0$ and $P_{\pi_k'}$ is nonnegative, we have $P_{\pi_k'}(v_{\pi_k} - v_k) \geq 0$ and hence $v_{\pi_{k+1}} - v_{k+1} \geq 0$.

Therefore, we can show by induction that $v_k \leq v_{\pi_k} \leq v^*$ for any $k \geq 0$. Since v_k converges to v^*, v_{π_k} also converges to v^*.

4.2.2 Elementwise form and implementation

To implement the policy iteration algorithm, we need to study its elementwise form.

⋄ First, the policy evaluation step solves v_{π_k} from $v_{\pi_k} = r_{\pi_k} + \gamma P_{\pi_k} v_{\pi_k}$ by using the

---

Algorithm 4.2: Policy iteration algorithm

Initialization: The system model, $p(r|s,a)$ and $p(s'|s,a)$ for all (s,a), is known. Initial guess π_0.

Goal: Search for the optimal state value and an optimal policy.

While v_{π_k} has not converged, for the kth iteration, do

    Policy evaluation:

    Initialization: an arbitrary initial guess $v_{\pi_k}^{(0)}$

    While $v_{\pi_k}^{(j)}$ has not converged, for the jth iteration, do

        For every state $s \in \mathcal{S}$, do

$$v_{\pi_k}^{(j+1)}(s) = \sum_a \pi_k(a|s) \left[\sum_r p(r|s,a)r + \gamma \sum_{s'} p(s'|s,a)v_{\pi_k}^{(j)}(s') \right]$$

    Policy improvement:

    For every state $s \in \mathcal{S}$, do

        For every action $a \in \mathcal{A}$, do

$$q_{\pi_k}(s,a) = \sum_r p(r|s,a)r + \gamma \sum_{s'} p(s'|s,a)v_{\pi_k}(s')$$

        $a_k^*(s) = \arg\max_a q_{\pi_k}(s,a)$

        $\pi_{k+1}(a|s) = 1$ if $a = a_k^*$, and $\pi_{k+1}(a|s) = 0$ otherwise

---

iterative algorithm in (4.4). The elementwise form of this algorithm is

$$v_{\pi_k}^{(j+1)}(s) = \sum_a \pi_k(a|s) \left(\sum_r p(r|s,a)r + \gamma \sum_{s'} p(s'|s,a)v_{\pi_k}^{(j)}(s') \right), \quad s \in \mathcal{S},$$

where $j = 0, 1, 2, \ldots$.

⋄ Second, the policy improvement step solves $\pi_{k+1} = \arg\max_\pi (r_\pi + \gamma P_\pi v_{\pi_k})$. The elementwise form of this equation is

$$\pi_{k+1}(s) = \arg\max_\pi \sum_a \pi(a|s) \underbrace{\left(\sum_r p(r|s,a)r + \gamma \sum_{s'} p(s'|s,a)v_{\pi_k}(s') \right)}_{q_{\pi_k}(s,a)}, \quad s \in \mathcal{S},$$

where $q_{\pi_k}(s,a)$ is the action value under policy π_k. Let $a_k^*(s) = \arg\max_a q_{\pi_k}(s,a)$. Then, the greedy optimal policy is

$$\pi_{k+1}(a|s) = \begin{cases} 1, & a = a_k^*(s), \\ 0, & a \neq a_k^*(s). \end{cases}$$

The implementation details are summarized in Algorithm 4.2.

4.2.3 Illustrative examples

A simple example

Consider a simple example shown in Figure 4.3. There are two states with three possible actions: $\mathcal{A} = \{a_\ell, a_0, a_r\}$. The three actions represent moving leftward, staying unchanged, and moving rightward. The reward settings are $r_{\text{boundary}} = -1$ and $r_{\text{target}} = 1$. The discount rate is $\gamma = 0.9$.

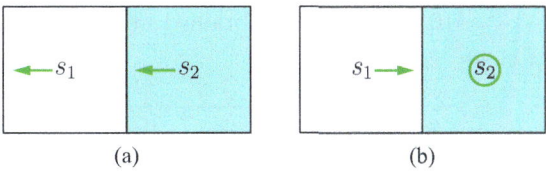

(a)   (b)

Figure 4.3: An example for illustrating the implementation of the policy iteration algorithm.

We next present the implementation of the policy iteration algorithm in a step-by-step manner. When $k = 0$, we start with the initial policy shown in Figure 4.3(a). This policy is not good because it does not move toward the target area. We next show how to apply the policy iteration algorithm to obtain an optimal policy.

◇ First, in the policy evaluation step, we need to solve the Bellman equation:

$$v_{\pi_0}(s_1) = -1 + \gamma v_{\pi_0}(s_1),$$
$$v_{\pi_0}(s_2) = 0 + \gamma v_{\pi_0}(s_1).$$

Since the equation is simple, it can be manually solved that

$$v_{\pi_0}(s_1) = -10, \quad v_{\pi_0}(s_2) = -9.$$

In practice, the equation can be solved by the iterative algorithm in (4.4). For example, select the initial state values as $v_{\pi_0}^{(0)}(s_1) = v_{\pi_0}^{(0)}(s_2) = 0$. It follows from (4.4) that

$$\begin{cases} v_{\pi_0}^{(1)}(s_1) = -1 + \gamma v_{\pi_0}^{(0)}(s_1) = -1, \\ v_{\pi_0}^{(1)}(s_2) = 0 + \gamma v_{\pi_0}^{(0)}(s_1) = 0, \end{cases}$$

$$\begin{cases} v_{\pi_0}^{(2)}(s_1) = -1 + \gamma v_{\pi_0}^{(1)}(s_1) = -1.9, \\ v_{\pi_0}^{(2)}(s_2) = 0 + \gamma v_{\pi_0}^{(1)}(s_1) = -0.9, \end{cases}$$

$$\begin{cases} v_{\pi_0}^{(3)}(s_1) = -1 + \gamma v_{\pi_0}^{(2)}(s_1) = -2.71, \\ v_{\pi_0}^{(3)}(s_2) = 0 + \gamma v_{\pi_0}^{(2)}(s_1) = -1.71, \end{cases}$$

$$\vdots$$

With more iterations, we can see the trend: $v_{\pi_0}^{(j)}(s_1) \to v_{\pi_0}(s_1) = -10$ and $v_{\pi_0}^{(j)}(s_2) \to v_{\pi_0}(s_2) = -9$ as j increases.

◇ Second, in the policy improvement step, the key is to calculate $q_{\pi_0}(s, a)$ for each state-action pair. The following q-table can be used to demonstrate such a process:

$q_{\pi_k}(s, a)$	a_ℓ	a_0	a_r
s_1	$-1 + \gamma v_{\pi_k}(s_1)$	$0 + \gamma v_{\pi_k}(s_1)$	$1 + \gamma v_{\pi_k}(s_2)$
s_2	$0 + \gamma v_{\pi_k}(s_1)$	$1 + \gamma v_{\pi_k}(s_2)$	$-1 + \gamma v_{\pi_k}(s_2)$

Table 4.4: The expression of $q_{\pi_k}(s, a)$ for the example in Figure 4.3.

Substituting $v_{\pi_0}(s_1) = -10, v_{\pi_0}(s_2) = -9$ obtained in the previous policy evaluation step into Table 4.4 yields Table 4.5.

$q_{\pi_0}(s, a)$	a_ℓ	a_0	a_r
s_1	-10	-9	-7.1
s_2	-9	-7.1	-9.1

Table 4.5: The value of $q_{\pi_k}(s, a)$ when $k = 0$.

By seeking the greatest value of q_{π_0}, the improved policy π_1 can be obtained as

$$\pi_1(a_r|s_1) = 1, \quad \pi_1(a_0|s_2) = 1.$$

This policy is illustrated in Figure 4.3(b). It is clear that this policy is optimal.

The above process shows that a single iteration is sufficient for finding the optimal policy in this simple example. More iterations are required for more complex examples.

A more complicated example

We next demonstrate the policy iteration algorithm using a more complicated example shown in Figure 4.4. The reward settings are $r_{\text{boundary}} = -1$, $r_{\text{forbidden}} = -10$, and $r_{\text{target}} = 1$. The discount rate is $\gamma = 0.9$. The policy iteration algorithm can converge to the optimal policy (Figure 4.4(h)) when starting from a random initial policy (Figure 4.4(a)). Two interesting phenomena are observed during the iteration process.

◇ First, if we observe how the policy evolves, an interesting pattern is that the states that are close to the target area find the optimal policies earlier than those far away. Only if the close states can find trajectories to the target first, can the farther states find trajectories passing through the close states to reach the target.

◇ Second, the spatial distribution of the state values exhibits an interesting pattern: the states that are located closer to the target have greater state values. The reason for this pattern is that an agent starting from a farther state must travel for many steps to obtain a positive reward. Such rewards would be severely discounted and hence relatively small.

Figure 4.4: The evolution processes of the policies generated by the policy iteration algorithm.

4.3 Truncated policy iteration

We next introduce a more general algorithm called *truncated policy iteration*. We will see that the value iteration and policy iteration algorithms are two special cases of the truncated policy iteration algorithm.

4.3.1 Comparing value iteration and policy iteration

First of all, we compare the value iteration and policy iteration algorithms by listing their steps as follows.

⬦ Policy iteration: Select an arbitrary initial policy π_0. In the kth iteration, do the following two steps.

- Step 1: Policy evaluation (PE). Given π_k, solve v_{π_k} from

$$v_{\pi_k} = r_{\pi_k} + \gamma P_{\pi_k} v_{\pi_k}.$$

- Step 2: Policy improvement (PI). Given v_{π_k}, solve π_{k+1} from

$$\pi_{k+1} = \arg\max_{\pi}(r_\pi + \gamma P_\pi v_{\pi_k}).$$

⬦ Value iteration: Select an arbitrary initial value v_0. In the kth iteration, do the following two steps.

- Step 1: Policy update (PU). Given v_k, solve π_{k+1} from

$$\pi_{k+1} = \arg\max_{\pi}(r_\pi + \gamma P_\pi v_k).$$

- Step 2: Value update (VU). Given π_{k+1}, solve v_{k+1} from

$$v_{k+1} = r_{\pi_{k+1}} + \gamma P_{\pi_{k+1}} v_k.$$

The above steps of the two algorithms can be illustrated as

$$\text{Policy iteration: } \pi_0 \xrightarrow{PE} v_{\pi_0} \xrightarrow{PI} \pi_1 \xrightarrow{PE} v_{\pi_1} \xrightarrow{PI} \pi_2 \xrightarrow{PE} v_{\pi_2} \xrightarrow{PI} \dots$$
$$\text{Value iteration: } \qquad v_0 \xrightarrow{PU} \pi_1' \xrightarrow{VU} v_1 \xrightarrow{PU} \pi_2' \xrightarrow{VU} v_2 \xrightarrow{PU} \dots$$

It can be seen that the procedures of the two algorithms are very similar.

We examine their value steps more closely to see the difference between the two algorithms. In particular, let both algorithms start from the *same initial condition*: $v_0 = v_{\pi_0}$. The procedures of the two algorithms are listed in Table 4.6. In the first three steps, the two algorithms generate the same results since $v_0 = v_{\pi_0}$. They become

	Policy iteration algorithm	Value iteration algorithm	Comments
1) Policy:	π_0	N/A	
2) Value:	$v_{\pi_0} = r_{\pi_0} + \gamma P_{\pi_0} v_{\pi_0}$	$v_0 \doteq v_{\pi_0}$	
3) Policy:	$\pi_1 = \arg\max_\pi (r_\pi + \gamma P_\pi v_{\pi_0})$	$\pi_1 = \arg\max_\pi (r_\pi + \gamma P_\pi v_0)$	The two policies are the same
4) Value:	$v_{\pi_1} = r_{\pi_1} + \gamma P_{\pi_1} v_{\pi_1}$	$v_1 = r_{\pi_1} + \gamma P_{\pi_1} v_0$	$v_{\pi_1} \geq v_1$ since $v_{\pi_1} \geq v_{\pi_0}$
5) Policy:	$\pi_2 = \arg\max_\pi (r_\pi + \gamma P_\pi v_{\pi_1})$	$\pi_2' = \arg\max_\pi (r_\pi + \gamma P_\pi v_1)$	
$\vdots$	$\vdots$	$\vdots$	$\vdots$

Table 4.6: A comparison between the implementation steps of policy iteration and value iteration.

different in the fourth step. During the fourth step, the value iteration algorithm executes $v_1 = r_{\pi_1} + \gamma P_{\pi_1} v_0$, which is a one-step calculation, whereas the policy iteration algorithm solves $v_{\pi_1} = r_{\pi_1} + \gamma P_{\pi_1} v_{\pi_1}$, which requires an infinite number of iterations. If we explicitly write out the iterative process for solving $v_{\pi_1} = r_{\pi_1} + \gamma P_{\pi_1} v_{\pi_1}$ in the fourth step, everything becomes clear. By letting $v_{\pi_1}^{(0)} = v_0$, we have

$$v_{\pi_1}^{(0)} = v_0$$
$$\text{value iteration} \leftarrow v_1 \longleftarrow \quad v_{\pi_1}^{(1)} = r_{\pi_1} + \gamma P_{\pi_1} v_{\pi_1}^{(0)}$$
$$v_{\pi_1}^{(2)} = r_{\pi_1} + \gamma P_{\pi_1} v_{\pi_1}^{(1)}$$
$$\vdots$$
$$\text{truncated policy iteration} \leftarrow \bar{v}_1 \longleftarrow \quad v_{\pi_1}^{(j)} = r_{\pi_1} + \gamma P_{\pi_1} v_{\pi_1}^{(j-1)}$$
$$\vdots$$
$$\text{policy iteration} \leftarrow v_{\pi_1} \longleftarrow \quad v_{\pi_1}^{(\infty)} = r_{\pi_1} + \gamma P_{\pi_1} v_{\pi_1}^{(\infty)}$$

The following observations can be obtained from the above process.

⋄ If the iteration is run *only once*, then $v_{\pi_1}^{(1)}$ is actually v_1, as calculated in the value iteration algorithm.

⋄ If the iteration is run *an infinite number of times*, then $v_{\pi_1}^{(\infty)}$ is actually v_{π_1}, as calculated in the policy iteration algorithm.

⋄ If the iteration is run *a finite number of times* (denoted as j_{truncate}), then such an algorithm is called *truncated policy iteration*. It is called *truncated* because the remaining iterations from j_{truncate} to ∞ are truncated.

As a result, the value iteration and policy iteration algorithms can be viewed as two extreme cases of the truncated policy iteration algorithm: value iteration terminates

---

Algorithm 4.3: Truncated policy iteration algorithm

Initialization: The probability models $p(r|s,a)$ and $p(s'|s,a)$ for all (s,a) are known. Initial guess π_0.

Goal: Search for the optimal state value and an optimal policy.

While v_k has not converged, for the kth iteration, do

    Policy evaluation:

    Initialization: select the initial guess as $v_k^{(0)} = v_{k-1}$. The maximum number of iterations is set as j_{truncate}.

    While $j < j_{\text{truncate}}$, do

        For every state $s \in \mathcal{S}$, do

$$v_k^{(j+1)}(s) = \sum_a \pi_k(a|s) \left[\sum_r p(r|s,a)r + \gamma \sum_{s'} p(s'|s,a)v_k^{(j)}(s') \right]$$

    Set $v_k = v_k^{(j_{\text{truncate}})}$

    Policy improvement:

    For every state $s \in \mathcal{S}$, do

        For every action $a \in \mathcal{A}(s)$, do

$$q_k(s,a) = \sum_r p(r|s,a)r + \gamma \sum_{s'} p(s'|s,a)v_k(s')$$

        $a_k^*(s) = \arg\max_a q_k(s,a)$

        $\pi_{k+1}(a|s) = 1$ if $a = a_k^*$, and $\pi_{k+1}(a|s) = 0$ otherwise

---

at $j_{\text{truncate}} = 1$, and policy iteration terminates at $j_{\text{truncate}} = \infty$. It should be noted that, although the above comparison is illustrative, it is based on the condition that $v_{\pi_1}^{(0)} = v_0 = v_{\pi_0}$. The two algorithms cannot be directly compared without this condition.

4.3.2 Truncated policy iteration algorithm

In a nutshell, the truncated policy iteration algorithm is the same as the policy iteration algorithm except that it merely runs a finite number of iterations in the policy evaluation step. Its implementation details are summarized in Algorithm 4.3. It is notable that v_k and $v_k^{(j)}$ in the algorithm are not state values. Instead, they are approximations of the true state values because only a finite number of iterations are executed in the policy evaluation step.

If v_k does not equal v_{π_k}, will the algorithm still be able to find optimal policies? The answer is yes. Intuitively, truncated policy iteration is in between value iteration and policy iteration. On the one hand, it converges faster than the value iteration algorithm because it computes more than one iteration during the policy evaluation step. On the other hand, it converges slower than the policy iteration algorithm because it only computes a finite number of iterations. This intuition is illustrated in Figure 4.5. Such intuition is also supported by the following analysis.

Proposition 4.1 (Value improvement). *Consider the iterative algorithm in the policy*

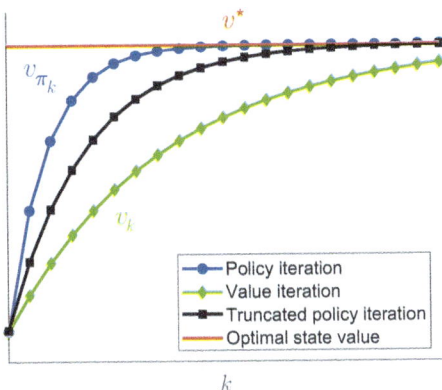

Figure 4.5: An illustration of the relationships between the value iteration, policy iteration, and truncated policy iteration algorithms.

evaluation step:

$$v_{\pi_k}^{(j+1)} = r_{\pi_k} + \gamma P_{\pi_k} v_{\pi_k}^{(j)}, \quad j = 0, 1, 2, \dots$$

If the initial guess is selected as $v_{\pi_k}^{(0)} = v_{\pi_{k-1}}$, it holds that

$$v_{\pi_k}^{(j+1)} \geq v_{\pi_k}^{(j)}$$

for $j = 0, 1, 2, \dots$.

Box 4.3: Proof of Proposition 4.1

First, since $v_{\pi_k}^{(j)} = r_{\pi_k} + \gamma P_{\pi_k} v_{\pi_k}^{(j-1)}$ and $v_{\pi_k}^{(j+1)} = r_{\pi_k} + \gamma P_{\pi_k} v_{\pi_k}^{(j)}$, we have

$$v_{\pi_k}^{(j+1)} - v_{\pi_k}^{(j)} = \gamma P_{\pi_k}(v_{\pi_k}^{(j)} - v_{\pi_k}^{(j-1)}) = \dots = \gamma^j P_{\pi_k}^j (v_{\pi_k}^{(1)} - v_{\pi_k}^{(0)}). \qquad (4.5)$$

Second, since $v_{\pi_k}^{(0)} = v_{\pi_{k-1}}$, we have

$$v_{\pi_k}^{(1)} = r_{\pi_k} + \gamma P_{\pi_k} v_{\pi_k}^{(0)} = r_{\pi_k} + \gamma P_{\pi_k} v_{\pi_{k-1}} \geq r_{\pi_{k-1}} + \gamma P_{\pi_{k-1}} v_{\pi_{k-1}} = v_{\pi_{k-1}} = v_{\pi_k}^{(0)},$$

where the inequality is due to $\pi_k = \arg\max_\pi (r_\pi + \gamma P_\pi v_{\pi_{k-1}})$. Substituting $v_{\pi_k}^{(1)} \geq v_{\pi_k}^{(0)}$ into (4.5) yields $v_{\pi_k}^{(j+1)} \geq v_{\pi_k}^{(j)}$.

Notably, Proposition 4.1 requires the assumption that $v_{\pi_k}^{(0)} = v_{\pi_{k-1}}$. However, $v_{\pi_{k-1}}$ is unavailable in practice, and only v_{k-1} is available. Nevertheless, Proposition 4.1 still sheds light on the convergence of the truncated policy iteration algorithm. A more in-depth discussion of this topic can be found in [2, Section 6.5].

Up to now, the advantages of truncated policy iteration are clear. Compared to the policy iteration algorithm, the truncated one merely requires a finite number of iterations in the policy evaluation step and hence is more computationally efficient. Compared to value iteration, the truncated policy iteration algorithm can speed up its convergence rate by running for a few more iterations in the policy evaluation step.

4.4 Summary

This chapter introduced three algorithms that can be used to find optimal policies.

◇ Value iteration: The value iteration algorithm is the same as the algorithm suggested by the contraction mapping theorem for solving the Bellman optimality equation. It can be decomposed into two steps: value update and policy update.

◇ Policy iteration: The policy iteration algorithm is slightly more complicated than the value iteration algorithm. It also contains two steps: policy evaluation and policy improvement.

◇ Truncated policy iteration: The value iteration and policy iteration algorithms can be viewed as two extreme cases of the truncated policy iteration algorithm.

A common property of the three algorithms is that every iteration has two steps. One step is to update the value, and the other step is to update the policy. The idea of interaction between value and policy updates widely exists in reinforcement learning algorithms. This idea is also called *generalized policy iteration* [3].

Finally, the algorithms introduced in this chapter require the system model. Starting in Chapter 5, we will study model-free reinforcement learning algorithms. We will see that the model-free can be obtained by extending the algorithms introduced in this chapter.

4.5 Q&A

◇ Q: Is the value iteration algorithm guaranteed to find optimal policies?

A: Yes. This is because value iteration is exactly the algorithm suggested by the contraction mapping theorem for solving the Bellman optimality equation in the last chapter. The convergence of this algorithm is guaranteed by the contraction mapping theorem.

◇ Q: Are the intermediate values generated by the value iteration algorithm state values?

A: No. These values are not guaranteed to satisfy the Bellman equation of any policy.

◇ Q: What steps are included in the policy iteration algorithm?

A: Each iteration of the policy iteration algorithm contains two steps: policy evaluation and policy improvement. In the policy evaluation step, the algorithm aims to solve the Bellman equation to obtain the state value of the current policy. In the policy improvement step, the algorithm aims to update the policy so that the newly generated policy has greater state values.

⋄ Q: Is another iterative algorithm embedded in the policy iteration algorithm?

A: Yes. In the policy evaluation step of the policy iteration algorithm, an iterative algorithm is required to solve the Bellman equation of the current policy.

⋄ Q: Are the intermediate values generated by the policy iteration algorithm state values?

A: Yes. This is because these values are the solutions of the Bellman equation of the current policy.

⋄ Q: Is the policy iteration algorithm guaranteed to find optimal policies?

A: Yes. We have presented a rigorous proof of its convergence in this chapter.

⋄ Q: What is the relationship between the truncated policy iteration and policy iteration algorithms?

A: As its name suggests, the truncated policy iteration algorithm can be obtained from the policy iteration algorithm by simply executing a finite number of iterations during the policy evaluation step.

⋄ Q: What is the relationship between truncated policy iteration and value iteration?

A: Value iteration can be viewed as an extreme case of truncated policy iteration, where a single iteration is run during the policy evaluation step.

⋄ Q: Are the intermediate values generated by the truncated policy iteration algorithm state values?

A: No. Only if we run an infinite number of iterations in the policy evaluation step, can we obtain true state values. If we run a finite number of iterations, we can only obtain approximates of the true state values.

⋄ Q: How many iterations should we run in the policy evaluation step of the truncated policy iteration algorithm?

A: The general guideline is to run a few iterations but not too many. The use of a few iterations in the policy evaluation step can speed up the overall convergence rate, but running too many iterations would not significantly speed up the convergence rate.

⋄ Q: What is generalized policy iteration?

A: Generalized policy iteration is not a specific algorithm. Instead, it refers to the general idea of the interaction between value and policy updates. This idea is rooted in the policy iteration algorithm. Most of the reinforcement learning algorithms introduced in this book fall into the scope of generalized policy iteration.

⋄ Q: What are model-based and model-free reinforcement learning?

A: Although the algorithms introduced in this chapter can find optimal policies, they are usually called dynamic programming algorithms rather than reinforcement learning algorithms because they require the system model. Reinforcement learning algorithms can be classified into two categories: model-based and model-free. Here, "model-based" does not refer to the requirement of the system model. Instead, model-based reinforcement learning uses data to estimate the system model and uses this model during the learning process. By contrast, model-free reinforcement learning does not involve model estimation during the learning process. More information about model-based reinforcement learning can be found in [13–16].

Chapter 5

Monte Carlo Methods

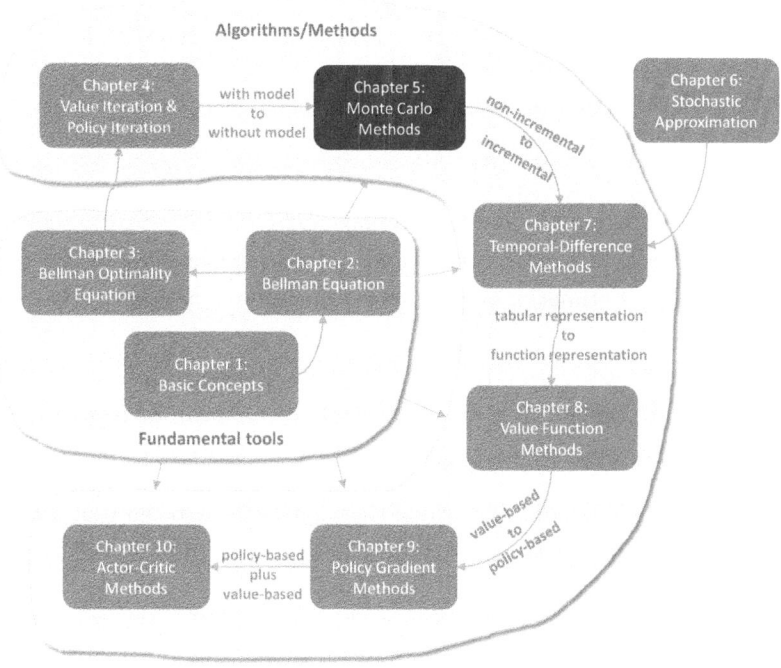

Figure 5.1: Where we are in this book.

In the previous chapter, we introduced algorithms that can find optimal policies based on the system model. In this chapter, we start introducing *model-free* reinforcement learning algorithms that do not presume system models.

While this is the first time we introduce model-free algorithms in this book, we must fill a knowledge gap: how can we find optimal policies without models? The philosophy is simple: If we do not have a model, we must have some data. If we do not have data, we must have a model. If we have neither, then we are not able to find optimal policies. The "data" in reinforcement learning usually refers to the agent's interaction experiences with the environment.

S. Zhao, *Mathematical Foundations of Reinforcement Learning*, https://doi.org/10.1007/978-981-97-3944-8_5

To demonstrate how to learn from data rather than a model, we start this chapter by introducing the *mean estimation* problem, where the expected value of a random variable is estimated from some samples. Understanding this problem is crucial for understanding the fundamental idea of *learning from data*.

Then, we introduce three algorithms based on Monte Carlo (MC) methods. These algorithms can learn optimal policies from experience samples. The first and simplest algorithm is called MC Basic, which can be readily obtained by modifying the policy iteration algorithm introduced in the last chapter. Understanding this algorithm is important for grasping the fundamental idea of MC-based reinforcement learning. By extending this algorithm, we further introduce another two algorithms that are more complicated but more efficient.

5.1 Motivating example: Mean estimation

We next introduce the *mean estimation* problem to demonstrate how to learn from data rather than a model. We will see that mean estimation can be achieved based on *Monte Carlo* methods, which refer to a broad class of techniques that use stochastic samples to solve estimation problems. The reader may wonder why we care about the mean estimation problem. It is simply because state and action values are both defined as the means of returns. Estimating a state or action value is actually a mean estimation problem.

Consider a random variable X that can take values from a finite set of real numbers denoted as $\mathcal{X}$. Suppose that our task is to calculate the mean or expected value of X: $\mathbb{E}[X]$. Two approaches can be used to calculate $\mathbb{E}[X]$.

⬦ The first approach is *model-based*. Here, the model refers to the probability distribution of X. If the model is known, then the mean can be directly calculated based on the definition of the expected value:

$$\mathbb{E}[X] = \sum_{x \in \mathcal{X}} p(x)x.$$

In this book, we use the terms *expected value*, *mean*, and *average* interchangeably.

⬦ The second approach is *model-free*. When the probability distribution (i.e., the model) of X is unknown, suppose that we have some samples $\{x_1, x_2, \ldots, x_n\}$ of X. Then, the mean can be approximated as

$$\mathbb{E}[X] \approx \bar{x} = \frac{1}{n} \sum_{j=1}^{n} x_j.$$

When n is small, this approximation may not be accurate. However, as n increases, the approximation becomes increasingly accurate. When $n \to \infty$, we have $\bar{x} \to \mathbb{E}[X]$.

This is guaranteed by the *law of large numbers*: the average of a large number of samples is close to the expected value. The law of large numbers is introduced in Box 5.1.

The following example illustrates the two approaches described above. Consider a coin flipping game. Let random variable X denote which side is showing when the coin lands. X has two possible values: $X = 1$ when the head is showing, and $X = -1$ when the tail is showing. Suppose that the true probability distribution (i.e., the model) of X is

$$p(X = 1) = 0.5, \quad p(X = -1) = 0.5.$$

If the probability distribution is known in advance, we can directly calculate the mean as

$$\mathbb{E}[X] = 0.5 \cdot 1 + 0.5 \cdot (-1) = 0.$$

If the probability distribution is unknown, then we can flip the coin many times and record the sampling results $\{x_i\}_{i=1}^n$. By calculating the average of the samples, we can obtain an estimate of the mean. As shown in Figure 5.2, the estimated mean becomes increasingly accurate as the number of samples increases.

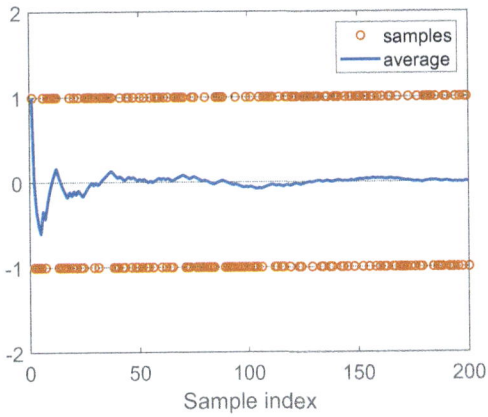

Figure 5.2: An example for demonstrating the law of large numbers. Here, the samples are drawn from $\{+1, -1\}$ following a uniform distribution. The average of the samples gradually converges to zero, which is the true expected value, as the number of samples increases.

It is worth mentioning that the samples used for mean estimation must be *independent and identically distributed* (i.i.d. or iid). Otherwise, if the sampling values correlate, it may be impossible to correctly estimate the expected value. An extreme case is that all the sampling values are the same as the first one, whatever the first one is. In this case, the average of the samples is always equal to the first sample, no matter how many samples we use.

Box 5.1: Law of large numbers

For a random variable X, suppose that $\{x_i\}_{i=1}^n$ are some i.i.d. samples. Let $\bar{x} = \frac{1}{n}\sum_{i=1}^n x_i$ be the average of the samples. Then,

$$\mathbb{E}[\bar{x}] = \mathbb{E}[X],$$

$$\text{var}[\bar{x}] = \frac{1}{n}\text{var}[X].$$

The above two equations indicate that $\bar{x}$ is an unbiased estimate of $\mathbb{E}[X]$, and its variance decreases to zero as n increases to infinity.

The proof is given below.

First, $\mathbb{E}[\bar{x}] = \mathbb{E}\left[\sum_{i=1}^n x_i/n\right] = \sum_{i=1}^n \mathbb{E}[x_i]/n = \mathbb{E}[X]$, where the last equability is due to the fact that the samples are *identically distributed* (that is, $\mathbb{E}[x_i] = \mathbb{E}[X]$).

Second, $\text{var}(\bar{x}) = \text{var}\left[\sum_{i=1}^n x_i/n\right] = \sum_{i=1}^n \text{var}[x_i]/n^2 = (n \cdot \text{var}[X])/n^2 = \text{var}[X]/n$, where the second equality is due to the fact that the samples are *independent*, and the third equability is a result of the samples being *identically distributed* (that is, $\text{var}[x_i] = \text{var}[X]$).

5.2 MC Basic: The simplest MC-based algorithm

This section introduces the first and the simplest MC-based reinforcement learning algorithm. This algorithm is obtained by replacing the *model-based policy evaluation step* in the policy iteration algorithm introduced in Section 4.2 with a *model-free MC estimation step*.

5.2.1 Converting policy iteration to be model-free

There are two steps in every iteration of the policy iteration algorithm (see Section 4.2). The first step is *policy evaluation*, which aims to compute v_{π_k} by solving $v_{\pi_k} = r_{\pi_k} + \gamma P_{\pi_k} v_{\pi_k}$. The second step is *policy improvement*, which aims to compute the greedy policy $\pi_{k+1} = \arg\max_\pi (r_\pi + \gamma P_\pi v_{\pi_k})$. The elementwise form of the policy improvement step is

$$\pi_{k+1}(s) = \arg\max_\pi \sum_a \pi(a|s)\left[\sum_r p(r|s,a)r + \gamma \sum_{s'} p(s'|s,a)v_{\pi_k}(s')\right]$$
$$= \arg\max_\pi \sum_a \pi(a|s)q_{\pi_k}(s,a), \quad s \in \mathcal{S}.$$

It must be noted that the action values lie in the *core* of these two steps. Specifically, in the first step, the state values are calculated for the purpose of calculating the action values. In the second step, the new policy is generated based on the calculated action values. Let us reconsider how we can calculate the action values. Two approaches are available.

◇ The first is a *model-based* approach. This is the approach adopted by the policy iteration algorithm. In particular, we can first calculate the state value v_{π_k} by solving the Bellman equation. Then, we can calculate the action values by using

$$q_{\pi_k}(s, a) = \sum_r p(r|s, a)r + \gamma \sum_{s'} p(s'|s, a)v_{\pi_k}(s'). \tag{5.1}$$

This approach requires the system model $\{p(r|s, a), p(s'|s, a)\}$ to be known.

◇ The second is a *model-free* approach. Recall that the definition of an action value is

$$q_{\pi_k}(s, a) = \mathbb{E}[G_t|S_t = s, A_t = a]$$
$$= \mathbb{E}[R_{t+1} + \gamma R_{t+2} + \gamma^2 R_{t+3} + \ldots |S_t = s, A_t = a],$$

which is the expected return obtained when starting from (s, a). Since $q_{\pi_k}(s, a)$ is an expectation, it can be estimated by MC methods as demonstrated in Section 5.1. To do that, starting from (s, a), the agent can interact with the environment by following policy π_k and then obtain a certain number of episodes. Suppose that there are n episodes and that the return of the ith episode is $g_{\pi_k}^{(i)}(s, a)$. Then, $q_{\pi_k}(s, a)$ can be approximated as

$$q_{\pi_k}(s, a) = \mathbb{E}[G_t|S_t = s, A_t = a] \approx \frac{1}{n} \sum_{i=1}^{n} g_{\pi_k}^{(i)}(s, a). \tag{5.2}$$

We already know that, if the number of episodes n is sufficiently large, the approximation will be sufficiently accurate according to the law of large numbers.

The fundamental idea of MC-based reinforcement learning is to use a model-free method for estimating action values, as shown in (5.2), to replace the model-based method in the policy iteration algorithm.

5.2.2 The MC Basic algorithm

We are now ready to present the first MC-based reinforcement learning algorithm. Starting from an initial policy π_0, the algorithm has two steps in the kth iteration ($k = 0, 1, 2, \ldots$).

◇ *Step 1: Policy evaluation.* This step is used to estimate $q_{\pi_k}(s, a)$ for all (s, a). Specifically, for every (s, a), we collect sufficiently many episodes and use the average of the returns, denoted as $q_k(s, a)$, to approximate $q_{\pi_k}(s, a)$.

Algorithm 5.1: MC Basic (a model-free variant of policy iteration)

Initialization: Initial guess π_0.
Goal: Search for an optimal policy.

For the kth iteration ($k = 0, 1, 2, \dots$), do
    For every state $s \in \mathcal{S}$, do
        For every action $a \in \mathcal{A}(s)$, do
            Collect sufficiently many episodes starting from (s, a) by following π_k
            Policy evaluation:
            $q_{\pi_k}(s, a) \approx q_k(s, a)$ = the average return of all the episodes starting from (s, a)
        Policy improvement:
        $a_k^*(s) = \arg\max_a q_k(s, a)$
        $\pi_{k+1}(a|s) = 1$ if $a = a_k^*$, and $\pi_{k+1}(a|s) = 0$ otherwise

---

$\diamond$ *Step 2: Policy improvement.* This step solves $\pi_{k+1}(s) = \arg\max_\pi \sum_a \pi(a|s) q_k(s, a)$ for all $s \in \mathcal{S}$. The greedy optimal policy is $\pi_{k+1}(a_k^*|s) = 1$ where $a_k^* = \arg\max_a q_k(s, a)$.

This is the simplest MC-based reinforcement learning algorithm, which is called *MC Basic* in this book. The pseudocode of the MC Basic algorithm is given in Algorithm 5.1. As can be seen, it is very similar to the policy iteration algorithm. The only difference is that it calculates action values directly from experience samples, whereas policy iteration calculates state values first and then calculates the action values based on the system model. It should be noted that the model-free algorithm directly estimates action values. Otherwise, if it estimates state values instead, we still need to calculate action values from these state values using the system model, as shown in (5.1).

Since policy iteration is convergent, MC Basic is also convergent when given sufficient samples. That is, for every (s, a), suppose that there are sufficiently many episodes starting from (s, a). Then, the average of the returns of these episodes can accurately approximate the action value of (s, a). In practice, we usually do not have sufficient episodes for every (s, a). As a result, the approximation of the action values may not be accurate. Nevertheless, the algorithm usually can still work. This is similar to the truncated policy iteration algorithm, where the action values are neither accurately calculated.

Finally, MC Basic is too simple to be practical due to its low sample efficiency. The reason why we introduce this algorithm is to let readers grasp the core idea of MC-based reinforcement learning. It is important to understand this algorithm well before studying more complex algorithms introduced later in this chapter. We will see that more complex and sample-efficient algorithms can be readily obtained by extending the MC Basic algorithm.

5.2.3 Illustrative examples

A simple example: A step-by-step implementation

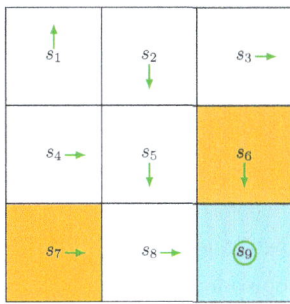

Figure 5.3: An example for illustrating the MC Basic algorithm.

We next use an example to demonstrate the implementation details of the MC Basic algorithm. The reward settings are $r_{\text{boundary}} = r_{\text{forbidden}} = -1$ and $r_{\text{target}} = 1$. The discount rate is $\gamma = 0.9$. The initial policy π_0 is shown in Figure 5.3. This initial policy is not optimal for s_1 or s_3.

While all the action values should be calculated, we merely present those of s_1 due to space limitations. At s_1, there are five possible actions. For each action, we need to collect many episodes that are sufficiently long to effectively approximate the action value. However, since this example is deterministic in terms of both the policy and model, running multiple times would generate the same trajectory. As a result, the estimation of each action value merely requires a single episode.

Following π_0, we can obtain the following episodes by respectively starting from $(s_1, a_1), (s_1, a_2), \ldots, (s_1, a_5)$.

⋄ Starting from (s_1, a_1), the episode is $s_1 \xrightarrow{a_1} s_1 \xrightarrow{a_1} s_1 \xrightarrow{a_1} \ldots$. The action value equals the discounted return of the episode:

$$q_{\pi_0}(s_1, a_1) = -1 + \gamma(-1) + \gamma^2(-1) + \cdots = \frac{-1}{1 - \gamma}.$$

⋄ Starting from (s_1, a_2), the episode is $s_1 \xrightarrow{a_2} s_2 \xrightarrow{a_3} s_5 \xrightarrow{a_3} \ldots$. The action value equals the discounted return of the episode:

$$q_{\pi_0}(s_1, a_2) = 0 + \gamma 0 + \gamma^2 0 + \gamma^3(1) + \gamma^4(1) + \cdots = \frac{\gamma^3}{1 - \gamma}.$$

⋄ Starting from (s_1, a_3), the episode is $s_1 \xrightarrow{a_3} s_4 \xrightarrow{a_2} s_5 \xrightarrow{a_3} \ldots$. The action value equals the discounted return of the episode:

$$q_{\pi_0}(s_1, a_3) = 0 + \gamma 0 + \gamma^2 0 + \gamma^3(1) + \gamma^4(1) + \cdots = \frac{\gamma^3}{1 - \gamma}.$$

⋄ Starting from (s_1, a_4), the episode is $s_1 \xrightarrow{a_4} s_1 \xrightarrow{a_1} s_1 \xrightarrow{a_1} \ldots$. The action value equals the discounted return of the episode:

$$q_{\pi_0}(s_1, a_4) = -1 + \gamma(-1) + \gamma^2(-1) + \cdots = \frac{-1}{1 - \gamma}.$$

⋄ Starting from (s_1, a_5), the episode is $s_1 \xrightarrow{a_5} s_1 \xrightarrow{a_1} s_1 \xrightarrow{a_1} \ldots$. The action value equals the discounted return of the episode:

$$q_{\pi_0}(s_1, a_5) = 0 + \gamma(-1) + \gamma^2(-1) + \cdots = \frac{-\gamma}{1 - \gamma}.$$

By comparing the five action values, we see that

$$q_{\pi_0}(s_1, a_2) = q_{\pi_0}(s_1, a_3) = \frac{\gamma^3}{1 - \gamma} > 0$$

are the maximum values. As a result, the new policy can be obtained as

$$\pi_1(a_2|s_1) = 1 \quad \text{or} \quad \pi_1(a_3|s_1) = 1.$$

It is intuitive that the improved policy, which takes either a_2 or a_3 at s_1, is optimal. Therefore, we can successfully obtain an optimal policy by using merely one iteration for this simple example. In this simple example, the initial policy is already optimal for all the states except s_1 and s_3. Therefore, the policy can become optimal after merely a single iteration. When the policy is nonoptimal for other states, more iterations are needed.

A comprehensive example: Episode length and sparse rewards

We next discuss some interesting properties of the MC Basic algorithm by examining a more comprehensive example. The example is a 5-by-5 grid world (Figure 5.4). The reward settings are $r_{\text{boundary}} = -1$, $r_{\text{forbidden}} = -10$, and $r_{\text{target}} = 1$. The discount rate is $\gamma = 0.9$.

First, we demonstrate that the *episode length* greatly impacts the final optimal policies. In particular, Figure 5.4 shows the final results generated by the MC Basic algorithm with different episode lengths. When the length of each episode is too short, neither the policy nor the value estimate is optimal (see Figures 5.4(a)-(d)). In the extreme case where the episode length is one, only the states that are adjacent to the target have

Figure 5.4: The policies and state values obtained by the MC Basic algorithm when given different episode lengths. Only if the length of each episode is sufficiently long, can the state values be accurately estimated.

nonzero values, and all the other states have zero values since each episode is too short to reach the target or get positive rewards (see Figure 5.4(a)). As the episode length increases, the policy and value estimates gradually approach the optimal ones (see Figure 5.4(h)).

As the episode length increases, an interesting spatial pattern emerges. That is, the states that are closer to the target possess nonzero values earlier than those that are farther away. The reason for this phenomenon is as follows. Starting from a state, the agent must travel at least a certain number of steps to reach the target state and then receive positive rewards. If the length of an episode is less than the minimum desired number of steps, it is certain that the return is zero, and so is the estimated state value. In this example, the episode length must be no less than 15, which is the minimum number of steps required to reach the target when starting from the bottom-left state.

While the above analysis suggests that each episode must be sufficiently long, the episodes are not necessarily infinitely long. As shown in Figure 5.4(g), when the length is 30, the algorithm can find an optimal policy, although the value estimate is not yet optimal.

The above analysis is related to an important reward design problem, *sparse reward*, which refers to the scenario in which no positive rewards can be obtained unless the target is reached. The sparse reward setting requires long episodes that can reach the target. This requirement is challenging to satisfy when the state space is large. As a result, the sparse reward problem downgrades the learning efficiency. One simple technique for solving this problem is to design *nonsparse rewards*. For instance, in the above grid world example, we can redesign the reward setting so that the agent can obtain a small positive reward when reaching the states near the target. In this way, an "attractive field" can be formed around the target so that the agent can find the target more easily. More information about sparse reward problems can be found in [17–19].

5.3 MC Exploring Starts

We next extend the MC Basic algorithm to obtain another MC-based reinforcement learning algorithm that is slightly more complicated but more sample-efficient.

5.3.1 Utilizing samples more efficiently

An important aspect of MC-based reinforcement learning is how to use samples more efficiently. Specifically, suppose that we have an episode of samples obtained by following a policy π:

$$s_1 \xrightarrow{a_2} s_2 \xrightarrow{a_4} s_1 \xrightarrow{a_2} s_2 \xrightarrow{a_3} s_5 \xrightarrow{a_1} \ldots \quad (5.3)$$

where the subscripts refer to the state or action indexes rather than time steps. Every time a state-action pair appears in an episode, it is called a *visit* of that state-action pair. Different strategies can be employed to utilize the visits.

The first and simplest strategy is to use the *initial visit*. That is, an episode is only used to estimate the action value of the initial state-action pair that the episode starts from. For the example in (5.3), the initial-visit strategy merely estimates the action value of (s_1, a_2). The MC Basic algorithm utilizes the initial-visit strategy. However, this strategy is *not sample-efficient* because the episode also visits many other state-action pairs such as (s_2, a_4), (s_2, a_3), and (s_5, a_1). These visits can also be used to estimate the corresponding action values. In particular, we can decompose the episode in (5.3) into multiple subepisodes:

$$s_1 \xrightarrow{a_2} s_2 \xrightarrow{a_4} s_1 \xrightarrow{a_2} s_2 \xrightarrow{a_3} s_5 \xrightarrow{a_1} \ldots \quad \text{[original episode]}$$
$$s_2 \xrightarrow{a_4} s_1 \xrightarrow{a_2} s_2 \xrightarrow{a_3} s_5 \xrightarrow{a_1} \ldots \quad \text{[subepisode starting from } (s_2, a_4)\text{]}$$
$$s_1 \xrightarrow{a_2} s_2 \xrightarrow{a_3} s_5 \xrightarrow{a_1} \ldots \quad \text{[subepisode starting from } (s_1, a_2)\text{]}$$
$$s_2 \xrightarrow{a_3} s_5 \xrightarrow{a_1} \ldots \quad \text{[subepisode starting from } (s_2, a_3)\text{]}$$
$$s_5 \xrightarrow{a_1} \ldots \quad \text{[subepisode starting from } (s_5, a_1)\text{]}$$

The trajectory generated after the visit of a state-action pair can be viewed as a new episode. These new episodes can be used to estimate more action values. In this way, the samples in the episode can be utilized more efficiently.

Moreover, a state-action pair may be visited multiple times in an episode. For example, (s_1, a_2) is visited twice in the episode in (5.3). If we only count the first-time visit, this is called a *first-visit* strategy. If we count every visit of a state-action pair, such a strategy is called *every-visit* [20].

In terms of sample usage efficiency, the every-visit strategy is the best. If an episode is sufficiently long such that it can visit all the state-action pairs many times, then this single episode may be sufficient for estimating all the action values using the every-visit strategy. However, the samples obtained by the every-visit strategy are correlated because the trajectory starting from the second visit is merely a subset of the trajectory starting from the first visit. Nevertheless, the correlation would not be strong if the two visits are far away from each other in the trajectory.

5.3.2   Updating policies more efficiently

Another aspect of MC-based reinforcement learning is when to update the policy. Two strategies are available.

⬦   The first strategy is, in the policy evaluation step, to collect all the episodes starting from the same state-action pair and then approximate the action value using the *average return of these episodes*. This strategy is adopted in the MC Basic algorithm.

Algorithm 5.2: MC Exploring Starts (an efficient variant of MC Basic)

Initialization: Initial policy $\pi_0(a|s)$ and initial value $q(s,a)$ for all (s,a). Returns$(s,a) = 0$ and Num$(s,a) = 0$ for all (s,a).

Goal: Search for an optimal policy.

For each episode, do

    Episode generation: Select a starting state-action pair (s_0, a_0) and ensure that all pairs can be possibly selected (this is the exploring-starts condition). Following the current policy, generate an episode of length T: $s_0, a_0, r_1, \ldots, s_{T-1}, a_{T-1}, r_T$.

    Initialization for each episode: $g \leftarrow 0$

    For each step of the episode, $t = T-1, T-2, \ldots, 0$, do

        $g \leftarrow \gamma g + r_{t+1}$

        Returns$(s_t, a_t) \leftarrow$ Returns$(s_t, a_t) + g$

        Num$(s_t, a_t) \leftarrow$ Num$(s_t, a_t) + 1$

        Policy evaluation:

        $q(s_t, a_t) \leftarrow$ Returns$(s_t, a_t)/$Num(s_t, a_t)

        Policy improvement:

        $\pi(a|s_t) = 1$ if $a = \arg\max_a q(s_t, a)$ and $\pi(a|s_t) = 0$ otherwise

The drawback of this strategy is that the agent must wait until all the episodes have been collected before the estimate can be updated.

$\diamond$ The second strategy, which can overcome this drawback, is to use the *return of a single episode* to approximate the corresponding action value. In this way, we can immediately obtain a rough estimate when we receive an episode. Then, the policy can be improved in an episode-by-episode fashion.

Since the return of a single episode cannot accurately approximate the corresponding action value, one may wonder whether the second strategy is good. In fact, this strategy falls into the scope of *generalized policy iteration* introduced in the last chapter. That is, we can still update the policy even if the value estimate is not sufficiently accurate.

5.3.3 Algorithm description

We can use the techniques introduced in Sections 5.3.1 and 5.3.2 to enhance the efficiency of the MC Basic algorithm. Then, a new algorithm called *MC Exploring Starts* can be obtained.

The details of MC Exploring Starts are given in Algorithm 5.2. This algorithm uses the every-visit strategy. Interestingly, when calculating the discounted return obtained by starting from each state-action pair, the procedure starts from the ending states and travels back to the starting state. Such techniques can make the algorithm more efficient, but it also makes the algorithm more complex. This is why the MC Basic algorithm, which is free of such techniques, is introduced first to reveal the core idea of MC-based reinforcement learning.

The *exploring starts* condition requires sufficiently many episodes starting from *every* state-action pair. Only if every state-action pair is well explored, can we accurately estimate their action values (according to the law of large numbers) and hence successfully find optimal policies. Otherwise, if an action is not well explored, its action value may be inaccurately estimated, and this action may not be selected by the policy even though it is indeed the best action. Both MC Basic and MC Exploring Starts require this condition. However, this condition is difficult to meet in many applications, especially those involving physical interactions with environments. Can we remove the exploring starts requirement? The answer is yes, as shown in the next section.

5.4 MC ϵ-Greedy: Learning without exploring starts

We next extend the MC Exploring Starts algorithm by removing the exploring starts condition. This condition actually requires that every state-action pair can be visited sufficiently many times, which can also be achieved based on soft policies.

5.4.1 ϵ-greedy policies

A policy is *soft* if it has a positive probability of taking any action at any state. Consider an extreme case in which we only have a single episode. With a soft policy, a single episode that is sufficiently long can visit *every* state-action pair many times (see the examples in Figure 5.8). Thus, we do not need to generate a large number of episodes starting from different state-action pairs, and then the exploring starts requirement can be removed.

One type of common soft policies is ϵ-*greedy* policies. An ϵ-greedy policy is a stochastic policy that has a higher chance of choosing the *greedy action* and the same nonzero probability of taking any other action. Here, the greedy action refers to the action with the greatest action value. In particular, suppose that $\epsilon \in [0, 1]$. The corresponding ϵ-greedy policy has the following form:

$$\pi(a|s) = \begin{cases} 1 - \dfrac{\epsilon}{|\mathcal{A}(s)|}(|\mathcal{A}(s)| - 1), & \text{for the greedy action,} \\[2mm] \dfrac{\epsilon}{|\mathcal{A}(s)|}, & \text{for the other } |\mathcal{A}(s)| - 1 \text{ actions,} \end{cases}$$

where $|\mathcal{A}(s)|$ denotes the number of actions associated with s.

When $\epsilon = 0$, ϵ-greedy becomes greedy. When $\epsilon = 1$, the probability of taking any action equals $\frac{1}{|\mathcal{A}(s)|}$.

The probability of taking the greedy action is always greater than that of taking any other action because

$$1 - \frac{\epsilon}{|\mathcal{A}(s)|}(|\mathcal{A}(s)| - 1) = 1 - \epsilon + \frac{\epsilon}{|\mathcal{A}(s)|} \geq \frac{\epsilon}{|\mathcal{A}(s)|}$$

for any $\epsilon \in [0, 1]$.

While an ϵ-greedy policy is stochastic, how can we select an action by following such a policy? We can first generate a random number x in $[0, 1]$ by following a uniform distribution. If $x \geq \epsilon$, then we select the greedy action. If $x < \epsilon$, then we randomly select an action in $\mathcal{A}(s)$ with the probability of $\frac{1}{|\mathcal{A}(s)|}$ (we may select the greedy action again). In this way, the total probability of selecting the greedy action is $1 - \epsilon + \frac{\epsilon}{|\mathcal{A}(s)|}$, and the probability of selecting any other action is $\frac{\epsilon}{|\mathcal{A}(s)|}$.

5.4.2 Algorithm description

To integrate ϵ-greedy policies into MC learning, we only need to change the policy improvement step from greedy to ϵ-greedy.

In particular, the policy improvement step in MC Basic or MC Exploring Starts aims to solve

$$\pi_{k+1}(s) = \arg\max_{\pi \in \Pi} \sum_a \pi(a|s) q_{\pi_k}(s, a), \qquad (5.4)$$

where Π denotes *the set of all possible policies*. We know that the solution of (5.4) is a greedy policy:

$$\pi_{k+1}(a|s) = \begin{cases} 1, & a = a_k^*, \\ 0, & a \neq a_k^*, \end{cases}$$

where $a_k^* = \arg\max_a q_{\pi_k}(s, a)$.

Now, the policy improvement step is changed to solve

$$\pi_{k+1}(s) = \arg\max_{\pi \in \Pi_\epsilon} \sum_a \pi(a|s) q_{\pi_k}(s, a), \qquad (5.5)$$

where Π_ϵ denotes *the set of all ϵ-greedy policies* with a given value of ϵ. In this way, we force the policy to be ϵ-greedy. The solution of (5.5) is

$$\pi_{k+1}(a|s) = \begin{cases} 1 - \frac{|\mathcal{A}(s)| - 1}{|\mathcal{A}(s)|}\epsilon, & a = a_k^*, \\ \frac{1}{|\mathcal{A}(s)|}\epsilon, & a \neq a_k^*, \end{cases}$$

where $a_k^* = \arg\max_a q_{\pi_k}(s, a)$. With the above change, we obtain another algorithm called *MC ϵ-Greedy*. The details of this algorithm are given in Algorithm 5.3. Here, the every-visit strategy is employed to better utilize the samples.

Algorithm 5.3: MC ϵ-Greedy (a variant of MC Exploring Starts)

Initialization: Initial policy $\pi_0(a|s)$ and initial value $q(s,a)$ for all (s,a). Returns$(s,a) = 0$ and Num$(s,a) = 0$ for all (s,a). $\epsilon \in (0,1]$
Goal: Search for an optimal policy.

For each episode, do
    Episode generation: Select a starting state-action pair (s_0, a_0) (the exploring starts condition is not required). Following the current policy, generate an episode of length T: $s_0, a_0, r_1, \ldots, s_{T-1}, a_{T-1}, r_T$.
    Initialization for each episode: $g \leftarrow 0$
    For each step of the episode, $t = T-1, T-2, \ldots, 0$, do
        $g \leftarrow \gamma g + r_{t+1}$
        Returns$(s_t, a_t) \leftarrow$ Returns$(s_t, a_t) + g$
        Num$(s_t, a_t) \leftarrow$ Num$(s_t, a_t) + 1$
        Policy evaluation:
        $q(s_t, a_t) \leftarrow$ Returns$(s_t, a_t)/$Num(s_t, a_t)
        Policy improvement:
        Let $a^* = \arg\max_a q(s_t, a)$ and

$$\pi(a|s_t) = \begin{cases} 1 - \frac{|\mathcal{A}(s_t)|-1}{|\mathcal{A}(s_t)|}\epsilon, & a = a^* \\ \frac{1}{|\mathcal{A}(s_t)|}\epsilon, & a \neq a^* \end{cases}$$

If greedy policies are replaced by ϵ-greedy policies in the policy improvement step, can we still guarantee to obtain optimal policies? The answer is both yes and no. By yes, we mean that, when given sufficient samples, the algorithm can converge to an ϵ-greedy policy that is optimal in the set Π_ϵ. By no, we mean that the policy is merely optimal in Π_ϵ but may not be optimal in Π. However, if ϵ is sufficiently small, the optimal policies in Π_ϵ are close to those in Π.

5.4.3 Illustrative examples

Consider the grid world example shown in Figure 5.5. The aim is to find the optimal policy for every state. A single episode with one million steps is generated in every iteration of the MC ϵ-Greedy algorithm. Here, we deliberately consider the extreme case with merely one single episode. We set $r_{\text{boundary}} = r_{\text{forbidden}} = -1$, $r_{\text{target}} = 1$, and $\gamma = 0.9$.

The initial policy is a uniform policy that has the same probability 0.2 of taking any action, as shown in Figure 5.5. The optimal ϵ-greedy policy with $\epsilon = 0.5$ can be obtained after two iterations. Although each iteration merely uses a single episode, the policy gradually improves because all the state-action pairs can be visited and hence their values can be accurately estimated.

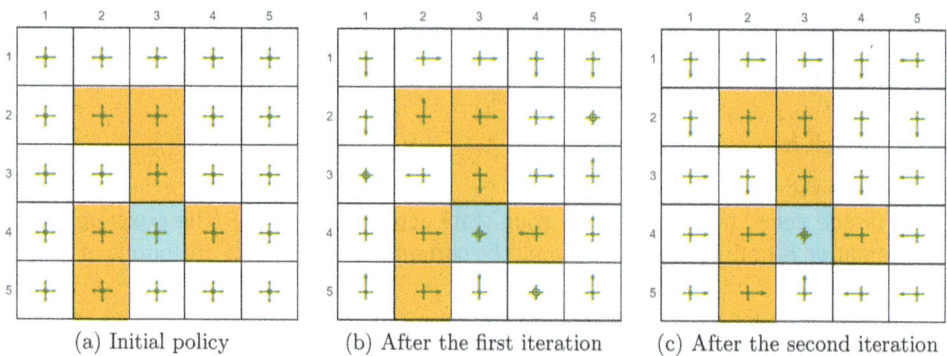

(a) Initial policy   (b) After the first iteration   (c) After the second iteration

Figure 5.5: The evolution process of the MC ϵ-Greedy algorithm based on single episodes.

5.5   Exploration and exploitation of ϵ-greedy policies

Exploration and *exploitation* constitute a fundamental tradeoff in reinforcement learning. Here, exploration means that the policy can possibly take as many actions as possible. In this way, all the actions can be visited and evaluated well. Exploitation means that the improved policy should take the greedy action that has the greatest action value. However, since the action values obtained at the current moment may not be accurate due to insufficient exploration, we should keep exploring while conducting exploitation to avoid missing optimal actions.

ϵ-greedy policies provide one way to balance exploration and exploitation. On the one hand, an ϵ-greedy policy has a higher probability of taking the greedy action so that it can exploit the estimated values. On the other hand, the ϵ-greedy policy also has a chance to take other actions so that it can keep exploring. ϵ-greedy policies are used not only in MC-based reinforcement learning but also in other reinforcement learning algorithms such as temporal-difference learning as introduced in Chapter 7.

Exploitation is related to *optimality* because optimal policies should be greedy. The fundamental idea of ϵ-greedy policies is to enhance exploration by sacrificing optimality/exploitation. If we would like to enhance exploitation and optimality, we need to reduce the value of ϵ. However, if we would like to enhance exploration, we need to increase the value of ϵ.

We next discuss this tradeoff based on some interesting examples. The reinforcement learning task here is a 5-by-5 grid world. The reward settings are $r_{\text{boundary}} = -1$, $r_{\text{forbidden}} = -10$, and $r_{\text{target}} = 1$. The discount rate is $\gamma = 0.9$.

Optimality of ϵ-greedy policies

We next show that the optimality of ϵ-greedy policies becomes worse when ϵ increases.

$\diamond$ First, a greedy optimal policy and the corresponding optimal state values are shown in Figure 5.6(a). The state values of some *consistent* ϵ-greedy policies are shown in

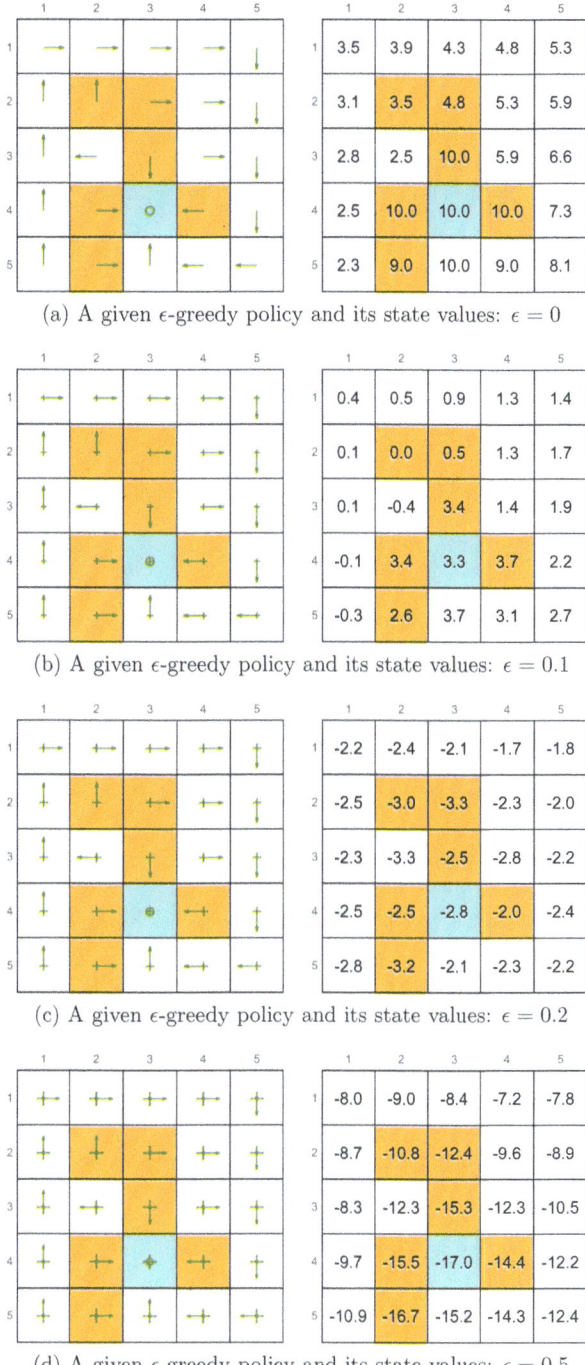

(a) A given ϵ-greedy policy and its state values: $\epsilon = 0$

(b) A given ϵ-greedy policy and its state values: $\epsilon = 0.1$

(c) A given ϵ-greedy policy and its state values: $\epsilon = 0.2$

(d) A given ϵ-greedy policy and its state values: $\epsilon = 0.5$

Figure 5.6: The state values of some ϵ-greedy policies. These ϵ-greedy policies are consistent with each other in the sense that the actions with the greatest probabilities are the same. It can be seen that, when the value of ϵ increases, the state values of the ϵ-greedy policies decrease and hence their optimality becomes worse.

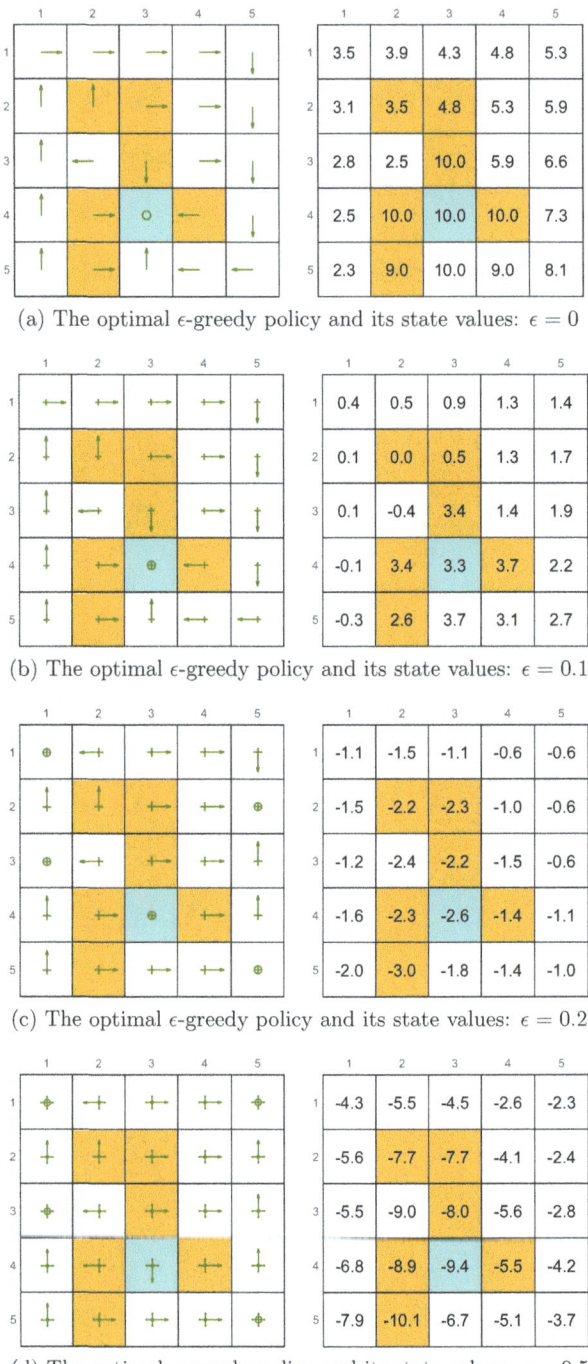

(a) The optimal ϵ-greedy policy and its state values: $\epsilon = 0$

(b) The optimal ϵ-greedy policy and its state values: $\epsilon = 0.1$

(c) The optimal ϵ-greedy policy and its state values: $\epsilon = 0.2$

(d) The optimal ϵ-greedy policy and its state values: $\epsilon = 0.5$

Figure 5.7: The optimal ϵ-greedy policies and their corresponding state values under different values of ϵ. Here, these ϵ-greedy policies are optimal among all ϵ-greedy ones (with the same value of ϵ). It can be seen that, when the value of ϵ increases, the optimal ϵ-greedy policies are no longer consistent with the optimal one as in (a).

Figures 5.6(b)-(d). Here, two ϵ-greedy policies are *consistent* if the actions with the greatest probabilities in the policies are the same.

As the value of ϵ increases, the state values of the ϵ-greedy policies decrease, indicating that the optimality of these ϵ-greedy policies becomes worse. Notably, the value of the target state becomes the smallest when ϵ is as large as 0.5. This is because, when ϵ is large, the agent starting from the target area may enter the surrounding forbidden areas and hence receive negative rewards with a higher probability.

⋄ Second, Figure 5.7 shows the optimal ϵ-greedy policies (they are optimal in Π_ϵ). When $\epsilon = 0$, the policy is greedy and optimal among all policies. When ϵ is as small as 0.1, the optimal ϵ-greedy policy is consistent with the optimal greedy one. However, when ϵ increases to, for example, 0.2, the obtained ϵ-greedy policies are not consistent with the optimal greedy one. Therefore, if we want to obtain ϵ-greedy policies that are consistent with the optimal greedy ones, the value of ϵ should be sufficiently small.

Why are the ϵ-greedy policies inconsistent with the optimal greedy one when ϵ is large? We can answer this question by considering the target state. In the greedy case, the optimal policy at the target state is to stay still to gain positive rewards. However, when ϵ is large, there is a high chance of entering the forbidden areas and receiving negative rewards. Therefore, the optimal policy at the target state in this case is to escape instead of staying still.

Exploration abilities of ϵ-greedy policies

We next illustrate that the exploration ability of an ϵ-greedy policy is strong when ϵ is large.

First, consider an ϵ-greedy policy with $\epsilon = 1$ (see Figure 5.5(a)). In this case, the exploration ability of the ϵ-greedy policy is strong since it has a 0.2 probability of taking any action at any state. Starting from (s_1, a_1), an episode generated by the ϵ-policy is given in Figures 5.8(a)-(c). It can be seen that this single episode can visit all the state-action pairs many times when the episode is sufficiently long due to the strong exploration ability of the policy. Moreover, the numbers of times that all the state-action pairs are visited are almost even, as shown in Figure 5.8(d).

Second, consider an ϵ-policy with $\epsilon = 0.5$ (see Figure 5.6(d)). In this case, the ϵ-greedy policy has a weaker exploration ability than the case of $\epsilon = 1$. Starting from (s_1, a_1), an episode generated by the ϵ-policy is given in Figures 5.8(e)-(g). Although every action can still be visited when the episode is sufficiently long, the distribution of the number of visits may be extremely uneven. For example, given an episode with one million steps, some actions are visited more than 250,000 times, while most actions are visited merely hundreds or even tens of times, as shown in Figure 5.8(h).

The above examples demonstrate that the exploration abilities of ϵ-greedy policies decrease when ϵ decreases. One useful technique is to initially set ϵ to be large to enhance

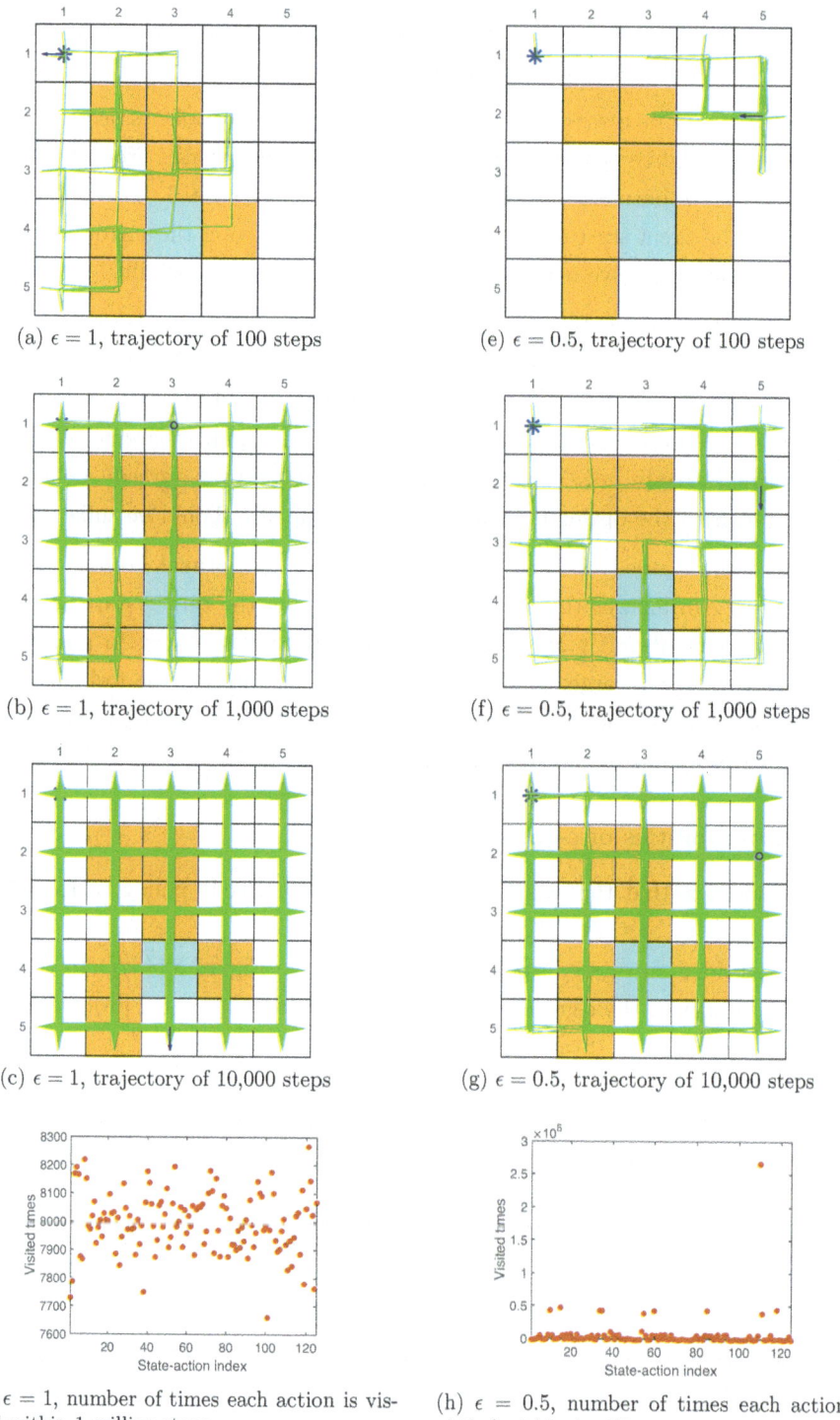

(a) $\epsilon = 1$, trajectory of 100 steps

(e) $\epsilon = 0.5$, trajectory of 100 steps

(b) $\epsilon = 1$, trajectory of 1,000 steps

(f) $\epsilon = 0.5$, trajectory of 1,000 steps

(c) $\epsilon = 1$, trajectory of 10,000 steps

(g) $\epsilon = 0.5$, trajectory of 10,000 steps

(d) $\epsilon = 1$, number of times each action is visited within 1 million steps

(h) $\epsilon = 0.5$, number of times each action is visited within 1 million steps

Figure 5.8: Exploration abilities of ϵ-greedy policies with different values of ϵ.

exploration and gradually reduce it to ensure the optimality of the final policy [21–23].

5.6 Summary

The algorithms in this chapter are the first model-free reinforcement learning algorithms ever introduced in this book. We first introduced the idea of MC estimation by examining an important mean estimation problem. Then, three MC-based algorithms were introduced.

⋄ MC Basic: This is the simplest MC-based reinforcement learning algorithm. This algorithm is obtained by replacing the model-based policy evaluation step in the policy iteration algorithm with a model-free MC-based estimation component. Given sufficient samples, it is guaranteed that this algorithm can converge to optimal policies and optimal state values.

⋄ MC Exploring Starts: This algorithm is a variant of MC Basic. It can be obtained from the MC Basic algorithm using the first-visit or every-visit strategy to use samples more efficiently.

⋄ MC ϵ-Greedy: This algorithm is a variant of MC Exploring Starts. Specifically, in the policy improvement step, it searches for the best ϵ-greedy policies instead of greedy policies. In this way, the exploration ability of the policy is enhanced and hence the condition of exploring starts can be removed.

Finally, a tradeoff between exploration and exploitation was introduced by examining the properties of ϵ-greedy policies. As the value of ϵ increases, the exploration ability of ϵ-greedy policies increases, and the exploitation of greedy actions decreases. On the other hand, if the value of ϵ decreases, we can better exploit the greedy actions, but the exploration ability is compromised.

5.7 Q&A

⋄ Q: What is Monte Carlo estimation?

A: Monte Carlo estimation refers to a broad class of techniques that use stochastic samples to solve approximation problems.

⋄ Q: What is the mean estimation problem?

A: The mean estimation problem refers to calculating the expected value of a random variable based on stochastic samples.

⋄ Q: How to solve the mean estimation problem?

A: There are two approaches: model-based and model-free. In particular, if the probability distribution of a random variable is known, the expected value can be calculated based on its definition. If the probability distribution is unknown, we can use Monte Carlo estimation to approximate the expected value. Such an approximation is accurate when the number of samples is large.

⋄ Q: Why is the mean estimation problem important for reinforcement learning?

A: Both state and action values are defined as expected values of returns. Hence, estimating state or action values is essentially a mean estimation problem.

⋄ Q: What is the core idea of model-free MC-based reinforcement learning?

A: The core idea is to convert the policy iteration algorithm to a model-free one. In particular, while the policy iteration algorithm aims to calculate values based on the system model, MC-based reinforcement learning replaces the model-based policy evaluation step in the policy iteration algorithm with a model-free MC-based policy evaluation step.

⋄ Q: What are initial-visit, first-visit, and every-visit strategies?

A: They are different strategies for utilizing the samples in an episode. An episode may visit many state-action pairs. The initial-visit strategy uses the entire episode to estimate the action value of the initial state-action pair. The every-visit and first-visit strategies can better utilize the given samples. If the rest of the episode is used to estimate the action value of a state-action pair every time it is visited, such a strategy is called every-visit. If we only count the first time a state-action pair is visited in the episode, such a strategy is called first-visit.

⋄ Q: What is exploring starts? Why is it important?

A: Exploring starts requires an infinite number of (or sufficiently many) episodes to be generated when starting from every state-action pair. In theory, the exploring starts condition is necessary to find optimal policies. That is, only if every action value is well explored, can we accurately evaluate all the actions and then correctly select the optimal ones.

⋄ Q: What is the idea used to avoid exploring starts?

A: The fundamental idea is to make policies soft. Soft policies are stochastic, enabling an episode to visit many state-action pairs. In this way, we do not need a large number of episodes starting from every state-action pair.

⋄ Q: Can an ϵ-greedy policy be optimal?

A: The answer is both yes and no. By yes, we mean that, if given sufficient samples, the MC ϵ-Greedy algorithm can converge to an optimal ϵ-greedy policy. By no, we mean that the converged policy is merely optimal among all ϵ-greedy policies (with the same value of ϵ).

⋄ Q: Is it possible to use one episode to visit all state-action pairs?

A: Yes, it is possible. If the policy is soft (e.g., ϵ-greedy) and the episode is sufficiently long.

◇ Q: What is the relationship between MC Basic, MC Exploring Starts, and MC ϵ-Greedy?

A: MC Basic is the simplest MC-based reinforcement learning algorithm. It is important because it reveals the fundamental idea of model-free MC-based reinforcement learning. MC Exploring Starts is a variant of MC Basic that adjusts the sample usage strategy. Furthermore, MC ϵ-Greedy is a variant of MC Exploring Starts that removes the exploring starts requirement. Therefore, while the basic idea is simple, complication appears when we want to achieve better performance. It is important to split the core idea from the complications that may be distracting for beginners.

Chapter 6

Stochastic Approximation

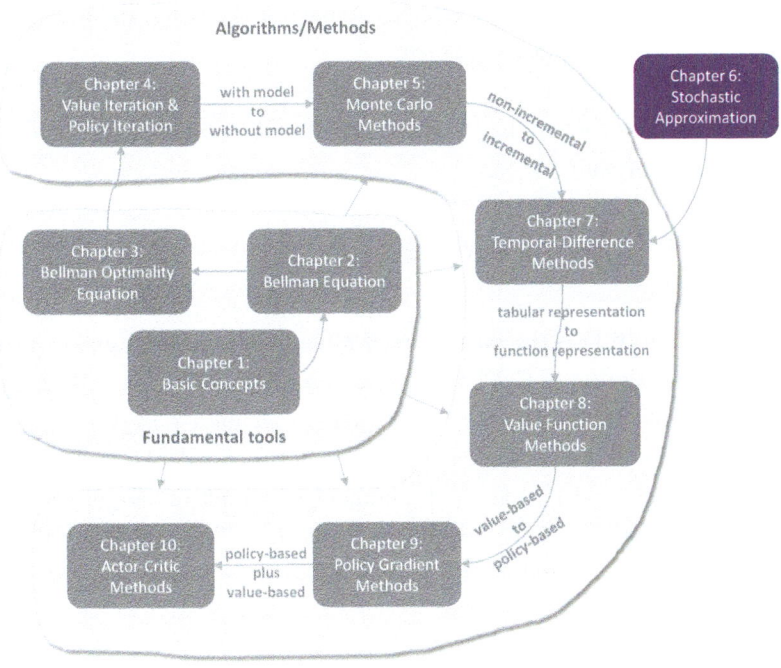

Figure 6.1: Where we are in this book.

Chapter 5 introduced the first class of model-free reinforcement learning algorithms based on Monte Carlo estimation. In the next chapter (Chapter 7), we will introduce another class of model-free reinforcement learning algorithms: temporal-difference learning. However, before proceeding to the next chapter, we need to press the pause button to better prepare ourselves. This is because temporal-difference algorithms are very different from the algorithms that we have studied so far. Many readers who see the temporal-difference algorithms for the first time often wonder how these algorithms were designed in the first place and why they can work effectively. In fact, there is a *knowledge gap* between the previous and subsequent chapters: the algorithms we have studied so far are *non-incremental*, but the algorithms that we will study in the subsequent chapters are *incremental*.

We use the present chapter to fill this knowledge gap by introducing the basics of stochastic approximation. Although this chapter does not introduce any specific reinforcement learning algorithms, it lays the necessary foundations for studying subsequent chapters. We will see in Chapter 7 that the temporal-difference algorithms can be viewed as special stochastic approximation algorithms. The well-known stochastic gradient descent algorithms widely used in machine learning are also introduced in the present chapter.

6.1 Motivating example: Mean estimation

We next demonstrate how to convert a non-incremental algorithm to an incremental one by examining the mean estimation problem.

Consider a random variable X that takes values from a finite set $\mathcal{X}$. Our goal is to estimate $\mathbb{E}[X]$. Suppose that we have a sequence of i.i.d. samples $\{x_i\}_{i=1}^n$. The expected value of X can be approximated by

$$\mathbb{E}[X] \approx \bar{x} \doteq \frac{1}{n} \sum_{i=1}^{n} x_i. \tag{6.1}$$

The approximation in (6.1) is the basic idea of Monte Carlo estimation, as introduced in Chapter 5. We know that $\bar{x} \to \mathbb{E}[X]$ as $n \to \infty$ according to the law of large numbers.

We next show that two methods can be used to calculate $\bar{x}$ in (6.1). The first *non-incremental* method collects all the samples first and then calculates the average. The drawback of such a method is that, if the number of samples is large, we may have to wait for a long time until all of the samples are collected. The second method can avoid this drawback because it calculates the average in an *incremental* manner. Specifically, suppose that

$$w_{k+1} \doteq \frac{1}{k} \sum_{i=1}^{k} x_i, \quad k = 1, 2, \dots$$

and hence

$$w_k = \frac{1}{k-1} \sum_{i=1}^{k-1} x_i, \quad k = 2, 3, \dots$$

Then, w_{k+1} can be expressed in terms of w_k as

$$w_{k+1} = \frac{1}{k} \sum_{i=1}^{k} x_i = \frac{1}{k} \left(\sum_{i=1}^{k-1} x_i + x_k \right) = \frac{1}{k}((k-1)w_k + x_k) = w_k - \frac{1}{k}(w_k - x_k).$$

Therefore, we obtain the following incremental algorithm:

$$w_{k+1} = w_k - \frac{1}{k}(w_k - x_k). \tag{6.2}$$

This algorithm can be used to calculate the mean $\bar{x}$ in an incremental manner. It can be verified that

$$w_1 = x_1,$$

$$w_2 = w_1 - \frac{1}{1}(w_1 - x_1) = x_1,$$

$$w_3 = w_2 - \frac{1}{2}(w_2 - x_2) = x_1 - \frac{1}{2}(x_1 - x_2) = \frac{1}{2}(x_1 + x_2),$$

$$w_4 = w_3 - \frac{1}{3}(w_3 - x_3) = \frac{1}{3}(x_1 + x_2 + x_3),$$

$$\vdots$$

$$w_{k+1} = \frac{1}{k}\sum_{i=1}^{k} x_i. \tag{6.3}$$

The advantage of (6.2) is that the average can be immediately calculated every time we receive a sample. This average can be used to approximate $\bar{x}$ and hence $\mathbb{E}[X]$. Notably, the approximation may not be accurate at the beginning due to insufficient samples. However, it is better than nothing. As more samples are obtained, the estimation accuracy can be gradually improved according to the law of large numbers. In addition, one can also define $w_{k+1} = \frac{1}{1+k}\sum_{i=1}^{k+1} x_i$ and $w_k = \frac{1}{k}\sum_{i=1}^{k} x_i$. Doing so would not make any significant difference. In this case, the corresponding iterative algorithm is $w_{k+1} = w_k - \frac{1}{1+k}(w_k - x_{k+1})$.

Furthermore, consider an algorithm with a more general expression:

$$w_{k+1} = w_k - \alpha_k(w_k - x_k). \tag{6.4}$$

This algorithm is important and frequently used in this chapter. It is the same as (6.2) except that the coefficient $1/k$ is replaced by $\alpha_k > 0$. Since the expression of α_k is not given, we are not able to obtain the explicit expression of w_k as in (6.3). However, we will show in the next section that, if $\{\alpha_k\}$ satisfies some mild conditions, $w_k \to \mathbb{E}[X]$ as $k \to \infty$. In Chapter 7, we will see that temporal-difference algorithms have similar (but more complex) expressions.

6.2 Robbins-Monro algorithm

Stochastic approximation refers to a broad class of stochastic iterative algorithms for solving root-finding or optimization problems [24]. Compared to many other root-finding algorithms such as gradient-based ones, stochastic approximation is powerful in the sense that it does not require the expression of the objective function or its derivative.

The Robbins-Monro (RM) algorithm is a pioneering work in the field of stochastic approximation [24–27]. The famous stochastic gradient descent algorithm is a special form of the RM algorithm, as shown in Section 6.4. We next introduce the details of the RM algorithm.

Suppose that we would like to find the root of the equation

$$g(w) = 0,$$

where $w \in \mathbb{R}$ is the unknown variable and $g : \mathbb{R} \to \mathbb{R}$ is a function. Many problems can be formulated as root-finding problems. For example, if $J(w)$ is an objective function to be optimized, this optimization problem can be converted to solving $g(w) \doteq \nabla_w J(w) = 0$. In addition, an equation such as $g(w) = c$, where c is a constant, can also be converted to the above equation by rewriting $g(w) - c$ as a new function.

If the expression of g or its derivative is known, there are many numerical algorithms that can be used. However, the problem we are facing is that the expression of the function g is unknown. For example, the function may be represented by an artificial neural network whose structure and parameters are unknown. Moreover, we can only obtain a noisy observation of $g(w)$:

$$\tilde{g}(w, \eta) = g(w) + \eta,$$

where $\eta \in \mathbb{R}$ is the observation error, which may or may not be Gaussian. In summary, it is a black-box system where only the input w and the noisy output $\tilde{g}(w, \eta)$ are known (see Figure 6.2). Our aim is to solve $g(w) = 0$ using w and $\tilde{g}$.

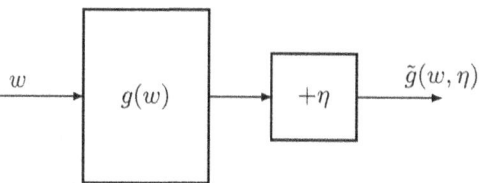

Figure 6.2: An illustration of the problem of solving $g(w) = 0$ from w and $\tilde{g}$.

The RM algorithm that can solve $g(w) = 0$ is

$$w_{k+1} = w_k - a_k \tilde{g}(w_k, \eta_k), \qquad k = 1, 2, 3, \dots \tag{6.5}$$

where w_k is the kth estimate of the root, $\tilde{g}(w_k, \eta_k)$ is the kth noisy observation, and a_k is a positive coefficient. As can be seen, the RM algorithm does not require any information about the function. It only requires the input and output.

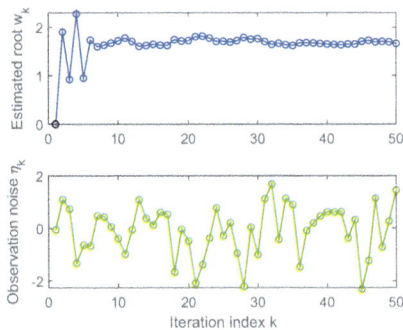

Figure 6.3: An illustrative example of the RM algorithm.

To illustrate the RM algorithm, consider an example in which $g(w) = w^3 - 5$. The true root is $5^{1/3} \approx 1.71$. Now, suppose that we can only observe the input w and the output $\tilde{g}(w) = g(w) + \eta$, where η is i.i.d. and obeys a standard normal distribution with a zero mean and a standard deviation of 1. The initial guess is $w_1 = 0$, and the coefficient is $a_k = 1/k$. The evolution process of w_k is shown in Figure 6.3. Even though the observation is corrupted by noise η_k, the estimate w_k can still converge to the true root. Note that the initial guess w_1 must be properly selected to ensure convergence for the specific function of $g(w) = w^3 - 5$. In the following subsection, we present the conditions under which the RM algorithm converges for any initial guesses.

6.2.1 Convergence properties

Why can the RM algorithm in (6.5) find the root of $g(w) = 0$? We next illustrate the idea with an example and then provide a rigorous convergence analysis.

Consider the example shown in Figure 6.4. In this example, $g(w) = \tanh(w - 1)$. The true root of $g(w) = 0$ is $w^* = 1$. We apply the RM algorithm with $w_1 = 3$ and $a_k = 1/k$. To better illustrate the reason for convergence, we simply set $\eta_k \equiv 0$, and consequently, $\tilde{g}(w_k, \eta_k) = g(w_k)$. The RM algorithm in this case is $w_{k+1} = w_k - a_k g(w_k)$. The resulting $\{w_k\}$ generated by the RM algorithm is shown in Figure 6.4. It can be seen that w_k converges to the true root $w^* = 1$.

This simple example can illustrate why the RM algorithm converges.

⋄ When $w_k > w^*$, we have $g(w_k) > 0$. Then, $w_{k+1} = w_k - a_k g(w_k) < w_k$. If $a_k g(w_k)$ is sufficiently small, we have $w^* < w_{k+1} < w_k$. As a result, w_{k+1} is closer to w^* than w_k.

⋄ When $w_k < w^*$, we have $g(w_k) < 0$. Then, $w_{k+1} = w_k - a_k g(w_k) > w_k$. If $|a_k g(w_k)|$ is sufficiently small, we have $w^* > w_{k+1} > w_k$. As a result, w_{k+1} is closer to w^* than w_k.

In either case, w_{k+1} is closer to w^* than w_k. Therefore, it is intuitive that w_k converges to w^*.

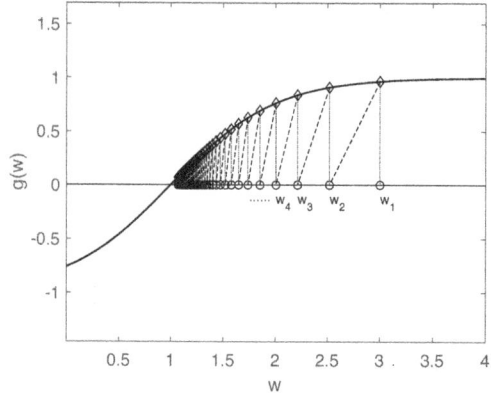

Figure 6.4: An example for illustrating the convergence of the RM algorithm.

The above example is simple since the observation error is assumed to be zero. It would be nontrivial to analyze the convergence in the presence of stochastic observation errors. A rigorous convergence result is given below.

Theorem 6.1 (Robbins-Monro theorem). *In the Robbins-Monro algorithm in (6.5), if*

(a) $0 < c_1 \leq \nabla_w g(w) \leq c_2$ *for all* w;

(b) $\sum_{k=1}^{\infty} a_k = \infty$ *and* $\sum_{k=1}^{\infty} a_k^2 < \infty$;

(c) $\mathbb{E}[\eta_k | \mathcal{H}_k] = 0$ *and* $\mathbb{E}[\eta_k^2 | \mathcal{H}_k] < \infty$;

where $\mathcal{H}_k = \{w_k, w_{k-1}, \dots\}$, *then* w_k *almost surely converges to the root* w^* *satisfying* $g(w^*) = 0$.

We postpone the proof of this theorem to Section 6.3.3. This theorem relies on the notion of *almost sure* convergence, which is introduced in Appendix B.

The three conditions in Theorem 6.1 are explained as follows.

⋄ In the first condition, $0 < c_1 \leq \nabla_w g(w)$ indicates that $g(w)$ is a monotonically increasing function. This condition ensures that the root of $g(w) = 0$ exists and is unique. If $g(w)$ is monotonically decreasing, we can simply treat $-g(w)$ as a new function that is monotonically increasing.

As an application, we can formulate an optimization problem in which the objective function is $J(w)$ as a root-finding problem: $g(w) \doteq \nabla_w J(w) = 0$. In this case, the condition that $g(w)$ is monotonically increasing indicates that $J(w)$ is *convex*, which is a commonly adopted assumption in optimization problems.

The inequality $\nabla_w g(w) \leq c_2$ indicates that the gradient of $g(w)$ is bounded from above. For example, $g(w) = \tanh(w - 1)$ satisfies this condition, but $g(w) = w^3 - 5$ does not.

106

◇ The second condition about $\{a_k\}$ is interesting. We often see conditions like this in reinforcement learning algorithms. In particular, the condition $\sum_{k=1}^{\infty} a_k^2 < \infty$ means that $\lim_{n \to \infty} \sum_{k=1}^{n} a_k^2$ is bounded from above. It requires that a_k converges to zero as $k \to \infty$. The condition $\sum_{k=1}^{\infty} a_k = \infty$ means that $\lim_{n \to \infty} \sum_{k=1}^{n} a_k$ is infinitely large. It requires that a_k should not converge to zero too fast. These conditions have interesting properties, which will be analyzed in detail shortly.

◇ The third condition is mild. It does not require the observation error η_k to be Gaussian. An important special case is that $\{\eta_k\}$ is an i.i.d. stochastic sequence satisfying $\mathbb{E}[\eta_k] = 0$ and $\mathbb{E}[\eta_k^2] < \infty$. In this case, the third condition is valid because η_k is independent of $\mathcal{H}_k$ and hence we have $\mathbb{E}[\eta_k|\mathcal{H}_k] = \mathbb{E}[\eta_k] = 0$ and $\mathbb{E}[\eta_k^2|\mathcal{H}_k] = \mathbb{E}[\eta_k^2]$.

We next examine the second condition about the coefficients $\{a_k\}$ more closely.

◇ Why is the second condition important for the convergence of the RM algorithm?

This question can naturally be answered when we present a rigorous proof of the above theorem later. Here, we would like to provide some insightful intuition.

First, $\sum_{k=1}^{\infty} a_k^2 < \infty$ indicates that $a_k \to 0$ as $k \to \infty$. Why is this condition important? Suppose that the observation $\tilde{g}(w_k, \eta_k)$ is always bounded. Since

$$w_{k+1} - w_k = -a_k \tilde{g}(w_k, \eta_k),$$

if $a_k \to 0$, then $a_k \tilde{g}(w_k, \eta_k) \to 0$ and hence $w_{k+1} - w_k \to 0$, indicating that w_{k+1} and w_k approach each other when $k \to \infty$. Otherwise, if a_k does not converge, then w_k may still fluctuate when $k \to \infty$.

Second, $\sum_{k=1}^{\infty} a_k = \infty$ indicates that a_k should not converge to zero too fast. Why is this condition important? Summarizing both sides of the equations of $w_2 - w_1 = -a_1 \tilde{g}(w_1, \eta_1)$, $w_3 - w_2 = -a_2 \tilde{g}(w_2, \eta_2)$, $w_4 - w_3 = -a_3 \tilde{g}(w_3, \eta_3)$, ... gives

$$w_1 - w_\infty = \sum_{k=1}^{\infty} a_k \tilde{g}(w_k, \eta_k).$$

If $\sum_{k=1}^{\infty} a_k < \infty$, then $|\sum_{k=1}^{\infty} a_k \tilde{g}(w_k, \eta_k)|$ is also bounded. Let b denote the finite upper bound such that

$$|w_1 - w_\infty| = \left| \sum_{k=1}^{\infty} a_k \tilde{g}(w_k, \eta_k) \right| \leq b. \tag{6.6}$$

If the initial guess w_1 is selected far away from w^* so that $|w_1 - w^*| > b$, then it is impossible to have $w_\infty = w^*$ according to (6.6). This suggests that the RM algorithm cannot find the true solution w^* in this case. Therefore, the condition $\sum_{k=1}^{\infty} a_k = \infty$ is necessary to ensure convergency given an *arbitrary* initial guess.

⬦ What kinds of sequences satisfy $\sum_{k=1}^{\infty} a_k = \infty$ and $\sum_{k=1}^{\infty} a_k^2 < \infty$?

One typical sequence is

$$a_k = \frac{1}{k}.$$

On the one hand, it holds that

$$\lim_{n \to \infty} \left(\sum_{k=1}^{n} \frac{1}{k} - \ln n \right) = \kappa,$$

where $\kappa \approx 0.577$ is called the Euler-Mascheroni constant (or Euler's constant) [28]. Since $\ln n \to \infty$ as $n \to \infty$, we have

$$\sum_{k=1}^{\infty} \frac{1}{k} = \infty.$$

In fact, $H_n = \sum_{k=1}^{n} \frac{1}{k}$ is called the harmonic number in number theory [29]. On the other hand, it holds that

$$\sum_{k=1}^{\infty} \frac{1}{k^2} = \frac{\pi^2}{6} < \infty.$$

Finding the value of $\sum_{k=1}^{\infty} \frac{1}{k^2}$ is known as the Basel problem [30].

In summary, the sequence $\{a_k = 1/k\}$ satisfies the second condition in Theorem 6.1. Notably, a slight modification, such as $a_k = 1/(k+1)$ or $a_k = c_k/k$ where c_k is bounded, also preserves this condition.

In the RM algorithm, a_k is often selected as a sufficiently small *constant* in many applications. Although the second condition is not satisfied anymore in this case because $\sum_{k=1}^{\infty} a_k^2 = \infty$ rather than $\sum_{k=1}^{\infty} a_k^2 < \infty$, the algorithm can still converge in a certain sense [24, Section 1.5]. In addition, $g(x) = x^3 - 5$ in the example shown in Figure 6.3 does not satisfy the first condition, but the RM algorithm can still find the root if the initial guess is adequately (not arbitrarily) selected.

6.2.2 Application to mean estimation

We next apply the Robbins-Monro theorem to analyze the mean estimation problem, which has been discussed in Section 6.1. Recall that

$$w_{k+1} = w_k + \alpha_k(x_k - w_k)$$

is the mean estimation algorithm in (6.4). When $\alpha_k = 1/k$, we can obtain the analytical expression of w_{k+1} as $w_{k+1} = 1/k \sum_{i=1}^{k} x_i$. However, we would not be able to obtain an analytical expression when given general values of α_k. In this case, the convergence analysis is nontrivial. We can show that the algorithm in this case is a special RM algorithm and hence its convergence naturally follows.

In particular, define a function as

$$g(w) \doteq w - \mathbb{E}[X].$$

The original problem is to obtain the value of $\mathbb{E}[X]$. This problem is formulated as a root-finding problem to solve $g(w) = 0$. Given a value of w, the noisy observation that we can obtain is $\tilde{g} \doteq w - x$, where x is a sample of X. Note that $\tilde{g}$ can be written as

$$\begin{aligned}
\tilde{g}(w, \eta) &= w - x \\
&= w - x + \mathbb{E}[X] - \mathbb{E}[X] \\
&= (w - \mathbb{E}[X]) + (\mathbb{E}[X] - x) \doteq g(w) + \eta,
\end{aligned}$$

where $\eta \doteq \mathbb{E}[X] - x$.

The RM algorithm for solving this problem is

$$w_{k+1} = w_k - \alpha_k \tilde{g}(w_k, \eta_k) = w_k - \alpha_k(w_k - x_k),$$

which is exactly the algorithm in (6.4). As a result, it is guaranteed by Theorem 6.1 that w_k converges to $\mathbb{E}[X]$ almost surely if $\sum_{k=1}^{\infty} \alpha_k = \infty$, $\sum_{k=1}^{\infty} \alpha_k^2 < \infty$, and $\{x_k\}$ is i.i.d. It is worth mentioning that the convergence property does not rely on any assumption regarding the distribution of X.

6.3 Dvoretzky's convergence theorem

Until now, the convergence of the RM algorithm has not yet been proven. To do that, we next introduce Dvoretzky's theorem [31, 32], which is a classic result in the field of stochastic approximation. This theorem can be used to analyze the convergence of the RM algorithm and many reinforcement learning algorithms.

This section is slightly mathematically intensive. Readers who are interested in the convergence analyses of stochastic algorithms are recommended to study this section. Otherwise, this section can be skipped.

Theorem 6.2 (Dvoretzky's theorem). *Consider a stochastic process*

$$\Delta_{k+1} = (1 - \alpha_k)\Delta_k + \beta_k \eta_k,$$

where $\{\alpha_k\}_{k=1}^{\infty}, \{\beta_k\}_{k=1}^{\infty}, \{\eta_k\}_{k=1}^{\infty}$ are stochastic sequences. Here $\alpha_k \geq 0, \beta_k \geq 0$ for all k. Then, Δ_k converges to zero almost surely if the following conditions are satisfied:

(a) $\sum_{k=1}^{\infty} \alpha_k = \infty$, $\sum_{k=1}^{\infty} \alpha_k^2 < \infty$, and $\sum_{k=1}^{\infty} \beta_k^2 < \infty$ uniformly almost surely;

(b) $\mathbb{E}[\eta_k | \mathcal{H}_k] = 0$ and $\mathbb{E}[\eta_k^2 | \mathcal{H}_k] \leq C$ almost surely;

where $\mathcal{H}_k = \{\Delta_k, \Delta_{k-1}, \ldots, \eta_{k-1}, \ldots, \alpha_{k-1}, \ldots, \beta_{k-1}, \ldots\}$.

Before presenting the proof of this theorem, we first clarify some issues.

⬦ In the RM algorithm, the coefficient sequence $\{\alpha_k\}$ is deterministic. However, Dvoretzky's theorem allows $\{\alpha_k\}, \{\beta_k\}$ to be random variables that depend on $\mathcal{H}_k$. Thus, it is more useful in cases where α_k or β_k is a function of Δ_k.

⬦ In the first condition, it is stated as "uniformly almost surely". This is because α_k and β_k may be random variables and hence the definition of their limits must be in the stochastic sense. In the second condition, it is also stated as "almost surely". This is because $\mathcal{H}_k$ is a sequence of random variables rather than specific values. As a result, $\mathbb{E}[\eta_k | \mathcal{H}_k]$ and $\mathbb{E}[\eta_k^2 | \mathcal{H}_k]$ are random variables. The definition of the conditional expectation in this case is in the "almost sure" sense (Appendix B).

⬦ The statement of Theorem 6.2 is slightly different from [32] in the sense that Theorem 6.2 does not require $\sum_{k=1}^{\infty} \beta_k = \infty$ in the first condition. When $\sum_{k=1}^{\infty} \beta_k < \infty$, especially in the extreme case where $\beta_k = 0$ for all k, the sequence can still converge.

6.3.1 Proof of Dvoretzky's theorem

The original proof of Dvoretzky's theorem was given in 1956 [31]. There are also other proofs. We next present a proof based on quasimartingales. With the convergence results of quasimartingales, the proof of Dvoretzky's theorem is straightforward. More information about quasimartingales can be found in Appendix C.

Proof of Dvoretzky's theorem. Let $h_k \doteq \Delta_k^2$. Then,

$$
\begin{aligned}
h_{k+1} - h_k &= \Delta_{k+1}^2 - \Delta_k^2 \\
&= (\Delta_{k+1} - \Delta_k)(\Delta_{k+1} + \Delta_k) \\
&= (-\alpha_k \Delta_k + \beta_k \eta_k)[(2 - \alpha_k)\Delta_k + \beta_k \eta_k] \\
&= -\alpha_k(2 - \alpha_k)\Delta_k^2 + \beta_k^2 \eta_k^2 + 2(1 - \alpha_k)\beta_k \eta_k \Delta_k.
\end{aligned}
$$

Taking expectations on both sides of the above equation yields

$$
\mathbb{E}[h_{k+1} - h_k | \mathcal{H}_k] = \mathbb{E}[-\alpha_k(2 - \alpha_k)\Delta_k^2 | \mathcal{H}_k] + \mathbb{E}[\beta_k^2 \eta_k^2 | \mathcal{H}_k] + \mathbb{E}[2(1 - \alpha_k)\beta_k \eta_k \Delta_k | \mathcal{H}_k].
$$

$$(6.7)$$

First, since Δ_k is included and hence determined by $\mathcal{H}_k$, it can be taken out from the expectation (see property (e) in Lemma B.1). Second, consider the simple case where α_k, β_k is determined by $\mathcal{H}_k$. This case is valid when, for example, $\{\alpha_k\}$ and $\{\beta_k\}$ are functions of Δ_k or deterministic sequences. Then, they can also be taken out of the expectation. Therefore, (6.7) becomes

$$\mathbb{E}[h_{k+1} - h_k|\mathcal{H}_k] = -\alpha_k(2 - \alpha_k)\Delta_k^2 + \beta_k^2\mathbb{E}[\eta_k^2|\mathcal{H}_k] + 2(1 - \alpha_k)\beta_k\Delta_k\mathbb{E}[\eta_k|\mathcal{H}_k]. \quad (6.8)$$

For the first term, since $\sum_{k=1}^{\infty} \alpha_k^2 < \infty$ implies $\alpha_k \to 0$ almost surely, there exists a finite n such that $\alpha_k \leq 1$ almost surely for all $k \geq n$. Without loss of generality, we next merely consider the case of $\alpha_k \leq 1$. Then, $-\alpha_k(2 - \alpha_k)\Delta_k^2 \leq 0$. For the second term, we have $\beta_k^2\mathbb{E}[\eta_k^2|\mathcal{H}_k] \leq \beta_k^2 C$ as assumed. The third term equals zero because $\mathbb{E}[\eta_k|\mathcal{H}_k] = 0$ as assumed. Therefore, (6.8) becomes

$$\mathbb{E}[h_{k+1} - h_k|\mathcal{H}_k] = -\alpha_k(2 - \alpha_k)\Delta_k^2 + \beta_k^2\mathbb{E}[\eta_k^2|\mathcal{H}_k] \leq \beta_k^2 C, \quad (6.9)$$

and hence

$$\sum_{k=1}^{\infty} \mathbb{E}[h_{k+1} - h_k|\mathcal{H}_k] \leq \sum_{k=1}^{\infty} \beta_k^2 C < \infty.$$

The last inequality is due to the condition $\sum_{k=1}^{\infty} \beta_k^2 < \infty$. Then, based on the quasimartingale convergence theorem in Appendix C, we conclude that h_k converges almost surely.

We next determine what value Δ_k converges to. It follows from (6.9) that

$$\sum_{k=1}^{\infty} \alpha_k(2 - \alpha_k)\Delta_k^2 = \sum_{k=1}^{\infty} \beta_k^2\mathbb{E}[\eta_k^2|\mathcal{H}_k] - \sum_{k=1}^{\infty} \mathbb{E}[h_{k+1} - h_k|\mathcal{H}_k].$$

The first term on the right-hand side is bounded as assumed. The second term is also bounded because h_k converges and hence $h_{k+1} - h_k$ is summable. Thus, $\sum_{k=1}^{\infty} \alpha_k(2 - \alpha_k)\Delta_k^2$ on the left-hand side is also bounded. Since we consider the case of $\alpha_k \leq 1$, we have

$$\infty > \sum_{k=1}^{\infty} \alpha_k(2 - \alpha_k)\Delta_k^2 \geq \sum_{k=1}^{\infty} \alpha_k\Delta_k^2 \geq 0.$$

Therefore, $\sum_{k=1}^{\infty} \alpha_k\Delta_k^2$ is bounded. Since $\sum_{k=1}^{\infty} \alpha_k = \infty$, we must have $\Delta_k \to 0$ almost surely. $\qquad\square$

6.3.2 Application to mean estimation

While the mean estimation algorithm, $w_{k+1} = w_k + \alpha_k(x_k - w_k)$, has been analyzed using the RM theorem, we next show that its convergence can also be directly proven by Dvoretzky's theorem.

Proof. Let $w^* = \mathbb{E}[X]$. The mean estimation algorithm $w_{k+1} = w_k + \alpha_k(x_k - w_k)$ can be rewritten as

$$w_{k+1} - w^* = w_k - w^* + \alpha_k(x_k - w^* + w^* - w_k).$$

Let $\Delta \doteq w - w^*$. Then, we have

$$\Delta_{k+1} = \Delta_k + \alpha_k(x_k - w^* - \Delta_k)$$
$$= (1 - \alpha_k)\Delta_k + \alpha_k \underbrace{(x_k - w^*)}_{\eta_k}.$$

Since $\{x_k\}$ is i.i.d., we have $\mathbb{E}[x_k|\mathcal{H}_k] = \mathbb{E}[x_k] = w^*$. As a result, $\mathbb{E}[\eta_k|\mathcal{H}_k] = \mathbb{E}[x_k - w^*|\mathcal{H}_k] = 0$ and $\mathbb{E}[\eta_k^2|\mathcal{H}_k] = \mathbb{E}[x_k^2|\mathcal{H}_k] - (w^*)^2 = \mathbb{E}[x_k^2] - (w^*)^2$ are bounded if the variance of x_k is finite. Following Dvoretzky's theorem, we conclude that Δ_k converges to zero and hence w_k converges to $w^* = \mathbb{E}[X]$ almost surely. □

6.3.3 Application to the Robbins-Monro theorem

We are now ready to prove the Robbins-Monro theorem using Dvoretzky's theorem.

Proof of the Robbins-Monro theorem. The RM algorithm aims to find the root of $g(w) = 0$. Suppose that the root is w^* such that $g(w^*) = 0$. The RM algorithm is

$$w_{k+1} = w_k - a_k \tilde{g}(w_k, \eta_k)$$
$$= w_k - a_k[g(w_k) + \eta_k].$$

Then, we have

$$w_{k+1} - w^* = w_k - w^* - a_k[g(w_k) - g(w^*) + \eta_k].$$

Due to the mean value theorem [7,8], we have $g(w_k) - g(w^*) = \nabla_w g(w_k')(w_k - w^*)$, where $w_k' \in [w_k, w^*]$. Let $\Delta_k \doteq w_k - w^*$. The above equation becomes

$$
\begin{aligned}
\Delta_{k+1} &= \Delta_k - a_k[\nabla_w g(w_k')(w_k - w^*) + \eta_k] \\
&= \Delta_k - a_k \nabla_w g(w_k')\Delta_k + a_k(-\eta_k) \\
&= [1 - \underbrace{a_k \nabla_w g(w_k')}_{\alpha_k}]\Delta_k + a_k(-\eta_k).
\end{aligned}
$$

Note that $\nabla_w g(w)$ is bounded as $0 < c_1 \leq \nabla_w g(w) \leq c_2$ as assumed. Since $\sum_{k=1}^{\infty} a_k = \infty$ and $\sum_{k=1}^{\infty} a_k^2 < \infty$ as assumed, we know $\sum_{k=1}^{\infty} \alpha_k = \infty$ and $\sum_{k=1}^{\infty} \alpha_k^2 < \infty$. Thus, all the conditions in Dvoretzky's theorem are satisfied and hence Δ_k converges to zero almost surely. $\qquad\square$

The proof of the RM theorem demonstrates the power of Dvoretzky's theorem. In particular, α_k in the proof is a stochastic sequence depending on w_k rather than a deterministic sequence. In this case, Dvoretzky's theorem is still applicable.

6.3.4   An extension of Dvoretzky's theorem

We next extend Dvoretzky's theorem to a more general theorem that can handle *multiple* variables. This general theorem, proposed by [32], can be used to analyze the convergence of stochastic iterative algorithms such as Q-learning.

Theorem 6.3. *Consider a finite set $\mathcal{S}$ of real numbers. For the stochastic process*

$$
\Delta_{k+1}(s) = (1 - \alpha_k(s))\Delta_k(s) + \beta_k(s)\eta_k(s),
$$

it holds that $\Delta_k(s)$ converges to zero almost surely for every $s \in \mathcal{S}$ if the following conditions are satisfied for $s \in \mathcal{S}$:

(a) $\sum_k \alpha_k(s) = \infty$, $\sum_k \alpha_k^2(s) < \infty$, $\sum_k \beta_k^2(s) < \infty$, *and* $\mathbb{E}[\beta_k(s)|\mathcal{H}_k] \leq \mathbb{E}[\alpha_k(s)|\mathcal{H}_k]$ *uniformly almost surely;*

(b) $\|\mathbb{E}[\eta_k(s)|\mathcal{H}_k]\|_\infty \leq \gamma \|\Delta_k\|_\infty$, *where* $\gamma \in (0,1)$;

(c) $\mathrm{var}[\eta_k(s)|\mathcal{H}_k] \leq C(1 + \|\Delta_k(s)\|_\infty)^2$, *where C is a constant.*

Here, $\mathcal{H}_k = \{\Delta_k, \Delta_{k-1}, \ldots, \eta_{k-1}, \ldots, \alpha_{k-1}, \ldots, \beta_{k-1}, \ldots\}$ represents the historical information. The term $\|\cdot\|_\infty$ refers to the maximum norm.

Proof. As an extension, this theorem can be proven based on Dvoretzky's theorem. Details can be found in [32] and are omitted here. $\qquad\square$

Some remarks about this theorem are given below.

⋄ We first clarify some notations in the theorem. The variable s can be viewed as an index. In the context of reinforcement learning, it indicates a state or a state-action pair. The *maximum norm* $\|\cdot\|_\infty$ is defined over a set. It is similar but different from the L^∞ norm of vectors. In particular, $\|\mathbb{E}[\eta_k(s)|\mathcal{H}_k]\|_\infty \doteq \max_{s\in\mathcal{S}}|\mathbb{E}[\eta_k(s)|\mathcal{H}_k]|$ and $\|\Delta_k(s)\|_\infty \doteq \max_{s\in\mathcal{S}}|\Delta_k(s)|$.

⋄ This theorem is more general than Dvoretzky's theorem. First, it can handle the case of multiple variables due to the maximum norm operations. This is important for a reinforcement learning problem where there are multiple states. Second, while Dvoretzky's theorem requires $\mathbb{E}[\eta_k(s)|\mathcal{H}_k] = 0$ and $\mathrm{var}[\eta_k(s)|\mathcal{H}_k] \leq C$, this theorem only requires that the expectation and variance are bounded by the error Δ_k.

⋄ It should be noted that the convergence of $\Delta(s)$ for all $s \in \mathcal{S}$ requires that the conditions are valid for every $s \in \mathcal{S}$. Therefore, when applying this theorem to prove the convergence of reinforcement learning algorithms, we need to show that the conditions are valid for every state (or state-action pair).

6.4 Stochastic gradient descent

This section introduces stochastic gradient descent (SGD) algorithms, which are widely used in the field of machine learning. We will see that SGD is a special RM algorithm, and the mean estimation algorithm is a special SGD algorithm.

Consider the following optimization problem:

$$\min_w J(w) = \mathbb{E}[f(w, X)], \tag{6.10}$$

where w is the parameter to be optimized, and X is a random variable. The expectation is calculated with respect to X. Here, w and X can be either scalars or vectors. The function $f(\cdot)$ is a scalar.

A straightforward method for solving (6.10) is *gradient descent*. In particular, the gradient of $\mathbb{E}[f(w, X)]$ is $\nabla_w \mathbb{E}[f(w, X)] = \mathbb{E}[\nabla_w f(w, X)]$. Then, the gradient descent algorithm is

$$w_{k+1} = w_k - \alpha_k \nabla_w J(w_k) = w_k - \alpha_k \mathbb{E}[\nabla_w f(w_k, X)]. \tag{6.11}$$

This gradient descent algorithm can find the optimal solution w^* under some mild conditions such as the convexity of f. Preliminaries about gradient descent algorithms can be found in Appendix D.

The gradient descent algorithm requires the expected value $\mathbb{E}[\nabla_w f(w_k, X)]$. One way to obtain the expected value is based on the probability distribution of X. The distribution is, however, often unknown in practice. Another way is to collect a large number of i.i.d. samples $\{x_i\}_{i=1}^n$ of X so that the expected value can be approximated as

$$\mathbb{E}[\nabla_w f(w_k, X)] \approx \frac{1}{n} \sum_{i=1}^n \nabla_w f(w_k, x_i).$$

Then, (6.11) becomes

$$w_{k+1} = w_k - \frac{\alpha_k}{n} \sum_{i=1}^n \nabla_w f(w_k, x_i). \tag{6.12}$$

One problem of the algorithm in (6.12) is that it requires all the samples in each iteration. In practice, if the samples are collected one by one, then it is favorable to update w every time a sample is collected. To that end, we can use the following algorithm:

$$w_{k+1} = w_k - \alpha_k \nabla_w f(w_k, x_k), \tag{6.13}$$

where x_k is the sample collected at time step k. This is the well-known *stochastic gradient descent* algorithm. This algorithm is called "stochastic" because it relies on stochastic samples $\{x_k\}$.

Compared to the gradient descent algorithm in (6.11), SGD replaces the true gradient $\mathbb{E}[\nabla_w f(w, X)]$ with the *stochastic gradient* $\nabla_w f(w_k, x_k)$. Since $\nabla_w f(w_k, x_k) \neq \mathbb{E}[\nabla_w f(w, X)]$, can such a replacement still ensure $w_k \to w^*$ as $k \to \infty$? The answer is yes. We next present an intuitive explanation and postpone the rigorous proof of the convergence to Section 6.4.5.

In particular, since

$$\nabla_w f(w_k, x_k) = \mathbb{E}[\nabla_w f(w_k, X)] + \left(\nabla_w f(w_k, x_k) - \mathbb{E}[\nabla_w f(w_k, X)] \right)$$
$$\doteq \mathbb{E}[\nabla_w f(w_k, X)] + \eta_k,$$

the SGD algorithm in (6.13) can be rewritten as

$$w_{k+1} = w_k - \alpha_k \mathbb{E}[\nabla_w f(w_k, X)] - \alpha_k \eta_k.$$

Therefore, the SGD algorithm is the same as the regular gradient descent algorithm except that it has a perturbation term $\alpha_k \eta_k$. Since $\{x_k\}$ is i.i.d., we have $\mathbb{E}_{x_k}[\nabla_w f(w_k, x_k)] = \mathbb{E}_X[\nabla_w f(w_k, X)]$. As a result,

$$\mathbb{E}[\eta_k] = \mathbb{E}\left[\nabla_w f(w_k, x_k) - \mathbb{E}[\nabla_w f(w_k, X)] \right] = \mathbb{E}_{x_k}[\nabla_w f(w_k, x_k)] - \mathbb{E}_X[\nabla_w f(w_k, X)] = 0.$$

Therefore, the perturbation term η_k has a zero mean, which intuitively suggests that it may not jeopardize the convergence property. A rigorous proof of the convergence of

SGD is given in Section 6.4.5.

6.4.1 Application to mean estimation

We next apply SGD to analyze the mean estimation problem and show that the mean estimation algorithm in (6.4) is a special SGD algorithm. To that end, we formulate the mean estimation problem as an optimization problem:

$$\min_w J(w) = \mathbb{E}\left[\frac{1}{2}\|w - X\|^2\right] \doteq \mathbb{E}[f(w, X)], \tag{6.14}$$

where $f(w, X) = \|w - X\|^2/2$ and the gradient is $\nabla_w f(w, X) = w - X$. It can be verified that the optimal solution is $w^* = \mathbb{E}[X]$ by solving $\nabla_w J(w) = 0$. Therefore, this optimization problem is equivalent to the mean estimation problem.

⋄ The gradient descent algorithm for solving (6.14) is

$$w_{k+1} = w_k - \alpha_k \nabla_w J(w_k)$$
$$= w_k - \alpha_k \mathbb{E}[\nabla_w f(w_k, X)]$$
$$= w_k - \alpha_k \mathbb{E}[w_k - X].$$

This gradient descent algorithm is not applicable since $\mathbb{E}[w_k - X]$ or $\mathbb{E}[X]$ on the right-hand side is unknown (in fact, it is what we need to solve).

⋄ The SGD algorithm for solving (6.14) is

$$w_{k+1} = w_k - \alpha_k \nabla_w f(w_k, x_k) = w_k - \alpha_k(w_k - x_k),$$

where x_k is a sample obtained at time step k. Notably, this SGD algorithm is the same as the iterative mean estimation algorithm in (6.4). Therefore, (6.4) is an SGD algorithm designed specifically for solving the mean estimation problem.

6.4.2 Convergence pattern of SGD

The idea of the SGD algorithm is to replace the true gradient with a stochastic gradient. However, since the stochastic gradient is random, one may ask whether the convergence speed of SGD is slow or random. Fortunately, SGD can converge efficiently in general. An interesting *convergence pattern* is that it behaves similarly to the regular gradient descent algorithm when the estimate w_k is far from the optimal solution w^*. Only when w_k is close to w^*, does the convergence of SGD exhibit more randomness.

An analysis of this pattern and an illustrative example are given below.

◇ Analysis: The *relative error* between the stochastic and true gradients is

$$\delta_k \doteq \frac{|\nabla_w f(w_k, x_k) - \mathbb{E}[\nabla_w f(w_k, X)]|}{|\mathbb{E}[\nabla_w f(w_k, X)]|}.$$

For the sake of simplicity, we consider the case where w and $\nabla_w f(w, x)$ are both *scalars*. Since w^* is the optimal solution, it holds that $\mathbb{E}[\nabla_w f(w^*, X)] = 0$. Then, the relative error can be rewritten as

$$\delta_k = \frac{|\nabla_w f(w_k, x_k) - \mathbb{E}[\nabla_w f(w_k, X)]|}{|\mathbb{E}[\nabla_w f(w_k, X)] - \mathbb{E}[\nabla_w f(w^*, X)]|} = \frac{|\nabla_w f(w_k, x_k) - \mathbb{E}[\nabla_w f(w_k, X)]|}{|\mathbb{E}[\nabla_w^2 f(\tilde{w}_k, X)(w_k - w^*)]|}, \quad (6.15)$$

where the last equality is due to the mean value theorem [7, 8] and $\tilde{w}_k \in [w_k, w^*]$. Suppose that f is strictly convex such that $\nabla_w^2 f \geq c > 0$ for all w, X. Then, the denominator in (6.15) becomes

$$\left| \mathbb{E}[\nabla_w^2 f(\tilde{w}_k, X)(w_k - w^*)] \right| = \left| \mathbb{E}[\nabla_w^2 f(\tilde{w}_k, X)] \right| \left| (w_k - w^*) \right|$$
$$\geq c|w_k - w^*|.$$

Substituting the above inequality into (6.15) yields

$$\delta_k \leq \frac{|\overbrace{\nabla_w f(w_k, x_k)}^{\text{stochastic gradient}} - \overbrace{\mathbb{E}[\nabla_w f(w_k, X)]}^{\text{true gradient}}|}{\underbrace{c|w_k - w^*|}_{\text{distance to the optimal solution}}}.$$

The above inequality suggests an interesting convergence pattern of SGD: the relative error δ_k is inversely proportional to $|w_k - w^*|$. As a result, when $|w_k - w^*|$ is large, δ_k is small. In this case, the SGD algorithm behaves like the gradient descent algorithm and hence w_k quickly converges to w^*. When w_k is close to w^*, the relative error δ_k may be large, and the convergence exhibits more randomness.

◇ Example: A good example for demonstrating the above analysis is the mean estimation problem. Consider the mean estimation problem in (6.14). When w and X are both scalar, we have $f(w, X) = |w - X|^2/2$ and hence

$$\nabla_w f(w, x_k) = w - x_k,$$
$$\mathbb{E}[\nabla_w f(w, x_k)] = w - \mathbb{E}[X] = w - w^*.$$

Thus, the relative error is

$$\delta_k = \frac{|\nabla_w f(w_k, x_k) - \mathbb{E}[\nabla_w f(w_k, X)]|}{|\mathbb{E}[\nabla_w f(w_k, X)]|} = \frac{|(w_k - x_k) - (w_k - \mathbb{E}[X])|}{|w_k - w^*|} = \frac{|\mathbb{E}[X] - x_k|}{|w_k - w^*|}.$$

The expression of the relative error clearly shows that δ_k is *inversely proportional* to

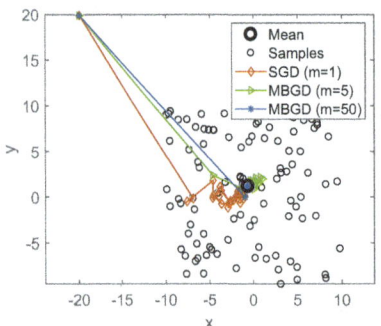 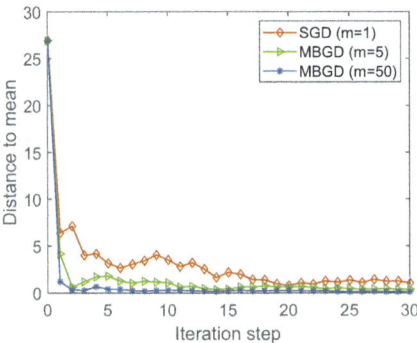

Figure 6.5: An example for demonstrating stochastic and mini-batch gradient descent algorithms. The distribution of $X \in \mathbb{R}^2$ is uniform in the square area centered at the origin with a side length as 20. The mean is $\mathbb{E}[X] = 0$. The mean estimation is based on 100 i.i.d. samples.

$|w_k - w^*|$. As a result, when w_k is far from w^*, the relative error is small, and SGD behaves like gradient descent. In addition, since δ_k is proportional to $|\mathbb{E}[X] - x_k|$, the mean of δ_k is proportional to the variance of X.

The simulation results are shown in Figure 6.5. Here, $X \in \mathbb{R}^2$ represents a random position in the plane. Its distribution is uniform in the square area centered at the origin and $\mathbb{E}[X] = 0$. The mean estimation is based on 100 i.i.d. samples. Although the initial guess of the mean is far away from the true value, it can be seen that the SGD estimate quickly approaches the neighborhood of the origin. When the estimate is close to the origin, the convergence process exhibits certain randomness.

6.4.3   A deterministic formulation of SGD

The formulation of SGD in (6.13) involves random variables. One may often encounter a deterministic formulation of SGD without involving any random variables.

In particular, consider a set of real numbers $\{x_i\}_{i=1}^n$, where x_i does not have to be a sample of any random variable. The optimization problem to be solved is to minimize the average:

$$\min_w J(w) = \frac{1}{n} \sum_{i=1}^n f(w, x_i),$$

where $f(w, x_i)$ is a parameterized function, and w is the parameter to be optimized. The gradient descent algorithm for solving this problem is

$$w_{k+1} = w_k - \alpha_k \nabla_w J(w_k) = w_k - \alpha_k \frac{1}{n} \sum_{i=1}^n \nabla_w f(w_k, x_i).$$

Suppose that the set $\{x_i\}_{i=1}^n$ is large and we can only fetch a single number each time.

In this case, it is favorable to update w_k in an incremental manner:

$$w_{k+1} = w_k - \alpha_k \nabla_w f(w_k, x_k). \tag{6.16}$$

It must be noted that x_k here is the number fetched at time step k instead of the kth element in the set $\{x_i\}_{i=1}^n$.

The algorithm in (6.16) is very similar to SGD, but its problem formulation is subtly different because it does not involve any random variables or expected values. Then, many questions arise. For example, is this algorithm SGD? How should we use the finite set of numbers $\{x_i\}_{i=1}^n$? Should we sort these numbers in a certain order and then use them one by one, or should we randomly sample a number from the set?

A quick answer to the above questions is that, although no random variables are involved in the above formulation, we can convert the *deterministic formulation* to the *stochastic formulation* by introducing a random variable. In particular, let X be a random variable defined on the set $\{x_i\}_{i=1}^n$. Suppose that its probability distribution is uniform such that $p(X = x_i) = 1/n$. Then, the deterministic optimization problem becomes a stochastic one:

$$\min_w J(w) = \frac{1}{n} \sum_{i=1}^n f(w, x_i) = \mathbb{E}[f(w, X)].$$

The last equality in the above equation is strict instead of approximate. Therefore, the algorithm in (6.16) is SGD, and the estimate converges if x_k is *uniformly* and independently sampled from $\{x_i\}_{i=1}^n$. Note that x_k may repeatedly take the same number in $\{x_i\}_{i=1}^n$ since it is sampled randomly.

6.4.4 BGD, SGD, and mini-batch GD

While SGD uses a single sample in every iteration, we next introduce *mini-batch gradient descent* (MBGD), which uses a few more samples in every iteration. When all samples are used in every iteration, the algorithm is called *batch gradient descent* (BGD).

In particular, suppose that we would like to find the optimal solution that can minimize $J(w) = \mathbb{E}[f(w, X)]$ given a set of random samples $\{x_i\}_{i=1}^n$ of X. The BGD, SGD, and MBGD algorithms for solving this problem are, respectively,

$$w_{k+1} = w_k - \alpha_k \frac{1}{n} \sum_{i=1}^n \nabla_w f(w_k, x_i), \qquad \text{(BGD)}$$

$$w_{k+1} = w_k - \alpha_k \frac{1}{m} \sum_{j \in \mathcal{I}_k} \nabla_w f(w_k, x_j), \qquad \text{(MBGD)}$$

$$w_{k+1} = w_k - \alpha_k \nabla_w f(w_k, x_k). \qquad \text{(SGD)}$$

In the BGD algorithm, all the samples are used in every iteration. When n is large, $(1/n) \sum_{i=1}^n \nabla_w f(w_k, x_i)$ is close to the true gradient $\mathbb{E}[\nabla_w f(w_k, X)]$. In the MBGD algorithm, $\mathcal{I}_k$ is a subset of $\{1,\ldots,n\}$ obtained at time k. The size of the set is $|\mathcal{I}_k| = m$. The samples in $\mathcal{I}_k$ are also assumed to be i.i.d. In the SGD algorithm, x_k is randomly sampled from $\{x_i\}_{i=1}^n$ at time k.

MBGD can be viewed as an intermediate version between SGD and BGD. Compared to SGD, MBGD has less randomness because it uses more samples instead of just one as in SGD. Compared to BGD, MBGD does not require using all the samples in every iteration, making it more flexible. If $m = 1$, then MBGD becomes SGD. However, if $m = n$, MBGD may *not* become BGD. This is because MBGD uses n randomly fetched samples, whereas BGD uses all n numbers. These n randomly fetched samples may contain the same number multiple times and hence may not cover all n numbers in $\{x_i\}_{i=1}^n$.

The convergence speed of MBGD is faster than that of SGD in general. This is because SGD uses $\nabla_w f(w_k, x_k)$ to approximate the true gradient, whereas MBGD uses $(1/m)\sum_{j \in \mathcal{I}_k} \nabla_w f(w_k, x_j)$, which is closer to the true gradient because the randomness is averaged out. The convergence of the MBGD algorithm can be proven similarly to the SGD case.

A good example for demonstrating the above analysis is the mean estimation problem. In particular, given some numbers $\{x_i\}_{i=1}^n$, our goal is to calculate the mean $\bar{x} = \sum_{i=1}^n x_i/n$. This problem can be equivalently stated as the following optimization problem:

$$\min_w J(w) = \frac{1}{2n}\sum_{i=1}^n \|w - x_i\|^2,$$

whose optimal solution is $w^* = \bar{x}$. The three algorithms for solving this problem are, respectively,

$$w_{k+1} = w_k - \alpha_k \frac{1}{n}\sum_{i=1}^n (w_k - x_i) = w_k - \alpha_k(w_k - \bar{x}), \qquad \text{(BGD)}$$

$$w_{k+1} = w_k - \alpha_k \frac{1}{m}\sum_{j \in \mathcal{I}_k}(w_k - x_j) = w_k - \alpha_k\left(w_k - \bar{x}_k^{(m)}\right), \qquad \text{(MBGD)}$$

$$w_{k+1} = w_k - \alpha_k(w_k - x_k), \qquad \text{(SGD)}$$

where $\bar{x}_k^{(m)} = \sum_{j \in \mathcal{I}_k} x_j/m$. Furthermore, if $\alpha_k - 1/k$, the above equations can be solved as follows:

$$w_{k+1} = \frac{1}{k} \sum_{j=1}^{k} \bar{x} = \bar{x}, \qquad \text{(BGD)}$$

$$w_{k+1} = \frac{1}{k} \sum_{j=1}^{k} \bar{x}_j^{(m)}, \qquad \text{(MBGD)}$$

$$w_{k+1} = \frac{1}{k} \sum_{j=1}^{k} x_j. \qquad \text{(SGD)}$$

The derivation of the above equations is similar to that of (6.3) and is omitted here. It can be seen that the estimate given by BGD at each step is exactly the optimal solution $w^* = \bar{x}$. MBGD converges to the mean faster than SGD because $\bar{x}_k^{(m)}$ is already an average.

A simulation example is given in Figure 6.5 to demonstrate the convergence of MBGD. Let $\alpha_k = 1/k$. It is shown that all MBGD algorithms with different mini-batch sizes can converge to the mean. The case with $m = 50$ converges the fastest, while SGD with $m = 1$ is the slowest. This is consistent with the above analysis. Nevertheless, the convergence rate of SGD is still fast, especially when w_k is far from w^*.

6.4.5 Convergence of SGD

The rigorous proof of the convergence of SGD is given as follows.

Theorem 6.4 (Convergence of SGD). *For the SGD algorithm in (6.13), if the following conditions are satisfied, then w_k converges to the root of $\nabla_w \mathbb{E}[f(w, X)] = 0$ almost surely.*

(a) $0 < c_1 \leq \nabla_w^2 f(w, X) \leq c_2$;

(b) $\sum_{k=1}^{\infty} a_k = \infty$ and $\sum_{k=1}^{\infty} a_k^2 < \infty$;

(c) $\{x_k\}_{k=1}^{\infty}$ are i.i.d.

The three conditions in Theorem 6.4 are discussed below.

⬦ Condition (a) is about the convexity of f. It requires the curvature of f to be bounded from above and below. Here, w is a scalar, and so is $\nabla_w^2 f(w, X)$. This condition can be generalized to the vector case. When w is a vector, $\nabla_w^2 f(w, X)$ is the well-known Hessian matrix.

⬦ Condition (b) is similar to that of the RM algorithm. In fact, the SGD algorithm is a special RM algorithm (as shown in the proof in Box 6.1). In practice, a_k is often selected as a sufficiently small *constant*. Although condition (b) is not satisfied in this case, the algorithm can still converge in a certain sense [24, Section 1.5].

⬦ Condition (c) is a common requirement.

Box 6.1: Proof of Theorem 6.4

We next show that the SGD algorithm is a special RM algorithm. Then, the convergence of SGD naturally follows from the RM theorem.

The problem to be solved by SGD is to minimize $J(w) = \mathbb{E}[f(w, X)]$. This problem can be converted to a root-finding problem. That is, finding the root of $\nabla_w J(w) = \mathbb{E}[\nabla_w f(w, X)] = 0$. Let

$$g(w) = \nabla_w J(w) = \mathbb{E}[\nabla_w f(w, X)].$$

Then, SGD aims to find the root of $g(w) = 0$. This is exactly the problem solved by the RM algorithm. The quantity that we can measure is $\tilde{g} = \nabla_w f(w, x)$, where x is a sample of X. Note that $\tilde{g}$ can be rewritten as

$$\begin{aligned}\tilde{g}(w, \eta) &= \nabla_w f(w, x) \\ &= \mathbb{E}[\nabla_w f(w, X)] + \underbrace{\nabla_w f(w, x) - \mathbb{E}[\nabla_w f(w, X)]}_{\eta}.\end{aligned}$$

Then, the RM algorithm for solving $g(w) = 0$ is

$$w_{k+1} = w_k - a_k \tilde{g}(w_k, \eta_k) = w_k - a_k \nabla_w f(w_k, x_k),$$

which is the same as the SGD algorithm in (6.13). As a result, the SGD algorithm is a special RM algorithm. We next show that the three conditions in Theorem 6.1 are satisfied. Then, the convergence of SGD naturally follows from Theorem 6.1.

⋄ Since $\nabla_w g(w) = \nabla_w \mathbb{E}[\nabla_w f(w, X)] = \mathbb{E}[\nabla_w^2 f(w, X)]$, it follows from $c_1 \leq \nabla_w^2 f(w, X) \leq c_2$ that $c_1 \leq \nabla_w g(w) \leq c_2$. Thus, the first condition in Theorem 6.1 is satisfied.

⋄ The second condition in Theorem 6.1 is the same as the second condition in this theorem.

⋄ The third condition in Theorem 6.1 requires $\mathbb{E}[\eta_k | \mathcal{H}_k] = 0$ and $\mathbb{E}[\eta_k^2 | \mathcal{H}_k] < \infty$. Since $\{x_k\}$ is i.i.d., we have $\mathbb{E}_{x_k}[\nabla_w f(w, x_k)] = \mathbb{E}[\nabla_w f(w, X)]$ for all k. Therefore,

$$\mathbb{E}[\eta_k | \mathcal{H}_k] = \mathbb{E}[\nabla_w f(w_k, x_k) - \mathbb{E}[\nabla_w f(w_k, X)] | \mathcal{H}_k].$$

Since $\mathcal{H}_k = \{w_k, w_{k-1}, \dots\}$ and x_k is independent of $\mathcal{H}_k$, the first term on the right-hand side becomes $\mathbb{E}[\nabla_w f(w_k, x_k) | \mathcal{H}_k] = \mathbb{E}_{x_k}[\nabla_w f(w_k, x_k)]$. The second term becomes $\mathbb{E}[\mathbb{E}[\nabla_w f(w_k, X)] | \mathcal{H}_k] = \mathbb{E}[\nabla_w f(w_k, X)]$ because $\mathbb{E}[\nabla_w f(w_k, X)]$ is a function of w_k. Therefore,

$$\mathbb{E}[\eta_k | \mathcal{H}_k] = \mathbb{E}_{x_k}[\nabla_w f(w_k, x_k)] - \mathbb{E}[\nabla_w f(w_k, X)] = 0.$$

Similarly, it can be proven that $\mathbb{E}[\eta_k^2 | \mathcal{H}_k] < \infty$ if $|\nabla_w f(w, x)| < \infty$ for all w given any x.

Since the three conditions in Theorem 6.1 are satisfied, the convergence of the SGD algorithm follows.

6.5 Summary

Instead of introducing new reinforcement learning algorithms, this chapter introduced the preliminaries of stochastic approximation such as the RM and SGD algorithms. Compared to many other root-finding algorithms, the RM algorithm does not require the expression of the objective function or its derivative. It has been shown that the SGD algorithm is a special RM algorithm. Moreover, an important problem frequently discussed throughout this chapter is mean estimation. The mean estimation algorithm (6.4) is the first stochastic iterative algorithm we have ever introduced in this book. We showed that it is a special SGD algorithm. We will see in Chapter 7 that temporal-difference learning algorithms have similar expressions. Finally, the name "stochastic approximation" was first used by Robbins and Monro in 1951 [25]. More information about stochastic approximation can be found in [24].

6.6 Q&A

⋄ Q: What is stochastic approximation?

A: Stochastic approximation refers to a broad class of stochastic iterative algorithms for solving root-finding or optimization problems.

⋄ Q: Why do we need to study stochastic approximation?

A: This is because the temporal-difference reinforcement learning algorithms that will be introduced in Chapter 7 can be viewed as stochastic approximation algorithms. With the knowledge introduced in this chapter, we can be better prepared, and it will not be abrupt for us to see these algorithms for the first time.

⋄ Q: Why do we frequently discuss the mean estimation problem in this chapter?

A: This is because the state and action values are defined as the means of random variables. The temporal-difference learning algorithms introduced in Chapter 7 are similar to stochastic approximation algorithms for mean estimation.

⋄ Q: What is the advantage of the RM algorithm over other root-finding algorithms?

A: Compared to many other root-finding algorithms, the RM algorithm is powerful in the sense that it does not require the expression of the objective function or its derivative. As a result, it is a black-box technique that only requires the input and output of the objective function. The famous SGD algorithm is a special form of the RM algorithm.

⋄ Q: What is the basic idea of SGD?

A: SGD aims to solve optimization problems involving random variables. When the probability distributions of the given random variables are not known, SGD can solve the optimization problems merely by using samples. Mathematically, the SGD algorithm can be obtained by replacing the true gradient expressed as an expectation in the gradient descent algorithm with a stochastic gradient.

⋄ Q: Can SGD converge quickly?

A: SGD has an interesting convergence pattern. That is, if the estimate is far from the optimal solution, then the convergence process is fast. When the estimate is close to the solution, the randomness of the stochastic gradient becomes influential, and the convergence rate decreases.

⋄ Q: What is MBGD? What are its advantages over SGD and BGD?

A: MBGD can be viewed as an intermediate version between SGD and BGD. Compared to SGD, it has less randomness because it uses more samples instead of just one as in SGD. Compared to BGD, it does not require the use of all the samples, making it more flexible.

Chapter 7

Temporal-Difference Methods

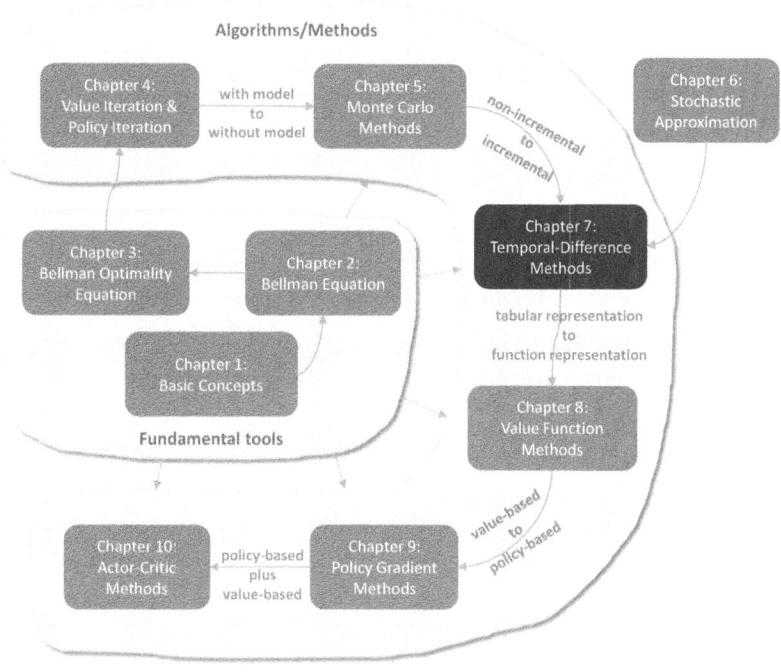

Figure 7.1: Where we are in this book.

This chapter introduces temporal-difference (TD) methods for reinforcement learning. Similar to Monte Carlo (MC) learning, TD learning is also model-free, but it has some advantages due to its incremental form. With the preparation in Chapter 6, readers will not feel alarmed when seeing TD learning algorithms. In fact, TD learning algorithms can be viewed as special stochastic algorithms for solving the Bellman or Bellman optimality equations.

Since this chapter introduces quite a few TD algorithms, we first overview these algorithms and clarify the relationships between them.

◇ Section 7.1 introduces the most basic TD algorithm, which can estimate the *state*

© The Author(s), under exclusive license to Springer Nature Singapore Pte Ltd. 2025
S. Zhao, *Mathematical Foundations of Reinforcement Learning*, https://doi.org/10.1007/978-981-97-3944-8_7

values of a given policy. It is important to understand this basic algorithm first before studying the other TD algorithms.

⬦ Section 7.2 introduces the Sarsa algorithm, which can estimate the *action values* of a given policy. This algorithm can be combined with a policy improvement step to find optimal policies. The Sarsa algorithm can be easily obtained from the TD algorithm in Section 7.1 by replacing state value estimation with action value estimation.

⬦ Section 7.3 introduces the *n*-step Sarsa algorithm, which is a generalization of the Sarsa algorithm. It will be shown that Sarsa and MC learning are two special cases of *n*-step Sarsa.

⬦ Section 7.4 introduces the Q-learning algorithm, which is one of the most classic reinforcement learning algorithms. While the other TD algorithms aim to solve the Bellman equation of a given policy, Q-learning aims to directly solve the Bellman optimality equation to obtain optimal policies.

⬦ Section 7.5 compares the TD algorithms introduced in this chapter and provides a unified point of view.

7.1 TD learning of state values

TD learning often refers to a broad class of reinforcement learning algorithms. For example, all the algorithms introduced in this chapter fall into the scope of TD learning. However, TD learning in this section specifically refers to a classic algorithm for estimating state values.

7.1.1 Algorithm description

Given a policy π, our goal is to estimate $v_\pi(s)$ for all $s \in \mathcal{S}$. Suppose that we have some experience samples $(s_0, r_1, s_1, \ldots, s_t, r_{t+1}, s_{t+1}, \ldots)$ generated following π. Here, t denotes the time step. The following TD algorithm can estimate the state values using these samples:

$$v_{t+1}(s_t) = v_t(s_t) - \alpha_t(s_t)\Big[v_t(s_t) - \big(r_{t+1} + \gamma v_t(s_{t+1})\big)\Big], \tag{7.1}$$

$$v_{t+1}(s) = v_t(s), \quad \text{for all } s \neq s_t, \tag{7.2}$$

where $t = 0, 1, 2, \ldots$. Here, $v_t(s_t)$ is the estimate of $v_\pi(s_t)$ at time t; $\alpha_t(s_t)$ is the learning rate for s_t at time t.

It should be noted that, at time t, only the value of the visited state s_t is updated. The values of the unvisited states $s \neq s_t$ remain unchanged as shown in (7.2). Equation (7.2) is often omitted for simplicity, but it should be kept in mind because the algorithm would be mathematically incomplete without this equation.

Readers who see the TD learning algorithm for the first time may wonder why it is designed like this. In fact, it can be viewed as a special stochastic approximation algorithm for solving the Bellman equation. To see that, first recall that the definition of the state value is

$$v_\pi(s) = \mathbb{E}\big[R_{t+1} + \gamma G_{t+1}|S_t = s\big], \quad s \in \mathcal{S}. \tag{7.3}$$

We can rewrite (7.3) as

$$v_\pi(s) = \mathbb{E}\big[R_{t+1} + \gamma v_\pi(S_{t+1})|S_t = s\big], \quad s \in \mathcal{S}. \tag{7.4}$$

That is because $\mathbb{E}[G_{t+1}|S_t = s] = \sum_a \pi(a|s) \sum_{s'} p(s'|s,a) v_\pi(s') = \mathbb{E}[v_\pi(S_{t+1})|S_t = s]$. Equation (7.4) is another expression of the Bellman equation. It is sometimes called the *Bellman expectation equation*.

The TD algorithm can be derived by applying the Robbins-Monro algorithm (Chapter 6) to solve the Bellman equation in (7.4). Interested readers can check the details in Box 7.1.

Box 7.1: Derivation of the TD algorithm

We next show that the TD algorithm in (7.1) can be obtained by applying the Robbins-Monro algorithm to solve (7.4).

For state s_t, we define a function as

$$g(v_\pi(s_t)) \doteq v_\pi(s_t) - \mathbb{E}\big[R_{t+1} + \gamma v_\pi(S_{t+1})|S_t = s_t\big].$$

Then, (7.4) is equivalent to

$$g(v_\pi(s_t)) = 0.$$

Our goal is to solve the above equation to obtain $v_\pi(s_t)$ using the Robbins-Monro algorithm. Since we can obtain r_{t+1} and s_{t+1}, which are the samples of R_{t+1} and S_{t+1}, the noisy observation of $g(v_\pi(s_t))$ that we can obtain is

$$\begin{aligned}
\tilde{g}(v_\pi(s_t)) &= v_\pi(s_t) - \big[r_{t+1} + \gamma v_\pi(s_{t+1})\big] \\
&= \underbrace{\left(v_\pi(s_t) - \mathbb{E}\big[R_{t+1} + \gamma v_\pi(S_{t+1})|S_t = s_t\big]\right)}_{g(v_\pi(s_t))} \\
&\quad + \underbrace{\left(\mathbb{E}\big[R_{t+1} + \gamma v_\pi(S_{t+1})|S_t = s_t\big] - \big[r_{t+1} + \gamma v_\pi(s_{t+1})\big]\right)}_{\eta}.
\end{aligned}$$

Therefore, the Robbins-Monro algorithm (Section 6.2) for solving $g(v_\pi(s_t)) = 0$ is

$$v_{t+1}(s_t) = v_t(s_t) - \alpha_t(s_t)\tilde{g}(v_t(s_t))$$
$$= v_t(s_t) - \alpha_t(s_t)\Big(v_t(s_t) - \big[r_{t+1} + \gamma v_\pi(s_{t+1})\big]\Big), \tag{7.5}$$

where $v_t(s_t)$ is the estimate of $v_\pi(s_t)$ at time t, and $\alpha_t(s_t)$ is the learning rate.

The algorithm in (7.5) has a similar expression to that of the TD algorithm in (7.1). The only difference is that the right-hand side of (7.5) contains $v_\pi(s_{t+1})$, whereas (7.1) contains $v_t(s_{t+1})$. That is because (7.5) is designed to merely estimate the state value of s_t by assuming that the state values of the other states are already known. If we would like to estimate the state values of all the states, then $v_\pi(s_{t+1})$ on the right-hand side should be replaced with $v_t(s_{t+1})$. Then, (7.5) is exactly the same as (7.1). However, can such a replacement still ensure convergence? The answer is yes, and it will be proven later in Theorem 7.1.

7.1.2 Property analysis

Some important properties of the TD algorithm are discussed as follows.

First, we examine the expression of the TD algorithm more closely. In particular, (7.1) can be described as

$$\underbrace{v_{t+1}(s_t)}_{\text{new estimate}} = \underbrace{v_t(s_t)}_{\text{current estimate}} - \alpha_t(s_t)\big[\overbrace{v_t(s_t) - \big(\underbrace{r_{t+1} + \gamma v_t(s_{t+1})}_{\text{TD target } \bar{v}_t}\big)}^{\text{TD error } \delta_t}\big], \tag{7.6}$$

where

$$\bar{v}_t \doteq r_{t+1} + \gamma v_t(s_{t+1})$$

is called the *TD target* and

$$\delta_t \doteq v(s_t) - \bar{v}_t = v_t(s_t) - (r_{t+1} + \gamma v_t(s_{t+1}))$$

is called the *TD error*. It can be seen that the new estimate $v_{t+1}(s_t)$ is a combination of the current estimate $v_t(s_t)$ and the TD error δ_t.

◇ Why is $\bar{v}_t$ called the TD target?

This is because $\bar{v}_t$ is the *target value* that the algorithm attempts to drive $v(s_t)$ to. To see that, subtracting $\bar{v}_t$ from both sides of (7.6) gives

$$v_{t+1}(s_t) - \bar{v}_t = \big[v_t(s_t) - \bar{v}_t\big] - \alpha_t(s_t)\big[v_t(s_t) - \bar{v}_t\big]$$
$$= [1 - \alpha_t(s_t)]\big[v_t(s_t) - \bar{v}_t\big].$$

Taking the absolute values of both sides of the above equation gives

$$|v_{t+1}(s_t) - \bar{v}_t| = |1 - \alpha_t(s_t)||v_t(s_t) - \bar{v}_t|.$$

Since $\alpha_t(s_t)$ is a small positive number, we have $0 < 1 - \alpha_t(s_t) < 1$. It then follows that

$$|v_{t+1}(s_t) - \bar{v}_t| < |v_t(s_t) - \bar{v}_t|.$$

The above inequality is important because it indicates that the new value $v_{t+1}(s_t)$ is closer to $\bar{v}_t$ than the old value $v_t(s_t)$. Therefore, this algorithm mathematically drives $v_t(s_t)$ toward $\bar{v}_t$. This is why $\bar{v}_t$ is called the TD target.

⋄ What is the interpretation of the TD error?

First, this error is called *temporal-difference* because $\delta_t = v_t(s_t) - (r_{t+1} + \gamma v_t(s_{t+1}))$ reflects the discrepancy between two time steps t and $t + 1$. Second, the TD error is zero in the expectation sense when the state value estimate is accurate. To see that, when $v_t = v_\pi$, the expected value of the TD error is

$$\mathbb{E}[\delta_t | S_t = s_t] = \mathbb{E}\big[v_\pi(S_t) - (R_{t+1} + \gamma v_\pi(S_{t+1}))|S_t = s_t\big]$$
$$= v_\pi(s_t) - \mathbb{E}\big[R_{t+1} + \gamma v_\pi(S_{t+1})|S_t = s_t\big]$$
$$= 0. \qquad \text{(due to (7.3))}$$

Therefore, the TD error reflects not only the discrepancy between two time steps but also, more importantly, the discrepancy between the estimate v_t and the true state value v_π.

On a more abstract level, the TD error can be interpreted as the *innovation*, which indicates new information obtained from the experience sample (s_t, r_{t+1}, s_{t+1}). The fundamental idea of TD learning is to correct our current estimate of the state value based on the newly obtained information. Innovation is fundamental in many estimation problems such as Kalman filtering [33, 34].

Second, the TD algorithm in (7.1) can only estimate the state values of a given policy. To find optimal policies, we still need to further calculate the action values and then conduct policy improvement. This will be introduced in Section 7.2. Nevertheless, the TD algorithm introduced in this section is very basic and important for understanding the other algorithms in this chapter.

Third, while both TD learning and MC learning are model-free, what are their advantages and disadvantages? The answers are summarized in Table 7.1.

TD learning	MC learning
Incremental: TD learning is incremental. It can update the state/action values immediately after receiving an experience sample.	*Non-incremental:* MC learning is non-incremental. It must wait until an episode has been completely collected. That is because it must calculate the discounted return of the episode.
Continuing tasks: Since TD learning is incremental, it can handle both episodic and continuing tasks. Continuing tasks may not have terminal states.	*Episodic tasks:* Since MC learning is non-incremental, it can only handle episodic tasks where the episodes terminate after a finite number of steps.
Bootstrapping: TD learning bootstraps because the update of a state/action value relies on the previous estimate of this value. As a result, TD learning requires an initial guess of the values.	*Non-bootstrapping:* MC is not bootstrapping because it can directly estimate state/action values without initial guesses.
Low estimation variance: The estimation variance of TD is lower than that of MC because it involves fewer random variables. For instance, to estimate an action value $q_\pi(s_t, a_t)$, Sarsa merely requires the samples of three random variables: $R_{t+1}, S_{t+1}, A_{t+1}$.	*High estimation variance:* The estimation variance of MC is higher since many random variables are involved. For example, to estimate $q_\pi(s_t, a_t)$, we need samples of $R_{t+1} + \gamma R_{t+2} + \gamma^2 R_{t+3} + \ldots$. Suppose that the length of each episode is L. Assume that each state has the same number of actions as $|\mathcal{A}|$. Then, there are $|\mathcal{A}|^L$ possible episodes following a soft policy. If we merely use a few episodes to estimate, it is not surprising that the estimation variance is high.

Table 7.1: A comparison between TD learning and MC learning.

7.1.3 Convergence analysis

The convergence analysis of the TD algorithm in (7.1) is given below.

Theorem 7.1 (Convergence of TD learning). *Given a policy π, by the TD algorithm in (7.1), $v_t(s)$ converges almost surely to $v_\pi(s)$ as $t \to \infty$ for all $s \in \mathcal{S}$ if $\sum_t \alpha_t(s) = \infty$ and $\sum_t \alpha_t^2(s) < \infty$ for all $s \in \mathcal{S}$.*

Some remarks about α_t are given below. First, the condition of $\sum_t \alpha_t(s) = \infty$ and $\sum_t \alpha_t^2(s) < \infty$ must be valid for all $s \in \mathcal{S}$. Note that, at time t, $\alpha_t(s) > 0$ if s is being visited and $\alpha_t(s) = 0$ otherwise. The condition $\sum_t \alpha_t(s) = \infty$ requires the state s to be visited an infinite (or sufficiently many) number of times. This requires either the condition of exploring starts or an exploratory policy so that every state-action pair can possibly be visited many times. Second, the learning rate α_t is often selected as a small positive constant in practice. In this case, the condition that $\sum_t \alpha_t^2(s) < \infty$ is no longer valid. When α is constant, it can still be shown that the algorithm converges in the sense of expectation [24, Section 1.5].

Box 7.2: Proof of Theorem 7.1

We prove the convergence based on Theorem 6.3 in Chapter 6. To do that, we need first to construct a stochastic process as that in Theorem 6.3. Consider an arbitrary state $s \in \mathcal{S}$. At time t, it follows from the TD algorithm in (7.1) that

$$v_{t+1}(s) = v_t(s) - \alpha_t(s)\Big(v_t(s) - (r_{t+1} + \gamma v_t(s_{t+1}))\Big), \quad \text{if } s = s_t, \qquad (7.7)$$

or

$$v_{t+1}(s) = v_t(s), \quad \text{if } s \neq s_t. \qquad (7.8)$$

The estimation error is defined as

$$\Delta_t(s) \doteq v_t(s) - v_\pi(s),$$

where $v_\pi(s)$ is the state value of s under policy π. Deducting $v_\pi(s)$ from both sides of (7.7) gives

$$\Delta_{t+1}(s) = (1 - \alpha_t(s))\Delta_t(s) + \alpha_t(s)\underbrace{(r_{t+1} + \gamma v_t(s_{t+1}) - v_\pi(s))}_{\eta_t(s)}$$

$$= (1 - \alpha_t(s))\Delta_t(s) + \alpha_t(s)\eta_t(s), \quad s = s_t. \qquad (7.9)$$

Deducting $v_\pi(s)$ from both sides of (7.8) gives

$$\Delta_{t+1}(s) = \Delta_t(s) = (1 - \alpha_t(s))\Delta_t(s) + \alpha_t(s)\eta_t(s), \quad s \neq s_t,$$

whose expression is the same as that of (7.9) except that $\alpha_t(s) = 0$ and $\eta_t(s) = 0$. Therefore, regardless of whether $s = s_t$, we obtain the following unified expression:

$$\Delta_{t+1}(s) = (1 - \alpha_t(s))\Delta_t(s) + \alpha_t(s)\eta_t(s).$$

This is the process in Theorem 6.3. Our goal is to show that the three conditions in Theorem 6.3 are satisfied and hence the process converges.

The first condition is valid as assumed in Theorem 7.1. We next show that the second condition is valid. That is, $\|\mathbb{E}[\eta_t(s)|\mathcal{H}_t]\|_\infty \leq \gamma\|\Delta_t(s)\|_\infty$ for all $s \in \mathcal{S}$. Here, $\mathcal{H}_t$ represents the historical information (see the definition in Theorem 6.3). Due to the Markovian property, $\eta_t(s) = r_{t+1} + \gamma v_t(s_{t+1}) - v_\pi(s)$ or $\eta_t(s) = 0$ does not depend on the historical information once s is given. As a result, we have $\mathbb{E}[\eta_t(s)|\mathcal{H}_t] = \mathbb{E}[\eta_t(s)]$. For $s \neq s_t$, we have $\eta_t(s) = 0$. Then, it is trivial to see that

$$|\mathbb{E}[\eta_t(s)]| = 0 \leq \gamma \|\Delta_t(s)\|_\infty. \tag{7.10}$$

For $s = s_t$, we have

$$\begin{aligned}
\mathbb{E}[\eta_t(s)] &= \mathbb{E}[\eta_t(s_t)] \\
&= \mathbb{E}[r_{t+1} + \gamma v_t(s_{t+1}) - v_\pi(s_t)|s_t] \\
&= \mathbb{E}[r_{t+1} + \gamma v_t(s_{t+1})|s_t] - v_\pi(s_t).
\end{aligned}$$

Since $v_\pi(s_t) = \mathbb{E}[r_{t+1} + \gamma v_\pi(s_{t+1})|s_t]$, the above equation implies that

$$\begin{aligned}
\mathbb{E}[\eta_t(s)] &= \gamma \mathbb{E}[v_t(s_{t+1}) - v_\pi(s_{t+1})|s_t] \\
&= \gamma \sum_{s' \in \mathcal{S}} p(s'|s_t)[v_t(s') - v_\pi(s')].
\end{aligned}$$

It follows that

$$\begin{aligned}
|\mathbb{E}[\eta_t(s)]| &= \gamma \left| \sum_{s' \in \mathcal{S}} p(s'|s_t)[v_t(s') - v_\pi(s')] \right| \\
&\leq \gamma \sum_{s' \in \mathcal{S}} p(s'|s_t) \max_{s' \in \mathcal{S}} |v_t(s') - v_\pi(s')| \\
&= \gamma \max_{s' \in \mathcal{S}} |v_t(s') - v_\pi(s')| \\
&= \gamma \|v_t(s') - v_\pi(s')\|_\infty \\
&= \gamma \|\Delta_t(s)\|_\infty. \tag{7.11}
\end{aligned}$$

Therefore, at time t, we know from (7.10) and (7.11) that $|\mathbb{E}[\eta_t(s)]| \leq \gamma \|\Delta_t(s)\|_\infty$ for all $s \in \mathcal{S}$ regardless of whether $s = s_t$. Thus,

$$\|\mathbb{E}[\eta_t(s)]\|_\infty \leq \gamma \|\Delta_t(s)\|_\infty,$$

which is the second condition in Theorem 6.3. Finally, regarding the third condition, we have $\text{var}[\eta_t(s)|\mathcal{H}_t] = \text{var}[r_{t+1} + \gamma v_t(s_{t+1}) - v_\pi(s_t)|s_t] = \text{var}[r_{t+1} + \gamma v_t(s_{t+1})|s_t]$ for $s = s_t$ and $\text{var}[\eta_t(s)|\mathcal{H}_t] = 0$ for $s \neq s_t$. Since r_{t+1} is bounded, the third condition can be proven without difficulty.

The above proof is inspired by [32].

7.2 TD learning of action values: Sarsa

The TD algorithm introduced in Section 7.1 can only estimate *state values*. This section introduces another TD algorithm called Sarsa that can directly estimate *action values*. Estimating action values is important because it can be combined with a policy improvement step to learn optimal policies.

7.2.1 Algorithm description

Given a policy π, our goal is to estimate the action values. Suppose that we have some experience samples generated following π: $(s_0, a_0, r_1, s_1, a_1, \ldots, s_t, a_t, r_{t+1}, s_{t+1}, a_{t+1}, \ldots)$. We can use the following *Sarsa* algorithm to estimate the action values:

$$q_{t+1}(s_t, a_t) = q_t(s_t, a_t) - \alpha_t(s_t, a_t)\Big[q_t(s_t, a_t) - (r_{t+1} + \gamma q_t(s_{t+1}, a_{t+1}))\Big], \qquad (7.12)$$

$$q_{t+1}(s, a) = q_t(s, a), \quad \text{for all } (s, a) \neq (s_t, a_t),$$

where $t = 0, 1, 2, \ldots$ and $\alpha_t(s_t, a_t)$ is the learning rate. Here, $q_t(s_t, a_t)$ is the estimate of $q_\pi(s_t, a_t)$. At time t, only the q-value of (s_t, a_t) is updated, whereas the q-values of the others remain the same.

Some important properties of the Sarsa algorithm are discussed as follows.

⋄ Why is this algorithm called "Sarsa"? That is because each iteration of the algorithm requires $(s_t, a_t, r_{t+1}, s_{t+1}, a_{t+1})$. Sarsa is an abbreviation for state-action-reward-state-action. The Sarsa algorithm was first proposed in [35] and its name was coined by [3].

⋄ Why is Sarsa designed in this way? One may have noticed that Sarsa is similar to the TD algorithm in (7.1). In fact, Sarsa can be easily obtained from the TD algorithm by replacing state value estimation with action value estimation.

⋄ What does Sarsa do mathematically? Similar to the TD algorithm in (7.1), Sarsa is a stochastic approximation algorithm for solving the Bellman equation of a given policy:

$$q_\pi(s, a) = \mathbb{E}\left[R + \gamma q_\pi(S', A')|s, a\right], \quad \text{for all } (s, a). \qquad (7.13)$$

Equation (7.13) is the Bellman equation expressed in terms of action values. A proof is given in Box 7.3.

Box 7.3: Showing that (7.13) is the Bellman equation

As introduced in Section 2.8.2, the Bellman equation expressed in terms of action values is

$$q_\pi(s, a) = \sum_r rp(r|s, a) + \gamma \sum_{s'} \sum_{a'} q_\pi(s', a')p(s'|s, a)\pi(a'|s')$$

$$= \sum_r rp(r|s, a) + \gamma \sum_{s'} p(s'|s, a) \sum_{a'} q_\pi(s', a')\pi(a'|s'). \quad (7.14)$$

This equation establishes the relationships among the action values. Since

$$p(s', a'|s, a) = p(s'|s, a)p(a'|s', s, a)$$

$$= p(s'|s, a)p(a'|s') \quad \text{(due to conditional independence)}$$

$$\doteq p(s'|s, a)\pi(a'|s'),$$

(7.14) can be rewritten as

$$q_\pi(s, a) = \sum_r rp(r|s, a) + \gamma \sum_{s'} \sum_{a'} q_\pi(s', a')p(s', a'|s, a).$$

By the definition of the expected value, the above equation is equivalent to (7.13). Hence, (7.13) is the Bellman equation.

◇ Is Sarsa convergent? Since Sarsa is the action-value version of the TD algorithm in (7.1), the convergence result is similar to Theorem 7.1 and given below.

Theorem 7.2 (Convergence of Sarsa). *Given a policy π, by the Sarsa algorithm in (7.12), $q_t(s, a)$ converges almost surely to the action value $q_\pi(s, a)$ as $t \to \infty$ for all (s, a) if $\sum_t \alpha_t(s, a) = \infty$ and $\sum_t \alpha_t^2(s, a) < \infty$ for all (s, a).*

The proof is similar to that of Theorem 7.1 and is omitted here. The condition of $\sum_t \alpha_t(s, a) = \infty$ and $\sum_t \alpha_t^2(s, a) < \infty$ should be valid for all (s, a). In particular, $\sum_t \alpha_t(s, a) = \infty$ requires that every state-action pair must be visited an infinite (or sufficiently many) number of times. At time t, if $(s, a) = (s_t, a_t)$, then $\alpha_t(s, a) > 0$; otherwise, $\alpha_t(s, a) = 0$.

7.2.2 Optimal policy learning via Sarsa

The Sarsa algorithm in (7.12) can only estimate the action values of a given policy. To find optimal policies, we can combine it with a policy improvement step. The combination is also often called Sarsa, and its implementation procedure is given in Algorithm 7.1.

Algorithm 7.1: Optimal policy learning by Sarsa

Initialization: $\alpha_t(s, a) = \alpha > 0$ for all (s, a) and all t. $\epsilon \in (0, 1)$. Initial $q_0(s, a)$ for all (s, a). Initial ϵ-greedy policy π_0 derived from q_0.

Goal: Learn an optimal policy that can lead the agent to the target state from an initial state s_0.

For each episode, do
    Generate a_0 at s_0 following $\pi_0(s_0)$
    If s_t $(t = 0, 1, 2, \dots)$ is not the target state, do
        Collect an experience sample $(r_{t+1}, s_{t+1}, a_{t+1})$ given (s_t, a_t): generate r_{t+1}, s_{t+1}
        by interacting with the environment; generate a_{t+1} following $\pi_t(s_{t+1})$.
        Update q-value for (s_t, a_t):
$$q_{t+1}(s_t, a_t) = q_t(s_t, a_t) - \alpha_t(s_t, a_t)\Big[q_t(s_t, a_t) - (r_{t+1} + \gamma q_t(s_{t+1}, a_{t+1}))\Big]$$
        Update policy for s_t:
$$\pi_{t+1}(a|s_t) = 1 - \frac{\epsilon}{|\mathcal{A}(s_t)|}(|\mathcal{A}(s_t)| - 1) \text{ if } a = \arg\max_a q_{t+1}(s_t, a)$$
$$\pi_{t+1}(a|s_t) = \frac{\epsilon}{|\mathcal{A}(s_t)|} \text{ otherwise}$$
    $s_t \leftarrow s_{t+1}$, $a_t \leftarrow a_{t+1}$

As shown in Algorithm 7.1, each iteration has two steps. The first step is to update the q-value of the visited state-action pair. The second step is to update the policy to an ϵ-greedy one. The q-value update step only updates the single state-action pair visited at time t. Afterward, the policy of s_t is immediately updated. Therefore, we do *not* evaluate a given policy sufficiently well before updating the policy. This is based on the idea of generalized policy iteration. Moreover, after the policy is updated, the policy is immediately used to generate the next experience sample. The policy here is ϵ-greedy so that it is exploratory.

A simulation example is shown in Figure 7.2 to demonstrate the Sarsa algorithm. Unlike all the tasks we have seen in this book, the task here aims to find an optimal path from a specific starting state to a target state. It does *not* aim to find the optimal policies for all states. This task is often encountered in practice where the starting state (e.g., home) and the target state (e.g., workplace) are fixed, and we only need to find an optimal path connecting them. This task is relatively simple because we only need to explore the states that are close to the path and do not need to explore all the states. However, if we do not explore all the states, the final path may be *locally* optimal rather than globally optimal.

The simulation setup and simulation results are discussed below.

◇ *Simulation setup:* In this example, all the episodes start from the top-left state and terminate at the target state. The reward settings are $r_{\text{target}} = 0$, $r_{\text{forbidden}} = r_{\text{boundary}} = -10$, and $r_{\text{other}} = -1$. Moreover, $\alpha_t(s, a) = 0.1$ for all t and $\epsilon = 0.1$. The initial guesses of the action values are $q_0(s, a) = 0$ for all (s, a). The initial policy has a

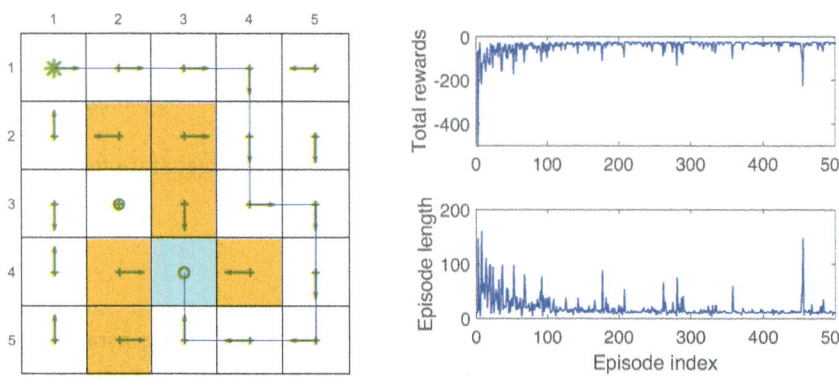

Figure 7.2: An example for demonstrating Sarsa. All the episodes start from the top-left state and terminate when reaching the target state (the blue cell). The goal is to find an optimal path from the starting state to the target state. The reward settings are $r_{\text{target}} = 0$, $r_{\text{forbidden}} = r_{\text{boundary}} = -10$, and $r_{\text{other}} = -1$. The learning rate is $\alpha = 0.1$ and the value of ϵ is 0.1. The left figure shows the final policy obtained by the algorithm. The right figures show the total reward and length of every episode.

uniform distribution: $\pi_0(a|s) = 0.2$ for all s, a.

⋄ *Learned policy:* The left figure in Figure 7.2 shows the final policy learned by Sarsa. As can be seen, this policy can successfully lead to the target state from the starting state. However, the policies of some other states may not be optimal. That is because the other states are not well explored.

⋄ *Total reward of each episode:* The top-right subfigure in Figure 7.2 shows the total reward of each episode. Here, the total reward is the non-discounted sum of all immediate rewards. As can be seen, the total reward of each episode increases gradually. That is because the initial policy is not good and hence negative rewards are frequently obtained. As the policy becomes better, the total reward increases.

⋄ *Length of each episode:* The bottom-right subfigure in Figure 7.2 shows that the length of each episode drops gradually. That is because the initial policy is not good and may take many detours before reaching the target. As the policy becomes better, the length of the trajectory becomes shorter. Notably, the length of an episode may increase abruptly (e.g., the 460th episode) and the corresponding total reward also drops sharply. That is because the policy is ϵ-greedy, and there is a chance for it to take non-optimal actions. One way to resolve this problem is to use decaying ϵ whose value converges to zero gradually.

Finally, Sarsa also has some variants such as Expected Sarsa. Interested readers may check Box 7.4.

Box 7.4: Expected Sarsa

Given a policy π, its action values can be evaluated by Expected Sarsa, which is a variant of Sarsa. The Expected Sarsa algorithm is

$$q_{t+1}(s_t, a_t) = q_t(s_t, a_t) - \alpha_t(s_t, a_t)\Big[q_t(s_t, a_t) - (r_{t+1} + \gamma\mathbb{E}[q_t(s_{t+1}, A)])\Big],$$

$$q_{t+1}(s, a) = q_t(s, a), \quad \text{for all } (s, a) \neq (s_t, a_t),$$

where

$$\mathbb{E}[q_t(s_{t+1}, A)] = \sum_a \pi_t(a|s_{t+1})q_t(s_{t+1}, a) \doteq v_t(s_{t+1})$$

is the expected value of $q_t(s_{t+1}, a)$ under policy π_t. The expression of the Expected Sarsa algorithm is very similar to that of Sarsa. They are different only in terms of their TD targets. In particular, the TD target in Expected Sarsa is $r_{t+1} + \gamma\mathbb{E}[q_t(s_{t+1}, A)]$, while that of Sarsa is $r_{t+1} + \gamma q_t(s_{t+1}, a_{t+1})$. Since the algorithm involves an expected value, it is called Expected Sarsa. Although calculating the expected value may increase the computational complexity slightly, it is beneficial in the sense that it reduces the estimation variances because it reduces the random variables in Sarsa from $\{s_t, a_t, r_{t+1}, s_{t+1}, a_{t+1}\}$ to $\{s_t, a_t, r_{t+1}, s_{t+1}\}$.

Similar to the TD learning algorithm in (7.1), Expected Sarsa can be viewed as a stochastic approximation algorithm for solving the following equation:

$$q_\pi(s, a) = \mathbb{E}\Big[R_{t+1} + \gamma\mathbb{E}[q_\pi(S_{t+1}, A_{t+1})|S_{t+1}]\Big|S_t = s, A_t = a\Big], \quad \text{for all } s, a. \quad (7.15)$$

The above equation may look strange at first glance. In fact, it is another expression of the Bellman equation. To see that, substituting

$$\mathbb{E}[q_\pi(S_{t+1}, A_{t+1})|S_{t+1}] = \sum_{A'} q_\pi(S_{t+1}, A')\pi(A'|S_{t+1}) = v_\pi(S_{t+1})$$

into (7.15) gives

$$q_\pi(s, a) = \mathbb{E}\Big[R_{t+1} + \gamma v_\pi(S_{t+1})|S_t = s, A_t = a\Big],$$

which is clearly the Bellman equation.

The implementation of Expected Sarsa is similar to that of Sarsa. More details can be found in [3, 36, 37].

7.3 TD learning of action values: n-step Sarsa

This section introduces *n-step Sarsa*, an extension of Sarsa. We will see that Sarsa and MC learning are two extreme cases of n-step Sarsa.

Recall that the definition of the action value is

$$q_\pi(s, a) = \mathbb{E}[G_t | S_t = s, A_t = a], \tag{7.16}$$

where G_t is the discounted return satisfying

$$G_t = R_{t+1} + \gamma R_{t+2} + \gamma^2 R_{t+3} + \ldots.$$

In fact, G_t can also be decomposed into different forms:

$$
\begin{aligned}
\text{Sarsa} \longleftarrow \quad G_t^{(1)} &= R_{t+1} + \gamma q_\pi(S_{t+1}, A_{t+1}), \\
G_t^{(2)} &= R_{t+1} + \gamma R_{t+2} + \gamma^2 q_\pi(S_{t+2}, A_{t+2}), \\
&\vdots \\
\text{n-step Sarsa} \longleftarrow \quad G_t^{(n)} &= R_{t+1} + \gamma R_{t+2} + \cdots + \gamma^n q_\pi(S_{t+n}, A_{t+n}), \\
&\vdots \\
\text{MC} \longleftarrow \quad G_t^{(\infty)} &= R_{t+1} + \gamma R_{t+2} + \gamma^2 R_{t+3} + \gamma^3 R_{t+4} \ldots
\end{aligned}
$$

It should be noted that $G_t = G_t^{(1)} = G_t^{(2)} = G_t^{(n)} = G_t^{(\infty)}$, where the superscripts merely indicate the different decomposition structures of G_t.

Substituting different decompositions of $G_t^{(n)}$ into $q_\pi(s, a)$ in (7.16) results in different algorithms.

◇ When $n = 1$, we have

$$q_\pi(s, a) = \mathbb{E}[G_t^{(1)} | s, a] = \mathbb{E}[R_{t+1} + \gamma q_\pi(S_{t+1}, A_{t+1}) | s, a].$$

The corresponding stochastic approximation algorithm for solving this equation is

$$q_{t+1}(s_t, a_t) = q_t(s_t, a_t) - \alpha_t(s_t, a_t)\Big[q_t(s_t, a_t) - (r_{t+1} + \gamma q_t(s_{t+1}, a_{t+1}))\Big],$$

which is the Sarsa algorithm in (7.12).

◇ When $n = \infty$, we have

$$q_\pi(s, a) = \mathbb{E}[G_t^{(\infty)} | s, a] = \mathbb{E}[R_{t+1} + \gamma R_{t+2} + \gamma^2 R_{t+3} + \ldots | s, a].$$

The corresponding algorithm for solving this equation is

$$q_{t+1}(s_t, a_t) = g_t \doteq r_{t+1} + \gamma r_{t+2} + \gamma^2 r_{t+3} + \cdots,$$

where g_t is a sample of G_t. In fact, this is the MC learning algorithm, which approximates the action value of (s_t, a_t) using the discounted return of an episode starting from (s_t, a_t).

⋄ For a general value of n, we have

$$q_\pi(s, a) = \mathbb{E}[G_t^{(n)}|s, a] = \mathbb{E}[R_{t+1} + \gamma R_{t+2} + \cdots + \gamma^n q_\pi(S_{t+n}, A_{t+n})|s, a].$$

The corresponding algorithm for solving the above equation is

$$q_{t+1}(s_t, a_t) = q_t(s_t, a_t)$$
$$- \alpha_t(s_t, a_t)\Big[q_t(s_t, a_t) - \big(r_{t+1} + \gamma r_{t+2} + \cdots + \gamma^n q_t(s_{t+n}, a_{t+n})\big)\Big]. \quad (7.17)$$

This algorithm is called *n-step Sarsa*.

In summary, n-step Sarsa is a more general algorithm because it becomes the (one-step) Sarsa algorithm when $n = 1$ and the MC learning algorithm when $n = \infty$ (by setting $\alpha_t = 1$).

To implement the n-step Sarsa algorithm in (7.17), we need the experience samples $(s_t, a_t, r_{t+1}, s_{t+1}, a_{t+1}, \ldots, r_{t+n}, s_{t+n}, a_{t+n})$. Since $(r_{t+n}, s_{t+n}, a_{t+n})$ has not been collected at time t, we have to wait until time $t + n$ to update the q-value of (s_t, a_t). To that end, (7.17) can be rewritten as

$$q_{t+n}(s_t, a_t) = q_{t+n-1}(s_t, a_t)$$
$$- \alpha_{t+n-1}(s_t, a_t)\Big[q_{t+n-1}(s_t, a_t) - \big(r_{t+1} + \gamma r_{t+2} + \cdots + \gamma^n q_{t+n-1}(s_{t+n}, a_{t+n})\big)\Big],$$

where $q_{t+n}(s_t, a_t)$ is the estimate of $q_\pi(s_t, a_t)$ at time $t + n$.

Since n-step Sarsa includes Sarsa and MC learning as two extreme cases, it is not surprising that the performance of n-step Sarsa is between that of Sarsa and MC learning. In particular, if n is selected as a large number, n-step Sarsa is close to MC learning: the estimate has a relatively high variance but a small bias. If n is selected to be small, n-step Sarsa is close to Sarsa: the estimate has a relatively large bias but a low variance. Finally, the n-step Sarsa algorithm presented here is merely used for policy evaluation. It must be combined with a policy improvement step to learn optimal policies. The implementation is similar to that of Sarsa and is omitted here. Interested readers may check [3, Chapter 7] for a detailed analysis of multi-step TD learning.

7.4 TD learning of optimal action values: Q-learning

In this section, we introduce the Q-learning algorithm, one of the most classic reinforcement learning algorithms [38,39]. Recall that Sarsa can only estimate the action values of a given policy, and it must be combined with a policy improvement step to find optimal policies. By contrast, Q-learning can directly estimate optimal action values and find optimal policies.

7.4.1 Algorithm description

The Q-learning algorithm is

$$q_{t+1}(s_t, a_t) = q_t(s_t, a_t) - \alpha_t(s_t, a_t) \left[q_t(s_t, a_t) - \left(r_{t+1} + \gamma \max_{a \in \mathcal{A}(s_{t+1})} q_t(s_{t+1}, a) \right) \right], \quad (7.18)$$

$$q_{t+1}(s, a) = q_t(s, a), \quad \text{for all } (s, a) \neq (s_t, a_t),$$

where $t = 0, 1, 2, \ldots$. Here, $q_t(s_t, a_t)$ is the estimate of the *optimal* action value of (s_t, a_t) and $\alpha_t(s_t, a_t)$ is the learning rate for (s_t, a_t).

The expression of Q-learning is similar to that of Sarsa. They are different only in terms of their TD targets: the TD target of Q-learning is $r_{t+1} + \gamma \max_a q_t(s_{t+1}, a)$, whereas that of Sarsa is $r_{t+1} + \gamma q_t(s_{t+1}, a_{t+1})$. Moreover, given (s_t, a_t), Sarsa requires $(r_{t+1}, s_{t+1}, a_{t+1})$ in every iteration, whereas Q-learning merely requires (r_{t+1}, s_{t+1}).

Why is Q-learning designed as the expression in (7.18), and what does it do mathematically? Q-learning is a stochastic approximation algorithm for solving the following equation:

$$q(s, a) = \mathbb{E} \left[R_{t+1} + \gamma \max_a q(S_{t+1}, a) \Big| S_t = s, A_t = a \right]. \quad (7.19)$$

This is the Bellman optimality equation expressed in terms of action values. The proof is given in Box 7.5. The convergence analysis of Q-learning is similar to Theorem 7.1 and omitted here. More information can be found in [32, 39].

Box 7.5: Showing that (7.19) is the Bellman optimality equation

By the definition of expectation, (7.19) can be rewritten as

$$q(s, a) = \sum_r p(r|s, a)r + \gamma \sum_{s'} p(s'|s, a) \max_{a \in \mathcal{A}(s')} q(s', a).$$

Taking the maximum of both sides of the equation gives

$$\max_{a \in \mathcal{A}(s)} q(s, a) = \max_{a \in \mathcal{A}(s)} \left[\sum_r p(r|s, a)r + \gamma \sum_{s'} p(s'|s, a) \max_{a \in \mathcal{A}(s')} q(s', a) \right].$$

By denoting $v(s) \doteq \max_{a \in \mathcal{A}(s)} q(s, a)$, we can rewrite the above equation as

$$v(s) = \max_{a \in \mathcal{A}(s)} \left[\sum_r p(r|s, a)r + \gamma \sum_{s'} p(s'|s, a)v(s') \right]$$

$$= \max_{\pi} \sum_{a \in \mathcal{A}(s)} \pi(a|s) \left[\sum_r p(r|s, a)r + \gamma \sum_{s'} p(s'|s, a)v(s') \right],$$

which is clearly the Bellman optimality equation in terms of state values as introduced in Chapter 3.

7.4.2   Off-policy vs on-policy

We next introduce two important concepts: *on-policy learning* and *off-policy learning*. What makes Q-learning slightly special compared to the other TD algorithms is that Q-learning is off-policy while the others are on-policy.

Two policies exist in any reinforcement learning task: a *behavior policy* and a *target policy*. The behavior policy is the one used to generate experience samples. The target policy is the one that is constantly updated to converge to an optimal policy. When the behavior policy is the same as the target policy, such a learning process is called *on-policy*. Otherwise, when they are different, the learning process is called *off-policy*.

The advantage of off-policy learning is that it can learn optimal policies based on the experience samples generated by other policies, which may be, for example, a policy executed by a human operator. As an important case, the behavior policy can be selected to be *exploratory*. For example, if we would like to estimate the action values of all state-action pairs, we must generate episodes visiting every state-action pair sufficiently many times. Although Sarsa uses ϵ-greedy policies to maintain certain exploration abilities, the value of ϵ is usually small and hence the exploration ability is limited. By contrast, if we can use a policy with a strong exploration ability to generate episodes and then use off-policy learning to learn optimal policies, the learning efficiency would be significantly increased.

To determine if an algorithm is on-policy or off-policy, we can examine two aspects. The first is the mathematical problem that the algorithm aims to solve. The second is the experience samples required by the algorithm.

⋄   Sarsa is on-policy.

The reason is as follows. Sarsa has two steps in every iteration. The first step is to evaluate a policy π by solving its Bellman equation. To do that, we need samples generated by π. Therefore, π is the behavior policy. The second step is to obtain an improved policy based on the estimated values of π. As a result, π is the target policy that is constantly updated and eventually converges to an optimal policy. Therefore, the behavior policy and the target policy are the same.

From another point of view, we can examine the samples required by the algorithm. The samples required by Sarsa in every iteration include $(s_t, a_t, r_{t+1}, s_{t+1}, a_{t+1})$. How these samples are generated is illustrated below:

$$s_t \xrightarrow{\pi_b} a_t \xrightarrow{\text{model}} r_{t+1}, s_{t+1} \xrightarrow{\pi_b} a_{t+1}$$

As can be seen, the behavior policy π_b is the one that generates a_t at s_t and a_{t+1} at s_{t+1}. The Sarsa algorithm aims to estimate the *action value* of (s_t, a_t) of a policy denoted as π_T, which is the target policy because it is improved in every iteration based on the estimated values. In fact, π_T is the same as π_b because the evaluation of π_T relies on the samples $(r_{t+1}, s_{t+1}, a_{t+1})$, where a_{t+1} is generated following π_b. In other words, the policy that Sarsa evaluates is the policy used to generate samples.

◇ Q-learning is off-policy.

The fundamental reason is that Q-learning is an algorithm for solving the *Bellman optimality equation*, whereas Sarsa is for solving the *Bellman equation* of a given policy. While solving the Bellman equation can evaluate the associated policy, solving the Bellman optimality equation can directly generate the optimal values and optimal policies.

In particular, the samples required by Q-learning in every iteration is $(s_t, a_t, r_{t+1}, s_{t+1})$. How these samples are generated is illustrated below:

$$s_t \xrightarrow{\pi_b} a_t \xrightarrow{\text{model}} r_{t+1}, s_{t+1}$$

As can be seen, the behavior policy π_b is the one that generates a_t at s_t. The Q-learning algorithm aims to estimate the *optimal action value* of (s_t, a_t). This estimation process relies on the samples (r_{t+1}, s_{t+1}). The process of generating (r_{t+1}, s_{t+1}) does not involve π_b because it is governed by the system model (or by interacting with the environment). Therefore, the estimation of the optimal action value of (s_t, a_t) does not involve π_b and we can use any π_b to generate a_t at s_t. Moreover, the target policy π_T here is the greedy policy obtained based on the estimated optimal values (Algorithm 7.3). The behavior policy does not have to be the same as π_T.

◇ MC learning is on-policy. The reason is similar to that of Sarsa. The target policy to be evaluated and improved is the same as the behavior policy that generates samples.

Another concept that may be confused with on-policy/off-policy is *online/offline*. Online learning refers to the case where the agent updates the values and policies while interacting with the environment. Offline learning refers to the case where the agent updates the values and policies using pre-collected experience data without interacting with the environment. If an algorithm is on-policy, then it can be implemented in an online fashion, but cannot use pre-collected data generated by other policies. If an algorithm is off-policy, then it can be implemented in either an online or offline fashion.

Algorithm 7.2: Optimal policy learning via Q-learning (on-policy version)

Initialization: $\alpha_t(s, a) = \alpha > 0$ for all (s, a) and all t. $\epsilon \in (0, 1)$. Initial $q_0(s, a)$ for all (s, a). Initial ϵ-greedy policy π_0 derived from q_0.
Goal: Learn an optimal path that can lead the agent to the target state from an initial state s_0.

For each episode, do
    If s_t $(t = 0, 1, 2, \dots)$ is not the target state, do
        Collect the experience sample (a_t, r_{t+1}, s_{t+1}) given s_t: generate a_t following $\pi_t(s_t)$; generate r_{t+1}, s_{t+1} by interacting with the environment.
        Update q-value for (s_t, a_t):
$$q_{t+1}(s_t, a_t) = q_t(s_t, a_t) - \alpha_t(s_t, a_t)\Big[q_t(s_t, a_t) - (r_{t+1} + \gamma \max_a q_t(s_{t+1}, a))\Big]$$
        Update policy for s_t:
$$\pi_{t+1}(a|s_t) = 1 - \tfrac{\epsilon}{|\mathcal{A}(s_t)|}(|\mathcal{A}(s_t)| - 1) \text{ if } a = \arg\max_a q_{t+1}(s_t, a)$$
$$\pi_{t+1}(a|s_t) = \tfrac{\epsilon}{|\mathcal{A}(s_t)|} \text{ otherwise}$$

Algorithm 7.3: Optimal policy learning via Q-learning (off-policy version)

Initialization: Initial guess $q_0(s, a)$ for all (s, a). Behavior policy $\pi_b(a|s)$ for all (s, a). $\alpha_t(s, a) = \alpha > 0$ for all (s, a) and all t.
Goal: Learn an optimal target policy π_T for all states from the experience samples generated by π_b.

For each episode $\{s_0, a_0, r_1, s_1, a_1, r_2, \dots\}$ generated by π_b, do
    For each step $t = 0, 1, 2, \dots$ of the episode, do
        Update q-value for (s_t, a_t):
$$q_{t+1}(s_t, a_t) = q_t(s_t, a_t) - \alpha_t(s_t, a_t)\Big[q_t(s_t, a_t) - (r_{t+1} + \gamma \max_a q_t(s_{t+1}, a))\Big]$$
        Update target policy for s_t:
$$\pi_{T,t+1}(a|s_t) = 1 \text{ if } a = \arg\max_a q_{t+1}(s_t, a)$$
$$\pi_{T,t+1}(a|s_t) = 0 \text{ otherwise}$$

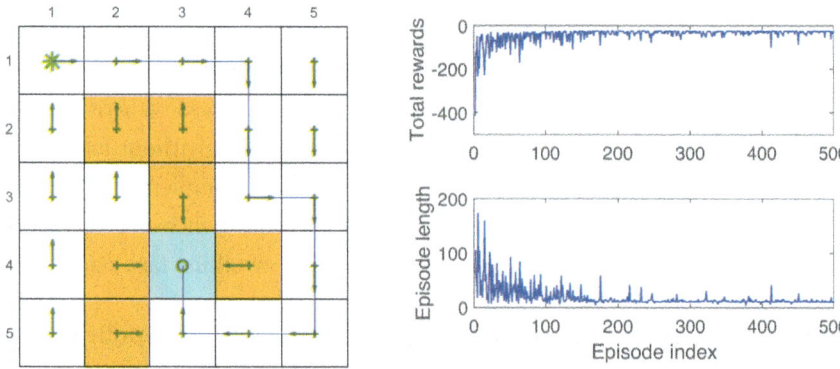

Figure 7.3: An example for demonstrating Q-learning. All the episodes start from the top-left state and terminate after reaching the target state. The aim is to find an optimal path from the starting state to the target state. The reward settings are $r_{\text{target}} = 0$, $r_{\text{forbidden}} = r_{\text{boundary}} = -10$, and $r_{\text{other}} = -1$. The learning rate is $\alpha = 0.1$ and the value of ϵ is 0.1. The left figure shows the final policy obtained by the algorithm. The right figure shows the total reward and length of every episode.

7.4.3 Implementation

Since Q-learning is off-policy, it can be implemented in either an on-policy or off-policy fashion.

The on-policy version of Q-learning is shown in Algorithm 7.2. This implementation is similar to the Sarsa one in Algorithm 7.1. Here, the behavior policy is the same as the target policy, which is an ϵ-greedy policy.

The off-policy version is shown in Algorithm 7.3. The behavior policy π_b can be any policy as long as it can generate sufficient experience samples. It is usually favorable when π_b is exploratory. Here, the target policy π_T is greedy rather than ϵ-greedy since it is not used to generate samples and hence is not required to be exploratory. Moreover, the off-policy version of Q-learning presented here is implemented offline: all the experience samples are collected first and then processed.

7.4.4 Illustrative examples

We next present examples to demonstrate Q-learning.

The first example is shown in Figure 7.3. It demonstrates on-policy Q-learning. The *goal* here is to find an optimal path from a starting state to the target state. The setup is given in the caption of Figure 7.3. As can be seen, Q-learning can eventually find an optimal path. During the learning process, the length of each episode decreases, whereas the total reward of each episode increases.

The second set of examples is shown in Figure 7.4 and Figure 7.5. They demonstrate off-policy Q-learning. The *goal* here is to find an optimal policy for all the states. The reward setting is $r_{\text{boundary}} = r_{\text{forbidden}} = -1$, and $r_{\text{target}} = 1$. The discount rate is $\gamma = 0.9$. The learning rate is $\alpha = 0.1$.

◇ *Ground truth:* To verify the effectiveness of Q-learning, we first need to know the ground truth of the optimal policies and optimal state values. Here, the ground truth is obtained by the model-based policy iteration algorithm. The ground truth is given in Figures 7.4(a) and (b).

◇ *Experience samples:* The behavior policy has a uniform distribution: the probability of taking any action at any state is 0.2 (Figure 7.4(c)). A single episode with 100,000 steps is generated (Figure 7.4(d)). Due to the good exploration ability of the behavior policy, the episode visits every state-action pair many times.

◇ *Learned results:* Based on the episode generated by the behavior policy, the final target policy learned by Q-learning is shown in Figure 7.4(e). This policy is optimal because the estimated state value error (root-mean-square error) converges to zero as shown in Figure 7.4(f). In addition, one may notice that the learned optimal policy is not exactly the same as that in Figure 7.4(a). In fact, there exist multiple optimal policies that have the same optimal state values.

◇ *Different initial values:* Since Q-learning bootstraps, the performance of the algorithm depends on the initial guess for the action values. As shown in Figure 7.4(g), when the initial guess is close to the true value, the estimate converges within approximately 10,000 steps. Otherwise, the convergence requires more steps (Figure 7.4(h)). Nevertheless, these figures demonstrate that Q-learning can still converge rapidly even though the initial value is not accurate.

◇ *Different behavior policies:* When the behavior policy is not exploratory, the learning performance drops significantly. For example, consider the behavior policies shown in Figure 7.5. They are ϵ-greedy policies with $\epsilon = 0.5$ or 0.1 (the uniform policy in Figure 7.4(c) can be viewed as ϵ-greedy with $\epsilon = 1$). It is shown that, when ϵ decreases from 1 to 0.5 and then to 0.1, the learning speed drops significantly. That is because the exploration ability of the policy is weak and hence the experience samples are insufficient.

7.5 A unified viewpoint

Up to now, we have introduced different TD algorithms such as Sarsa, n-step Sarsa, and Q-learning. In this section, we introduce a unified framework to accommodate all these algorithms and MC learning.

In particular, the TD algorithms (for action value estimation) can be expressed in a unified expression:

$$q_{t+1}(s_t, a_t) = q_t(s_t, a_t) - \alpha_t(s_t, a_t)[q_t(s_t, a_t) - \bar{q}_t], \tag{7.20}$$

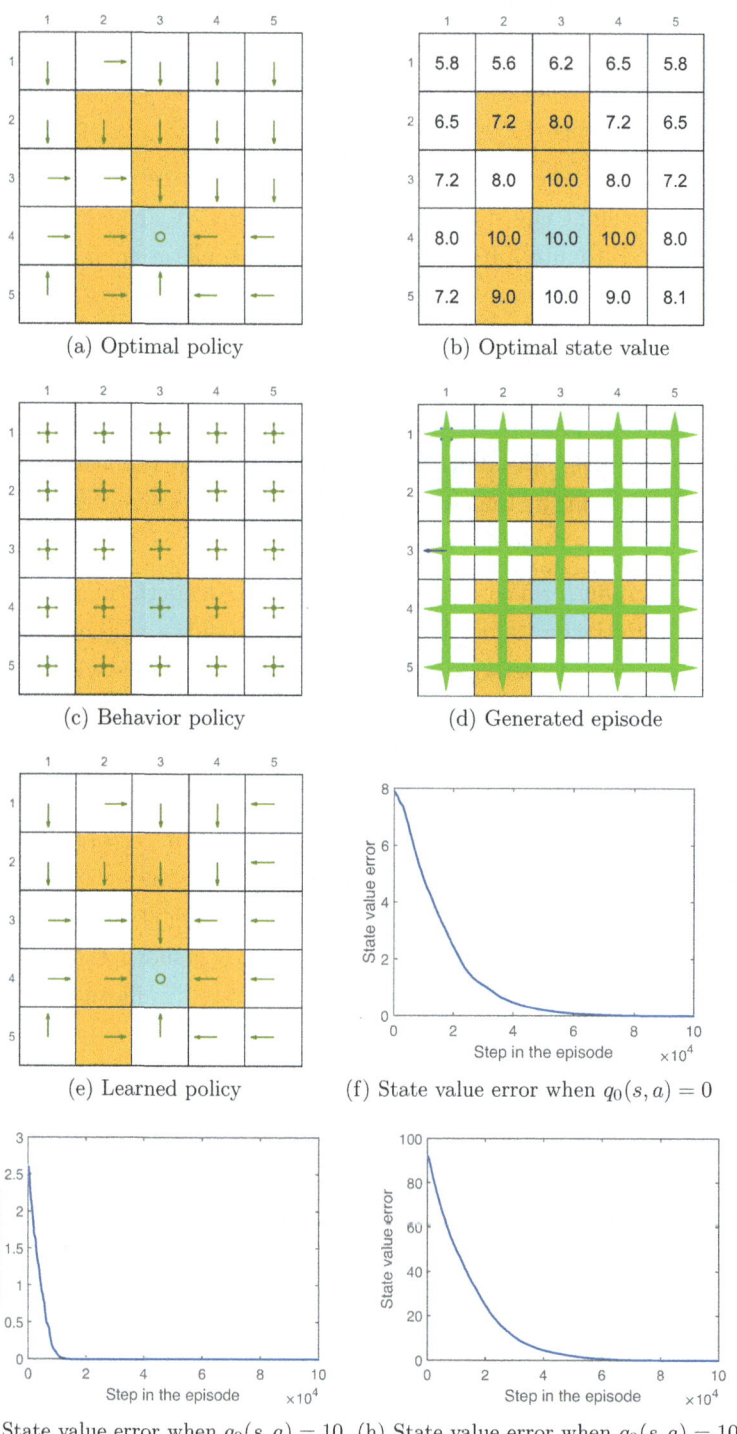

Figure 7.4: Examples for demonstrating off-policy learning via Q-learning. The optimal policy and optimal state values are shown in (a) and (b), respectively. The behavior policy and the generated episode are shown in (c) and (d), respectively. The estimated policy and the estimation error evolution are shown in (e) and (f), respectively. The cases with different initial values are shown in (g) and (h).

146

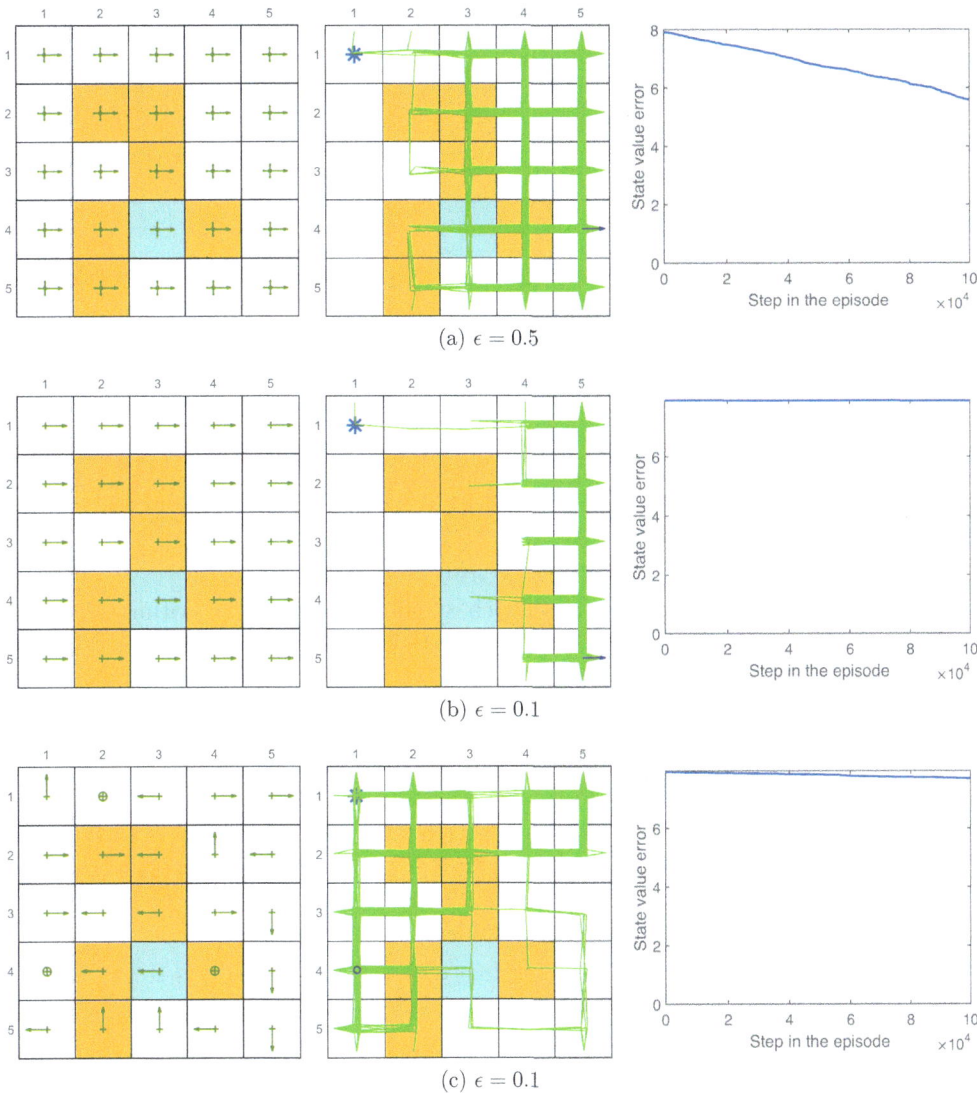

(a) $\epsilon = 0.5$

(b) $\epsilon = 0.1$

(c) $\epsilon = 0.1$

Figure 7.5: The performance of Q-learning drops when the behavior policy is not exploratory. The figures in the left column show the behavior policies. The figures in the middle column show the generated episodes following the corresponding behavior policies. The episode in each example has 100,000 steps. The figures in the right column show the evolution of the root-mean-square error of the estimated state values.

Algorithm	Expression of the TD target $\bar{q}_t$ in (7.20)
Sarsa	$\bar{q}_t = r_{t+1} + \gamma q_t(s_{t+1}, a_{t+1})$
n-step Sarsa	$\bar{q}_t = r_{t+1} + \gamma r_{t+2} + \cdots + \gamma^n q_t(s_{t+n}, a_{t+n})$
Q-learning	$\bar{q}_t = r_{t+1} + \gamma \max_a q_t(s_{t+1}, a)$
Monte Carlo	$\bar{q}_t = r_{t+1} + \gamma r_{t+2} + \gamma^2 r_{t+3} + \ldots$

Algorithm	Equation to be solved
Sarsa	BE: $q_\pi(s, a) = \mathbb{E}\left[R_{t+1} + \gamma q_\pi(S_{t+1}, A_{t+1})\vert S_t = s, A_t = a\right]$
n-step Sarsa	BE: $q_\pi(s, a) = \mathbb{E}[R_{t+1} + \gamma R_{t+2} + \cdots + \gamma^n q_\pi(S_{t+n}, A_{t+n})\vert S_t = s, A_t = a]$
Q-learning	BOE: $q(s, a) = \mathbb{E}\left[R_{t+1} + \gamma \max_a q(S_{t+1}, a)\big\vert S_t = s, A_t = a\right]$
Monte Carlo	BE: $q_\pi(s, a) = \mathbb{E}[R_{t+1} + \gamma R_{t+2} + \gamma^2 R_{t+3} + \ldots \vert S_t = s, A_t = a]$

Table 7.2: A unified point of view of TD algorithms. Here, BE and BOE denote the Bellman equation and Bellman optimality equation, respectively.

where $\bar{q}_t$ is the *TD target*. Different TD algorithms have different $\bar{q}_t$. See Table 7.2 for a summary. The MC learning algorithm can be viewed as a special case of (7.20): we can set $\alpha_t(s_t, a_t) = 1$ and then (7.20) becomes $q_{t+1}(s_t, a_t) = \bar{q}_t$.

Algorithm (7.20) can be viewed as a stochastic approximation algorithm for solving a unified equation: $q(s, a) = \mathbb{E}[\bar{q}_t \vert s, a]$. This equation has different expressions with different $\bar{q}_t$. These expressions are summarized in Table 7.2. As can be seen, all of the algorithms aim to solve the Bellman equation except Q-learning, which aims to solve the Bellman optimality equation.

7.6 Summary

This chapter introduced an important class of reinforcement learning algorithms called TD learning. The specific algorithms that we introduced include Sarsa, n-step Sarsa, and Q-learning. All these algorithms can be viewed as stochastic approximation algorithms for solving Bellman or Bellman optimality equations.

The TD algorithms introduced in this chapter, except Q-learning, are used to evaluate a given policy. That is to estimate a given policy's state/action values from some experience samples. Together with policy improvement, they can be used to learn optimal policies. Moreover, these algorithms are on-policy: the target policy is used as the behavior policy to generate experience samples.

Q-learning is slightly special compared to the other TD algorithms in the sense that it is off-policy. The target policy can be different from the behavior policy in Q-learning. The fundamental reason why Q-learning is off-policy is that Q-learning aims to solve the Bellman optimality equation rather than the Bellman equation of a given policy.

It is worth mentioning that there are some methods that can convert an on-policy algorithm to be off-policy. Importance sampling is a widely used one [3, 40] and will be introduced in Chapter 10. Finally, there are some variants and extensions of the TD algorithms introduced in this chapter [41–45]. For example, the TD(λ) method provides a more general and unified framework for TD learning. More information can be found in [3, 20, 46].

7.7 Q&A

⋄ Q: What does the term "TD" in TD learning mean?

A: Every TD algorithm has a TD error, which represents the discrepancy between the new sample and the current estimate. Since this discrepancy is calculated between different time steps, it is called temporal-difference.

⋄ Q: What does the term "learning" in TD learning mean?

A: From a mathematical point of view, "learning" simply means "estimation". That is to estimate state/action values from some samples and then obtain policies based on the estimated values.

⋄ Q: While Sarsa can estimate the action values of a given policy, how can it be used to learn optimal policies?

A: To obtain an optimal policy, the value estimation process should interact with the policy improvement process. That is, after a value is updated, the corresponding policy should be updated. Then, the updated policy generates new samples that can be used to estimate values again. This is the idea of generalized policy iteration.

⋄ Q: Why does Sarsa update policies to be ϵ-greedy?

A: That is because the policy is also used to generate samples for value estimation. Hence, it should be exploratory to generate sufficient experience samples.

⋄ Q: While Theorems 7.1 and 7.2 require that the learning rate α_t converges to zero gradually, why is it often set to be a small constant in practice?

A: The fundamental reason is that the policy to be evaluated keeps changing (or called nonstationary). In particular, a TD learning algorithm like Sarsa aims to estimate the action values of a given policy. If the policy is fixed, using a decaying learning rate is acceptable. However, in the optimal policy learning process, the policy that Sarsa aims to evaluate keeps *changing* after every iteration. We need a constant learning rate in this case; otherwise, a decaying learning rate may be too small to effectively evaluate policies. Although a drawback of constant learning rates is that the value estimate may fluctuate eventually, the fluctuation is neglectable as long as the constant learning rate is sufficiently small.

⋄ Q: Should we learn the optimal policies for all states or a subset of the states?

A: It depends on the task. One may notice that some tasks considered in this chapter (e.g., Figure 7.2) do *not* require finding the optimal policies for all states. Instead, they only need to find an optimal path from a given starting state to the target state. Such tasks are not demanding in terms of data because the agent does not need to visit every state-action pair sufficiently many times. It, however, must be noted that the obtained path is not guaranteed to be optimal. That is because better paths may be missed if not all state-action pairs are well explored. Nevertheless, given sufficient data, we can still find a good or locally optimal path.

⋄ Q: Why is Q-learning off-policy while all the other TD algorithms in this chapter are on-policy?

A: The fundamental reason is that Q-learning aims to solve the Bellman optimality equation, whereas the other TD algorithms aim to solve the Bellman equation of a given policy. Details can be found in Section 7.4.2.

⋄ Q: Why does the off-policy version of Q-learning update policies to be greedy instead of ϵ-greedy?

A: That is because the target policy is not required to generate experience samples. Hence, it is not required to be exploratory.

Chapter 8

Value Function Methods

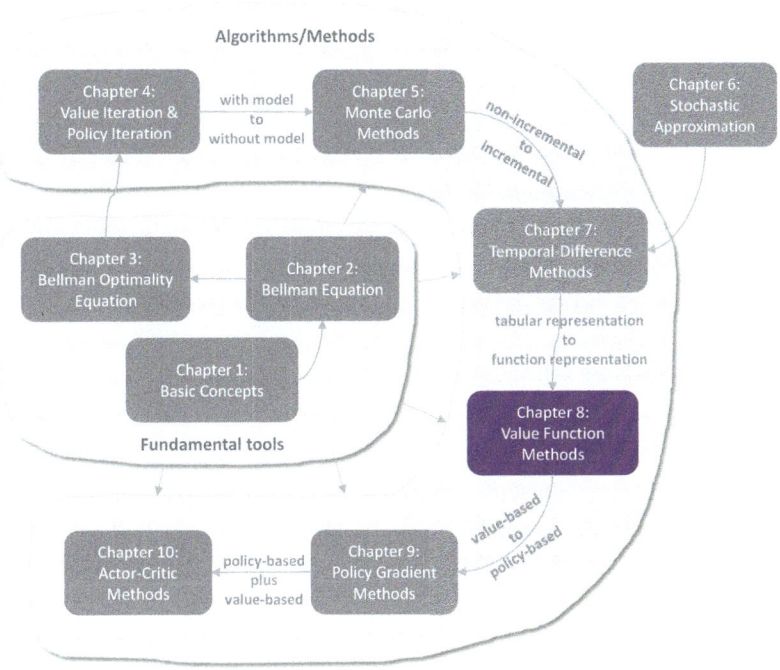

Figure 8.1: Where we are in this book.

In this chapter, we continue to study temporal-difference learning algorithms. However, a different method is used to represent state/action values. So far in this book, state/action values have been represented by *tables*. The tabular method is straightforward to understand, but it is inefficient for handling large state or action spaces. To solve this problem, this chapter introduces the *value function* method, which has become a standard way to represent values. It is also where artificial neural networks are incorporated into reinforcement learning as function approximators. The idea of value function can also be extended to *policy function*, as introduced in Chapter 9.

S. Zhao, *Mathematical Foundations of Reinforcement Learning*, https://doi.org/10.1007/978-981-97-3944-8_8

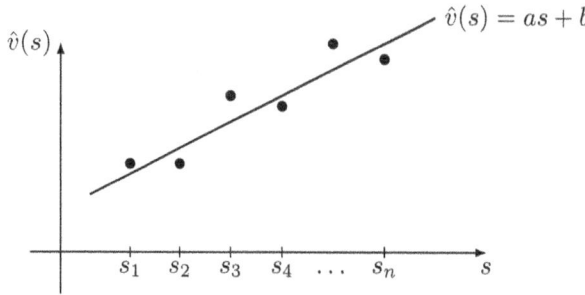

Figure 8.2: An illustration of the function approximation method. The x-axis and y-axis correspond to s and $\hat{v}(s)$, respectively.

8.1 Value representation: From table to function

We next use an example to demonstrate the difference between the tabular and function approximation methods.

Suppose that there are n states $\{s_i\}_{i=1}^n$, whose state values are $\{v_\pi(s_i)\}_{i=1}^n$. Here, π is a given policy. Let $\{\hat{v}(s_i)\}_{i=1}^n$ denote the estimates of the true state values. If we use the tabular method, the estimated values can be maintained in the following table. This table can be stored in memory as an array or a vector. To retrieve or update any value, we can directly read or rewrite the corresponding entry in the table.

State	s_1	s_2	$\cdots$	s_n
Estimated value	$\hat{v}(s_1)$	$\hat{v}(s_2)$	$\cdots$	$\hat{v}(s_n)$

We next show that the values in the above table can be approximated by a function. In particular, $\{(s_i, \hat{v}(s_i))\}_{i=1}^n$ are shown as n points in Figure 8.2. These points can be fitted or approximated by a curve. The simplest curve is a straight line, which can be described as

$$\hat{v}(s, w) = as + b = \underbrace{[s, 1]}_{\phi^T(s)} \begin{bmatrix} a \\ b \end{bmatrix}_w = \phi^T(s)w. \tag{8.1}$$

Here, $\hat{v}(s, w)$ is a function for approximating $v_\pi(s)$. It is determined jointly by the state s and the parameter vector $w \in \mathbb{R}^2$. $\hat{v}(s, w)$ is sometimes written as $\hat{v}_w(s)$. Here, $\phi(s) \in \mathbb{R}^2$ is called the *feature vector* of s.

The first notable difference between the tabular and function approximation methods concerns how they retrieve and update a value.

◇ How to *retrieve* a value: When the values are represented by a table, if we want to retrieve a value, we can directly read the corresponding entry in the table. However, when the values are represented by a function, it becomes slightly more complicated to retrieve a value. In particular, we need to input the state index s into the function and calculate the function value (Figure 8.3). For the example in (8.1), we first need to calculate the feature vector $\phi(s)$ and then calculate $\phi^T(s)w$. If the function is an artificial neural network, a forward propagation from the input to the output is needed.

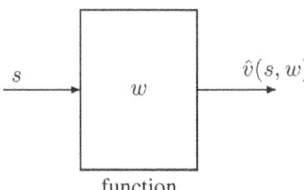

function

Figure 8.3: An illustration of the process for retrieving the value of s when using the function approximation method.

The function approximation method is more efficient in terms of storage due to the way in which the state values are retrieved. Specifically, while the tabular method needs to store n values, we now only need to store a lower dimensional parameter vector w. Thus, the storage efficiency can be significantly improved. Such a benefit is, however, *not* free. It comes with a cost: the state values may not be accurately represented by the function. For example, a straight line is not able to accurately fit the points in Figure 8.2. That is why this method is called approximation. From a fundamental point of view, some information will certainly be lost when we use a low-dimensional vector to represent a high-dimensional dataset. Therefore, the function approximation method enhances storage efficiency by sacrificing accuracy.

⋄ How to *update* a value: When the values are represented by a table, if we want to update one value, we can directly rewrite the corresponding entry in the table. However, when the values are represented by a function, the way to update a value is completely different. Specifically, we must update w to change the values indirectly. How to update w to find optimal state values will be addressed in detail later.

Thanks to the way in which the state values are updated, the function approximation method has another merit: its *generalization ability* is stronger than that of the tabular method. The reason is as follows. When using the tabular method, we can update a value if the corresponding state is visited in an episode. The values of the states that have not been visited cannot be updated. However, when using the function approximation method, we need to update w to update the value of a state. The update of w also affects the values of some other states even though these states have not been visited. Therefore, the experience sample for one state can generalize to help estimate the values of some other states.

The above analysis is illustrated in Figure 8.4, where there are three states $\{s_1, s_2, s_3\}$.

Suppose that we have an experience sample for s_3 and would like to update $\hat{v}(s_3)$. When using the tabular method, we can only update $\hat{v}(s_3)$ without changing $\hat{v}(s_1)$ or $\hat{v}(s_2)$, as shown in Figure 8.4(a). When using the function approximation method, updating w not only can update $\hat{v}(s_3)$ but also would change $\hat{v}(s_1)$ and $\hat{v}(s_2)$, as shown in Figure 8.4(b). Therefore, the experience sample of s_3 can help update the values of its neighboring states.

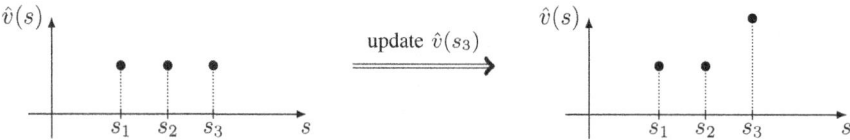

(a) Tabular method: when $\hat{v}(s_3)$ is updated, the other values remain the same.

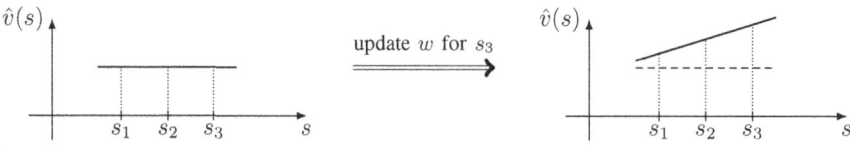

(b) Function approximation method: when we update $\hat{v}(s_3)$ by changing w, the values of the neighboring states are also changed.

Figure 8.4: An illustration of how to update the value of a state.

We can use more complex functions that have stronger approximation abilities than straight lines. For example, consider a second-order polynomial:

$$\hat{v}(s, w) = as^2 + bs + c = \underbrace{[s^2, s, 1]}_{\phi^T(s)} \underbrace{\begin{bmatrix} a \\ b \\ c \end{bmatrix}}_{w} = \phi^T(s)w. \tag{8.2}$$

We can use even higher-order polynomial curves to fit the points. As the order of the curve increases, the approximation accuracy can be improved, but the dimension of the parameter vector also increases, requiring more storage and computational resources.

Note that $\hat{v}(s, w)$ in either (8.1) or (8.2) is *linear* in w (though it may be nonlinear in s). This type of method is called *linear function approximation*, which is the simplest function approximation method. To realize linear function approximation, we need to select an appropriate feature vector $\phi(s)$. That is, we must decide, for example, whether we should use a first-order straight line or a second-order curve to fit the points. The selection of appropriate feature vectors is nontrivial. It requires prior knowledge of the given task: the better we understand the task, the better the feature vectors we can select. For instance, if we know that the points in Figure 8.2 are approximately located on a straight line, we can use a straight line to fit the points. However, such prior knowledge is usually unknown in practice. If we do not have any prior knowledge, a popular solution is to use artificial neural networks as nonlinear function approximations.

Another important problem is how to find the optimal parameter vector. If we know $\{v_\pi(s_i)\}_{i=1}^n$, this is a least-squares problem. The optimal parameter can be obtained by optimizing the following objective function:

$$J_1 = \sum_{i=1}^n \left(\hat{v}(s_i, w) - v_\pi(s_i) \right)^2 = \sum_{i=1}^n \left(\phi^T(s_i)w - v_\pi(s_i) \right)^2$$

$$= \left\| \begin{bmatrix} \phi^T(s_1) \\ \vdots \\ \phi^T(s_n) \end{bmatrix} w - \begin{bmatrix} v_\pi(s_1) \\ \vdots \\ v_\pi(s_n) \end{bmatrix} \right\|^2 \doteq \| \Phi w - v_\pi \|^2,$$

where

$$\Phi \doteq \begin{bmatrix} \phi^T(s_1) \\ \vdots \\ \phi^T(s_n) \end{bmatrix} \in \mathbb{R}^{n \times 2}, \qquad v_\pi \doteq \begin{bmatrix} v_\pi(s_1) \\ \vdots \\ v_\pi(s_n) \end{bmatrix} \in \mathbb{R}^n.$$

It can be verified that the optimal solution to this least-squares problem is

$$w^* = (\Phi^T \Phi)^{-1} \Phi v_\pi.$$

More information about least-squares problems can be found in [47, Section 3.3] and [48, Section 5.14].

The curve-fitting example presented in this section illustrates the basic idea of value function approximation. This idea will be formally introduced in the next section.

8.2 TD learning of state values based on function approximation

In this section, we show how to integrate the function approximation method into TD learning to estimate the state values of a given policy. This algorithm will be extended to learn action values and optimal policies in Section 8.3.

This section contains quite a few subsections and many coherent contents. It is better for us to review the contents first before diving into the details.

⋄ The function approximation method is formulated as an optimization problem. The objective function of this problem is introduced in Section 8.2.1. The TD learning algorithm for optimizing this objective function is introduced in Section 8.2.2.

⋄ To apply the TD learning algorithm, we need to select appropriate feature vectors. Section 8.2.3 discusses this problem.

⋄ Examples are given in Section 8.2.4 to demonstrate the TD algorithm and the impacts of different feature vectors.

⋄ A theoretical analysis of the TD algorithm is given in Section 8.2.5. This subsection is mathematically intensive. Readers may read it selectively based on their interests.

8.2.1 Objective function

Let $v_\pi(s)$ and $\hat{v}(s, w)$ be the true state value and approximated state value of $s \in \mathcal{S}$, respectively. The problem to be solved is to find an *optimal* w so that $\hat{v}(s, w)$ can best approximate $v_\pi(s)$ for every s. In particular, the objective function is

$$J(w) = \mathbb{E}[(v_\pi(S) - \hat{v}(S, w))^2], \tag{8.3}$$

where the expectation is calculated with respect to the random variable $S \in \mathcal{S}$. While S is a random variable, what is its probability distribution? This question is important for understanding this objective function. There are several ways to define the probability distribution of S.

⋄ The first way is to use a *uniform distribution*. That is to treat all the states as *equally important* by setting the probability of each state to $1/n$. In this case, the objective function in (8.3) becomes

$$J(w) = \frac{1}{n} \sum_{s \in \mathcal{S}} (v_\pi(s) - \hat{v}(s, w))^2, \tag{8.4}$$

which is the average value of the approximation errors of all the states. However, this way does not consider the real dynamics of the Markov process under the given policy. Since some states may be rarely visited by a policy, it may be unreasonable to treat all the states as equally important.

⋄ The second way, which is the focus of this chapter, is to use the *stationary distribution*. The stationary distribution describes the *long-term* behavior of a Markov decision process. More specifically, after the agent executes a given policy for a sufficiently long period, the probability of the agent being located at any state can be described by this stationary distribution. Interested readers may see the details in Box 8.1.

Let $\{d_\pi(s)\}_{s \in \mathcal{S}}$ denote the stationary distribution of the Markov process under policy π. That is, the probability for the agent visiting s after a long period of time is $d_\pi(s)$. By definition, $\sum_{s \in \mathcal{S}} d_\pi(s) = 1$. Then, the objective function in (8.3) can be rewritten as

$$J(w) = \sum_{s \in \mathcal{S}} d_\pi(s)(v_\pi(s) - \hat{v}(s, w))^2, \qquad (8.5)$$

which is a weighted average of the approximation errors. The states that have higher probabilities of being visited are given greater weights.

It is notable that the value of $d_\pi(s)$ is nontrivial to obtain because it requires knowing the state transition probability matrix P_π (see Box 8.1). Fortunately, we do not need to calculate the specific value of $d_\pi(s)$ to minimize this objective function as shown in the next subsection. In addition, it was assumed that the number of states was *finite* when we introduced (8.4) and (8.5). When the state space is continuous, we can replace the summations with integrals.

Box 8.1: Stationary distribution of a Markov decision process

The key tool for analyzing stationary distribution is $P_\pi \in \mathbb{R}^{n \times n}$, which is the probability transition matrix under the given policy π. If the states are indexed as $s_1, \ldots, s_n$, then $[P_\pi]_{ij}$ is defined as the probability for the agent moving from s_i to s_j. The definition of P_π can be found in Section 2.6.

◇ Interpretation of P_π^k ($k = 1, 2, 3, \ldots$).

First of all, it is necessary to examine the interpretation of the entries in P_π^k. The probability of the agent transitioning from s_i to s_j using exactly k steps is denoted as

$$p_{ij}^{(k)} = \Pr(S_{t_k} = j | S_{t_0} = i),$$

where t_0 and t_k are the initial and kth time steps, respectively. First, by the definition of P_π, we have

$$[P_\pi]_{ij} = p_{ij}^{(1)},$$

which means that $[P_\pi]_{ij}$ is the probability of transitioning from s_i to s_j using *a single step*. Second, consider P_π^2. It can be verified that

$$[P_\pi^2]_{ij} = [P_\pi P_\pi]_{ij} = \sum_{q=1}^{n} [P_\pi]_{iq}[P_\pi]_{qj}.$$

Since $[P_\pi]_{iq}[P_\pi]_{qj}$ is the joint probability of transitioning from s_i to s_q and then from s_q to s_j, we know that $[P_\pi^2]_{ij}$ is the probability of transitioning from s_i to s_j

using *exactly two steps*. That is

$$[P_\pi^2]_{ij} = p_{ij}^{(2)}.$$

Similarly, we know that

$$[P_\pi^k]_{ij} = p_{ij}^{(k)},$$

which means that $[P_\pi^k]_{ij}$ is the probability of transitioning from s_i to s_j using *exactly k steps*.

$\diamond$ Definition of stationary distributions.

Let $d_0 \in \mathbb{R}^n$ be a vector representing the probability distribution of the states at the initial time step. For example, if s is always selected as the starting state, then $d_0(s) = 1$ and the other entries of d_0 are 0. Let $d_k \in \mathbb{R}^n$ be the vector representing the probability distribution obtained after exactly k steps starting from d_0. Then, we have

$$d_k(s_i) = \sum_{j=1}^{n} d_0(s_j)[P_\pi^k]_{ji}, \quad i = 1, 2, \dots. \tag{8.6}$$

This equation indicates that the probability of the agent visiting s_i at step k equals the sum of the probabilities of the agent transitioning from $\{s_j\}_{j=1}^n$ to s_i using exactly k steps. The matrix-vector form of (8.6) is

$$d_k^T = d_0^T P_\pi^k. \tag{8.7}$$

When we consider the long-term behavior of the Markov process, it holds under certain conditions that

$$\lim_{k \to \infty} P_\pi^k = \mathbf{1}_n d_\pi^T, \tag{8.8}$$

where $\mathbf{1}_n = [1, \dots, 1]^T \in \mathbb{R}^n$ and $\mathbf{1}_n d_\pi^T$ is a constant matrix with all its rows equal to d_π^T. The conditions under which (8.8) is valid will be discussed later. Substituting (8.8) into (8.7) yields

$$\lim_{k \to \infty} d_k^T = d_0^T \lim_{k \to \infty} P_\pi^k = d_0^T \mathbf{1}_n d_\pi^T = d_\pi^T, \tag{8.9}$$

where the last equality is valid because $d_0^T \mathbf{1}_n = 1$.

Equation (8.9) means that the state distribution d_k converges to a constant value d_π, which is called the *limiting distribution*. The limiting distribution depends on the system model and the policy π. Interestingly, it is independent of the initial distribution d_0. That is, regardless of which state the agent starts from, the probability distribution of the agent after a sufficiently long period can always be described by the limiting distribution.

The value of d_π can be calculated in the following way. Taking the limit of both sides of $d_k^T = d_{k-1}^T P_\pi$ gives $\lim_{k\to\infty} d_k^T = \lim_{k\to\infty} d_{k-1}^T P_\pi$ and hence

$$d_\pi^T = d_\pi^T P_\pi. \tag{8.10}$$

As a result, d_π is the left eigenvector of P_π associated with the eigenvalue 1. The solution of (8.10) is called the stationary distribution. It holds that $\sum_{s\in\mathcal{S}} d_\pi(s) = 1$ and $d_\pi(s) > 0$ for all $s \in \mathcal{S}$. The reason why $d_\pi(s) > 0$ (not $d_\pi(s) \geq 0$) will be explained later.

◇ Conditions for the uniqueness of stationary distributions.

The solution d_π of (8.10) is usually called a stationary distribution, whereas the distribution d_π in (8.9) is usually called the limiting distribution. Note that (8.9) implies (8.10), but the converse may not be true. A general class of Markov processes that have unique stationary (or limiting) distributions is *irreducible* (or *regular*) Markov processes. Some necessary definitions are given below. More details can be found in [49, Chapter IV].

- State s_j is said to be *accessible* from state s_i if there exists a finite integer k so that $[P_\pi]_{ij}^k > 0$, which means that the agent starting from s_i can possibly reach s_j after a finite number of transitions.

- If two states s_i and s_j are mutually accessible, then the two states are said to *communicate*.

- A Markov process is called *irreducible* if all of its states communicate with each other. In other words, the agent starting from an arbitrary state can possibly reach any other state within a finite number of steps. Mathematically, it indicates that, for any s_i and s_j, there exists $k \geq 1$ such that $[P_\pi^k]_{ij} > 0$ (the value of k may vary for different i, j).

- A Markov process is called *regular* if there exists $k \geq 1$ such that $[P_\pi^k]_{ij} > 0$ for all i, j. Equivalently, there exists $k \geq 1$ such that $P_\pi^k > 0$, where $>$ is elementwise. As a result, it is possible that every state is reachable from any other state within at most k steps. A regular Markov process is also irreducible, but the converse is not true. However, if a Markov process is irreducible and there exists i such that $[P_\pi]_{ii} > 0$, then it is also regular. Moreover, if $P_\pi^k > 0$, then $P_\pi^{k'} > 0$ for any $k' \geq k$ since $P_\pi \geq 0$. It then follows from (8.9) that $d_\pi(s) > 0$ for every s.

⋄ Policies that may lead to unique stationary distributions.

Once the policy is given, a Markov decision process becomes a Markov process, whose long-term behavior is jointly determined by the given policy and the system model. Then, an important question is what kind of policies can lead to regular Markov processes? In general, the answer is *exploratory policies* such as ϵ-greedy policies. That is because an exploratory policy has a positive probability of taking any action at any state. As a result, the states can communicate with each other when the system model allows them to do so.

⋄ An example is given in Figure 8.5 to illustrate stationary distributions. The policy in this example is ϵ-greedy with $\epsilon = 0.5$. The states are indexed as s_1, s_2, s_3, s_4, which correspond to the top-left, top-right, bottom-left, and bottom-right cells in the grid, respectively.

We compare two methods to calculate the stationary distributions. The first method is to solve (8.10) to get the theoretical value of d_π. The second method is to estimate d_π numerically: we start from an arbitrary initial state and generate a sufficiently long episode by following the given policy. Then, d_π can be estimated by the ratio between the number of times each state is visited in the episode and the total length of the episode. The estimation result is more accurate when the episode is longer. We next compare the theoretical and estimated results.

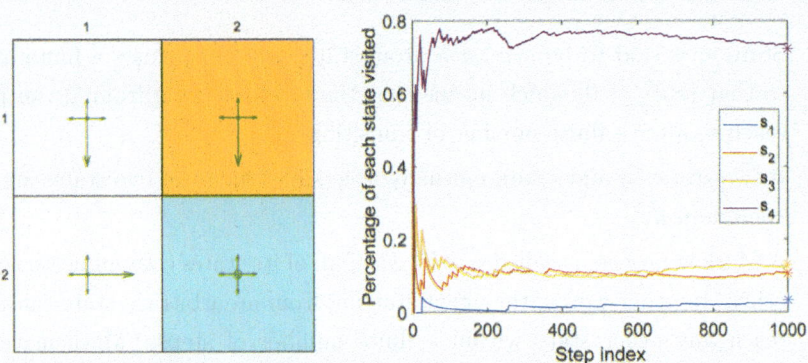

Figure 8.5: Long-term behavior of an ϵ-greedy policy with $\epsilon = 0.5$. The asterisks in the right figure represent the theoretical values of the elements of d_π.

- Theoretical value of d_π: It can be verified that the Markov process induced by the policy is both irreducible and regular. That is due to the following reasons. First, since all the states communicate, the resulting Markov process is irreducible. Second, since every state can transition to itself, the resulting

Markov process is regular. It can be seen from Figure 8.5 that

$$
P_\pi^T =
\begin{bmatrix}
0.3 & 0.1 & 0.1 & 0 \\
0.1 & 0.3 & 0 & 0.1 \\
0.6 & 0 & 0.3 & 0.1 \\
0 & 0.6 & 0.6 & 0.8
\end{bmatrix}.
$$

The eigenvalues of P_π^T can be calculated as $\{-0.0449, 0.3, 0.4449, 1\}$. The unit-length (right) eigenvector of P_π^T corresponding to the eigenvalue 1 is $[0.0463, 0.1455, 0.1785, 0.9720]^T$. After scaling this vector so that the sum of all its elements is equal to 1, we obtain the theoretical value of d_π as follows:

$$
d_\pi =
\begin{bmatrix}
0.0345 \\
0.1084 \\
0.1330 \\
0.7241
\end{bmatrix}.
$$

The ith element of d_π corresponds to the probability of the agent visiting s_i in the long run.

- Estimated value of d_π: We next verify the above theoretical value of d_π by executing the policy for sufficiently many steps in the simulation. Specifically, we select s_1 as the starting state and run 1,000 steps by following the policy. The proportion of the visits of each state during the process is shown in Figure 8.5. It can be seen that the proportions converge to the theoretical value of d_π after hundreds of steps.

8.2.2 Optimization algorithms

To minimize the objective function $J(w)$ in (8.3), we can use the gradient descent algorithm:

$$
w_{k+1} = w_k - \alpha_k \nabla_w J(w_k),
$$

where

$$
\begin{aligned}
\nabla_w J(w_k) &= \nabla_w \mathbb{E}[(v_\pi(S) - \hat{v}(S, w_k))^2] \\
&= \mathbb{E}[\nabla_w (v_\pi(S) - \hat{v}(S, w_k))^2] \\
&= 2\mathbb{E}[(v_\pi(S) - \hat{v}(S, w_k))(-\nabla_w \hat{v}(S, w_k))] \\
&= -2\mathbb{E}[(v_\pi(S) - \hat{v}(S, w_k))\nabla_w \hat{v}(S, w_k)].
\end{aligned}
$$

161

Therefore, the gradient descent algorithm is

$$w_{k+1} = w_k + 2\alpha_k \mathbb{E}[(v_\pi(S) - \hat{v}(S, w_k))\nabla_w \hat{v}(S, w_k)], \qquad (8.11)$$

where the coefficient 2 before α_k can be merged into α_k without loss of generality. The algorithm in (8.11) requires calculating the expectation. In the spirit of stochastic gradient descent, we can replace the true gradient with a stochastic gradient. Then, (8.11) becomes

$$w_{t+1} = w_t + \alpha_t \big(v_\pi(s_t) - \hat{v}(s_t, w_t)\big)\nabla_w \hat{v}(s_t, w_t), \qquad (8.12)$$

where s_t is a sample of S at time t.

Notably, (8.12) is *not* implementable because it requires the true state value v_π, which is unknown and must be estimated. We can replace $v_\pi(s_t)$ with an approximation to make the algorithm implementable. The following two methods can be used to do so.

⋄ Monte Carlo method: Suppose that we have an episode $(s_0, r_1, s_1, r_2, \dots)$. Let g_t be the discounted return starting from s_t. Then, g_t can be used as an approximation of $v_\pi(s_t)$. The algorithm in (8.12) becomes

$$w_{t+1} = w_t + \alpha_t \big(g_t - \hat{v}(s_t, w_t)\big)\nabla_w \hat{v}(s_t, w_t).$$

This is the algorithm of Monte Carlo learning with function approximation.

⋄ Temporal-difference method: In the spirit of TD learning, $r_{t+1} + \gamma\hat{v}(s_{t+1}, w_t)$ can be used as an approximation of $v_\pi(s_t)$. The algorithm in (8.12) becomes

$$w_{t+1} = w_t + \alpha_t \left[r_{t+1} + \gamma\hat{v}(s_{t+1}, w_t) - \hat{v}(s_t, w_t)\right]\nabla_w \hat{v}(s_t, w_t). \qquad (8.13)$$

This is the algorithm of TD learning with function approximation. This algorithm is summarized in Algorithm 8.1.

Understanding the TD algorithm in (8.13) is important for studying the other algorithms in this chapter. Notably, (8.13) can only learn the *state values* of a given policy. It will be extended to algorithms that can learn *action values* in Sections 8.3.1 and 8.3.2.

8.2.3   Selection of function approximators

To apply the TD algorithm in (8.13), we need to select appropriate $\hat{v}(s, w)$. There are two ways to do that. The first is to use an artificial neural network as a *nonlinear* function approximator. The input of the neural network is the state, the output is $\hat{v}(s, w)$, and the network parameter is w. The second is to simply use a *linear* function:

$$\hat{v}(s, w) = \phi^T(s)w,$$

Algorithm 8.1: TD learning of state values with function approximation

Initialization: A function $\hat{v}(s, w)$ that is differentiable in w. Initial parameter w_0.
Goal: Learn the true state values of a given policy π.

For each episode $\{(s_t, r_{t+1}, s_{t+1})\}_t$ generated by π, do
    For each sample (s_t, r_{t+1}, s_{t+1}), do
        In the general case, $w_{t+1} = w_t + \alpha_t \left[r_{t+1} + \gamma \hat{v}(s_{t+1}, w_t) - \hat{v}(s_t, w_t) \right] \nabla_w \hat{v}(s_t, w_t)$
        In the linear case, $w_{t+1} = w_t + \alpha_t \left[r_{t+1} + \gamma \phi^T(s_{t+1}) w_t - \phi^T(s_t) w_t \right] \phi(s_t)$

---

where $\phi(s) \in \mathbb{R}^m$ is the feature vector of s. The lengths of $\phi(s)$ and w are equal to m, which is usually much smaller than the number of states. In the linear case, the gradient is

$$\nabla_w \hat{v}(s, w) = \phi(s),$$

Substituting which into (8.13) yields

$$w_{t+1} = w_t + \alpha_t \left[r_{t+1} + \gamma \phi^T(s_{t+1}) w_t - \phi^T(s_t) w_t \right] \phi(s_t). \tag{8.14}$$

This is the algorithm of TD learning with linear function approximation. We call it *TD-Linear* for short.

The linear case is much better understood in theory than the nonlinear case. However, its approximation ability is limited. It is also nontrivial to select appropriate feature vectors for complex tasks. By contrast, artificial neural networks can approximate values as black-box universal nonlinear approximators, which are more friendly to use.

Nevertheless, it is still meaningful to study the linear case. A better understanding of the linear case can help readers better grasp the idea of the function approximation method. Moreover, the linear case is sufficient for solving the simple grid world tasks considered in this book. More importantly, the linear case is still powerful in the sense that the tabular method can be viewed as a special linear case. More information can be found in Box 8.2.

Box 8.2: Tabular TD learning is a special case of TD-Linear

We next show that the tabular TD algorithm in (7.1) in Chapter 7 is a special case of the TD-Linear algorithm in (8.14).

Consider the following special feature vector for any $s \in \mathcal{S}$:

$$\phi(s) = e_s \in \mathbb{R}^n,$$

where e_s is the vector with the entry corresponding to s equal to 1 and the other entries equal to 0. In this case,

$$\hat{v}(s, w) = e_s^T w = w(s),$$

where $w(s)$ is the entry in w that corresponds to s. Substituting the above equation into (8.14) yields

$$w_{t+1} = w_t + \alpha_t \big(r_{t+1} + \gamma w_t(s_{t+1}) - w_t(s_t)\big) e_{s_t}.$$

The above equation merely updates the entry $w_t(s_t)$ due to the definition of e_{s_t}. Motivated by this, multiplying $e_{s_t}^T$ on both sides of the equation yields

$$w_{t+1}(s_t) = w_t(s_t) + \alpha_t \big(r_{t+1} + \gamma w_t(s_{t+1}) - w_t(s_t)\big),$$

which is exactly the tabular TD algorithm in (7.1).

In summary, by selecting the feature vector as $\phi(s) = e_s$, the TD-Linear algorithm becomes the tabular TD algorithm.

8.2.4 Illustrative examples

We next present some examples for demonstrating how to use the TD-Linear algorithm in (8.14) to estimate the state values of a given policy. In the meantime, we demonstrate how to select feature vectors.

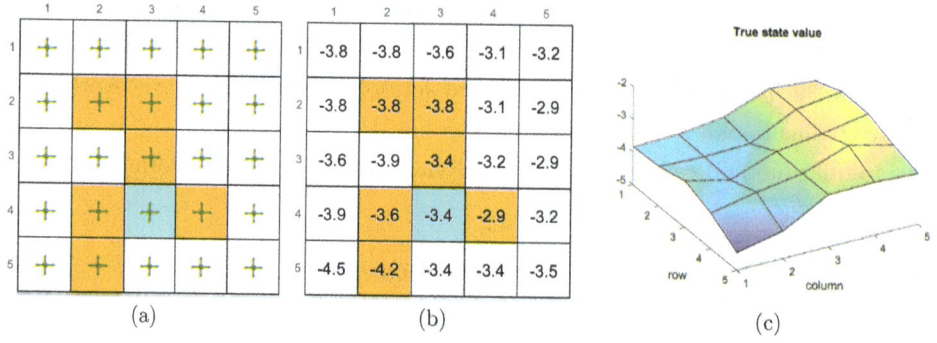

Figure 8.6: (a) The policy to be evaluated. (b) The true state values are represented as a table. (c) The true state values are represented as a 3D surface.

The grid world example is shown in Figure 8.6. The given policy takes any action at a state with a probability of 0.2. Our goal is to estimate the state values under this policy. There are 25 state values in total. The true state values are shown in Figure 8.6(b). The true state values are visualized as a three-dimensional surface in Figure 8.6(c).

We next show that we can use fewer than 25 parameters to approximate these state values. The simulation setup is as follows. Five hundred episodes are generated by the given policy. Each episode has 500 steps and starts from a randomly selected state-action pair following a uniform distribution. In addition, in each simulation trial, the parameter vector w is randomly initialized such that each element is drawn from a standard normal distribution with a zero mean and a standard deviation of 1. We set $r_{\text{forbidden}} = r_{\text{boundary}} = -1$, $r_{\text{target}} = 1$, and $\gamma = 0.9$.

To implement the TD-Linear algorithm, we need to select the feature vector $\phi(s)$ first. There are different ways to do that as shown below.

◇ The first type of feature vector is based on polynomials. In the grid world example, a state s corresponds to a 2D location. Let x and y denote the column and row indexes of s, respectively. To avoid numerical issues, we normalize x and y so that their values are within the interval of $[-1, +1]$. With a slight abuse of notation, the normalized values are also represented by x and y. Then, the simplest feature vector is

$$\phi(s) = \begin{bmatrix} x \\ y \end{bmatrix} \in \mathbb{R}^2.$$

In this case, we have

$$\hat{v}(s, w) = \phi^T(s)w = [x, y] \begin{bmatrix} w_1 \\ w_2 \end{bmatrix} = w_1 x + w_2 y.$$

When w is given, $\hat{v}(s, w) = w_1 x + w_2 y$ represents a 2D plane that passes through the origin. Since the surface of the state values may not pass through the origin, we need to introduce a bias to the 2D plane to better approximate the state values. To do that, we consider the following 3D feature vector:

$$\phi(s) = \begin{bmatrix} 1 \\ x \\ y \end{bmatrix} \in \mathbb{R}^3. \tag{8.15}$$

In this case, the approximated state value is

$$\hat{v}(s, w) = \phi^T(s)w = [1, x, y] \begin{bmatrix} w_1 \\ w_2 \\ w_3 \end{bmatrix} = w_1 + w_2 x + w_3 y.$$

When w is given, $\hat{v}(s, w)$ corresponds to a plane that may not pass through the origin. Notably, $\phi(s)$ can also be defined as $\phi(s) = [x, y, 1]^T$, where the order of the elements does not matter.

The estimation result when we use the feature vector in (8.15) is shown in Figure 8.7(a). It can be seen that the estimated state values form a 2D plane. Although the estimation error converges as more episodes are used, the error cannot decrease to zero due to the limited approximation ability of a 2D plane.

To enhance the approximation ability, we can increase the dimension of the feature vector. To that end, consider

$$\phi(s) = [1, x, y, x^2, y^2, xy]^T \in \mathbb{R}^6. \tag{8.16}$$

In this case, $\hat{v}(s, w) = \phi^T(s)w = w_1 + w_2x + w_3y + w_4x^2 + w_5y^2 + w_6xy$, which corresponds to a quadratic 3D surface. We can further increase the dimension of the feature vector:

$$\phi(s) = [1, x, y, x^2, y^2, xy, x^3, y^3, x^2y, xy^2]^T \in \mathbb{R}^{10}. \tag{8.17}$$

The estimation results when we use the feature vectors in (8.16) and (8.17) are shown in Figures 8.7(b)-(c). As can be seen, the longer the feature vector is, the more accurately the state values can be approximated. However, in all three cases, the estimation error cannot converge to zero because these linear approximators still have limited approximation abilities.

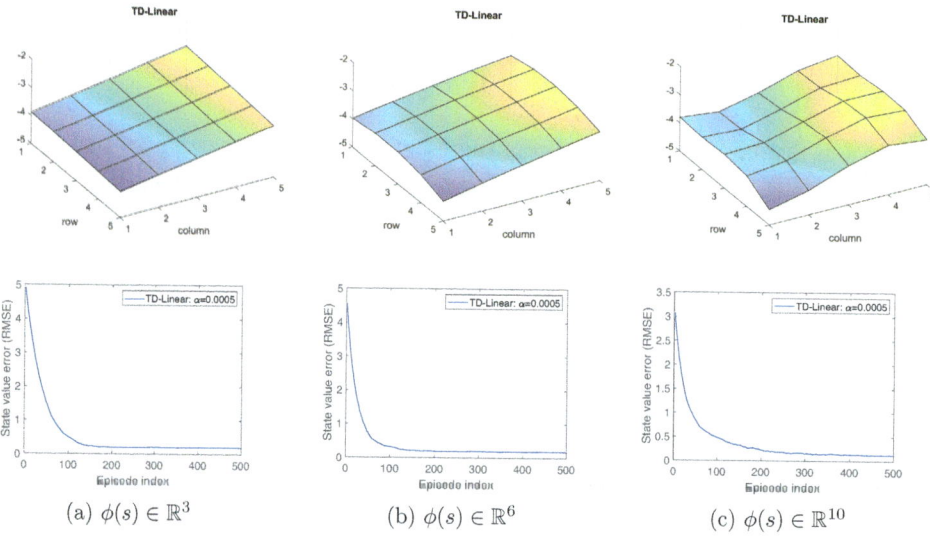

(a) $\phi(s) \in \mathbb{R}^3$ (b) $\phi(s) \in \mathbb{R}^6$ (c) $\phi(s) \in \mathbb{R}^{10}$

Figure 8.7: TD-Linear estimation results obtained with the polynomial features in (8.15), (8.16), and (8.17).

◇ In addition to polynomial feature vectors, many other types of features are available such as Fourier basis and tile coding [3, Chapter 9]. First, the values of x and y of

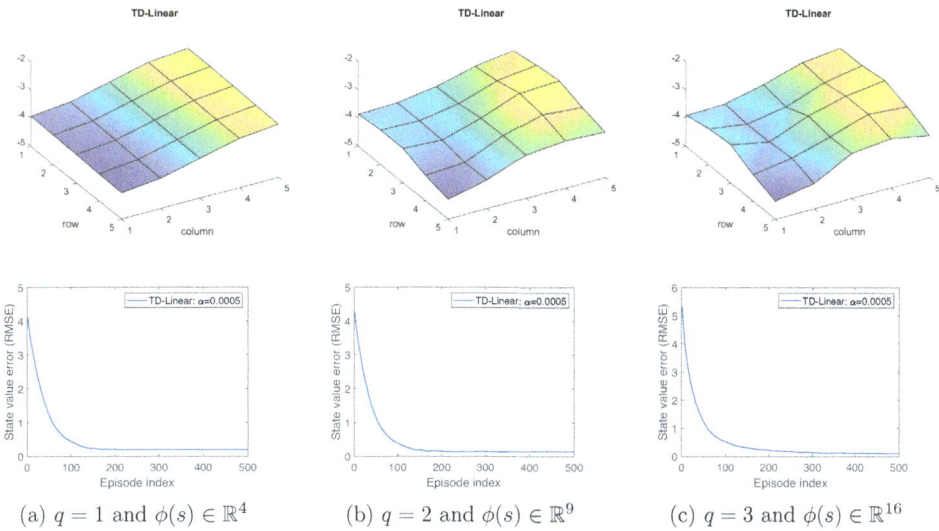

(a) $q = 1$ and $\phi(s) \in \mathbb{R}^4$ (b) $q = 2$ and $\phi(s) \in \mathbb{R}^9$ (c) $q = 3$ and $\phi(s) \in \mathbb{R}^{16}$

Figure 8.8: TD-Linear estimation results obtained with the Fourier features in (8.18).

each state are normalized to the interval of $[0, 1]$. The resulting feature vector is

$$
\phi(s) = \begin{bmatrix} \vdots \\ \cos\left(\pi(c_1 x + c_2 y)\right) \\ \vdots \end{bmatrix} \in \mathbb{R}^{(q+1)^2}, \tag{8.18}
$$

where π denotes the circumference ratio, which is $3.1415\ldots$, instead of a policy. Here, c_1 or c_2 can be set as any integers in $\{0, 1, \ldots, q\}$, where q is a user-specified integer. As a result, there are $(q + 1)^2$ possible values for the pair (c_1, c_2) to take. Hence, the dimension of $\phi(s)$ is $(q + 1)^2$. For example, in the case of $q = 1$, the feature vector is

$$
\phi(s) = \begin{bmatrix} \cos\left(\pi(0x + 0y)\right) \\ \cos\left(\pi(0x + 1y)\right) \\ \cos\left(\pi(1x + 0y)\right) \\ \cos\left(\pi(1x + 1y)\right) \end{bmatrix} = \begin{bmatrix} 1 \\ \cos(\pi y) \\ \cos(\pi x) \\ \cos(\pi(x + y)) \end{bmatrix} \in \mathbb{R}^4.
$$

The estimation results obtained when we use the Fourier features with $q = 1, 2, 3$ are shown in Figure 8.8. The dimensions of the feature vectors in the three cases are $4, 9, 16$, respectively. As can be seen, the higher the dimension of the feature vector is, the more accurately the state values can be approximated.

8.2.5   Theoretical analysis

Thus far, we have finished describing the story of TD learning with function approximation. This story started from the objective function in (8.3). To optimize this objective function, we introduced the stochastic algorithm in (8.12). Later, the true value function in the algorithm, which was unknown, was replaced by an approximation, leading to the TD algorithm in (8.13). Although this story is helpful for understanding the basic idea of value function approximation, it is not mathematically rigorous. For example, the algorithm in (8.13) actually does not minimize the objective function in (8.3).

We next present a theoretical analysis of the TD algorithm in (8.13) to reveal why the algorithm works effectively and what mathematical problems it solves. Since general nonlinear approximators are difficult to analyze, this part only considers the linear case. Readers are advised to read selectively based on their interests since this part is mathematically intensive.

Convergence analysis

To study the convergence property of (8.13), we first consider the following deterministic algorithm:

$$w_{t+1} = w_t + \alpha_t \mathbb{E}\left[\left(r_{t+1} + \gamma \phi^T(s_{t+1})w_t - \phi^T(s_t)w_t\right)\phi(s_t)\right], \tag{8.19}$$

where the expectation is calculated with respect to the random variables s_t, s_{t+1}, r_{t+1}. The distribution of s_t is assumed to be the stationary distribution d_π. The algorithm in (8.19) is deterministic because the random variables s_t, s_{t+1}, r_{t+1} all disappear after calculating the expectation.

Why would we consider this deterministic algorithm? First, the convergence of this deterministic algorithm is easier (though nontrivial) to analyze. Second and more importantly, the convergence of this deterministic algorithm implies the convergence of the stochastic TD algorithm in (8.13). That is because (8.13) can be viewed as a stochastic gradient descent (SGD) implementation of (8.19). Therefore, we only need to study the convergence property of the deterministic algorithm.

Although the expression of (8.19) may look complex at first glance, it can be greatly simplified. To do that, define

$$\Phi = \begin{bmatrix} \vdots \\ \phi^T(s) \\ \vdots \end{bmatrix} \in \mathbb{R}^{n \times m}, \quad D = \begin{bmatrix} \ddots & & \\ & d_\pi(s) & \\ & & \ddots \end{bmatrix} \in \mathbb{R}^{n \times n}, \tag{8.20}$$

where Φ is the matrix containing all the feature vectors, and D is a diagonal matrix with the stationary distribution in its diagonal entries. The two matrices will be frequently used.

Lemma 8.1. *The expectation in (8.19) can be rewritten as*

$$\mathbb{E}\left[\left(r_{t+1} + \gamma \phi^T(s_{t+1})w_t - \phi^T(s_t)w_t\right)\phi(s_t)\right] = b - Aw_t,$$

where

$$A \doteq \Phi^T D(I - \gamma P_\pi)\Phi \in \mathbb{R}^{m \times m},$$
$$b \doteq \Phi^T D r_\pi \in \mathbb{R}^m. \tag{8.21}$$

Here, P_π, r_π are the two terms in the Bellman equation $v_\pi = r_\pi + \gamma P_\pi v_\pi$, and I is the identity matrix with appropriate dimensions.

The proof is given in Box 8.3. With the expression in Lemma 8.1, the deterministic algorithm in (8.19) can be rewritten as

$$w_{t+1} = w_t + \alpha_t(b - Aw_t), \tag{8.22}$$

which is a simple deterministic process. Its convergence is analyzed below.

First, what is the converged value of w_t? Hypothetically, if w_t converges to a constant value w^* as $t \to \infty$, then (8.22) implies $w^* = w^* + \alpha_\infty(b - Aw^*)$, which suggests that $b - Aw^* = 0$ and hence

$$w^* = A^{-1}b.$$

Several remarks about this converged value are given below.

⋄ Is A invertible? The answer is yes. In fact, A is not only invertible but also positive definite. That is, for any nonzero vector x with appropriate dimensions, $x^T Ax > 0$. The proof is given in Box 8.4.

⋄ What is the interpretation of $w^* = A^{-1}b$? It is actually the optimal solution for minimizing the *projected Bellman error*. The details will be introduced in Section 8.2.5.

⋄ The tabular method is a special case. One interesting result is that, when the dimensionality of w equals $n = |\mathcal{S}|$ and $\phi(s) = [0, \ldots, 1, \ldots, 0]^T$, where the entry corresponding to s is 1, we have

$$w^* = A^{-1}b = v_\pi. \tag{8.23}$$

This equation indicates that the parameter vector to be learned is actually the true state value. This conclusion is consistent with the fact that the tabular TD algorithm is a special case of the TD-Linear algorithm, as introduced in Box 8.2. The proof of (8.23) is given below. It can be verified that $\Phi = I$ in this case and hence $A = \Phi^T D(I - \gamma P_\pi)\Phi = D(I - \gamma P_\pi)$ and $b = \Phi^T D r_\pi = D r_\pi$. Thus, $w^* = A^{-1}b = (I - \gamma P_\pi)^{-1}D^{-1}D r_\pi = (I - \gamma P_\pi)^{-1}r_\pi = v_\pi$.

Second, we prove that w_t in (8.22) converges to $w^* = A^{-1}b$ as $t \to \infty$. Since (8.22) is a simple deterministic process, it can be proven in many ways. We present two proofs as follows.

169

◇ Proof 1: Define the convergence error as $\delta_t \doteq w_t - w^*$. We only need to show that δ_t converges to zero. To do that, substituting $w_t = \delta_t + w^*$ into (8.22) gives

$$\delta_{t+1} = \delta_t - \alpha_t A \delta_t = (I - \alpha_t A)\delta_t.$$

It then follows that

$$\delta_{t+1} = (I - \alpha_t A) \cdots (I - \alpha_0 A)\delta_0.$$

Consider the simple case where $\alpha_t = \alpha$ for all t. Then, we have

$$\|\delta_{t+1}\|_2 \leq \|I - \alpha A\|_2^{t+1}\|\delta_0\|_2.$$

When $\alpha > 0$ is sufficiently small, we have that $\|I - \alpha A\|_2 < 1$ and hence $\delta_t \to 0$ as $t \to \infty$. The reason why $\|I - \alpha A\|_2 < 1$ holds is that A is positive definite and hence $x^T(I - \alpha A)x < 1$ for any x.

◇ Proof 2: Consider $g(w) \doteq b - Aw$. Since w^* is the root of $g(w) = 0$, the task is actually a root-finding problem. The algorithm in (8.22) is actually a Robbins-Monro (RM) algorithm. Although the original RM algorithm was designed for stochastic processes, it can also be applied to deterministic cases. The convergence of RM algorithms can shed light on the convergence of $w_{t+1} = w_t + \alpha_t(b - Aw_t)$. That is, w_t converges to w^* when $\sum_t \alpha_t = \infty$ and $\sum_t \alpha_t^2 < \infty$.

Box 8.3: Proof of Lemma 8.1

By using the law of total expectation, we have

$$\mathbb{E}\left[r_{t+1}\phi(s_t) + \phi(s_t)\left(\gamma\phi^T(s_{t+1}) - \phi^T(s_t)\right)w_t\right]$$
$$= \sum_{s \in \mathcal{S}} d_\pi(s)\mathbb{E}\left[r_{t+1}\phi(s_t) + \phi(s_t)\left(\gamma\phi^T(s_{t+1}) - \phi^T(s_t)\right)w_t \big| s_t = s\right]$$
$$= \sum_{s \in \mathcal{S}} d_\pi(s)\mathbb{E}\left[r_{t+1}\phi(s_t)\big| s_t = s\right] + \sum_{s \in \mathcal{S}} d_\pi(s)\mathbb{E}\left[\phi(s_t)\left(\gamma\phi^T(s_{t+1}) - \phi^T(s_t)\right)w_t\big| s_t = s\right].$$

$$(8.24)$$

Here, s_t is assumed to obey the stationary distribution d_π.

First, consider the first term in (8.24). Note that

$$\mathbb{E}\left[r_{t+1}\phi(s_t)\big| s_t = s\right] = \phi(s)\mathbb{E}\left[r_{t+1}\big| s_t = s\right] = \phi(s)r_\pi(s),$$

where $r_\pi(s) = \sum_a \pi(a|s)\sum_r rp(r|s,a)$. Then, the first term in (8.24) can be rewritten as

$$\sum_{s \in \mathcal{S}} d_\pi(s) \mathbb{E}\Big[r_{t+1}\phi(s_t)\big|s_t = s\Big] = \sum_{s \in \mathcal{S}} d_\pi(s)\phi(s)r_\pi(s) = \Phi^T D r_\pi, \qquad (8.25)$$

where $r_\pi = [\cdots, r_\pi(s), \cdots]^T \in \mathbb{R}^n$.

Second, consider the second term in (8.24). Since

$$\mathbb{E}\Big[\phi(s_t)\big(\gamma\phi^T(s_{t+1}) - \phi^T(s_t)\big)w_t\big|s_t = s\Big]$$
$$= -\mathbb{E}\Big[\phi(s_t)\phi^T(s_t)w_t\big|s_t = s\Big] + \mathbb{E}\Big[\gamma\phi(s_t)\phi^T(s_{t+1})w_t\big|s_t = s\Big]$$
$$= -\phi(s)\phi^T(s)w_t + \gamma\phi(s)\mathbb{E}\Big[\phi^T(s_{t+1})\big|s_t = s\Big]w_t$$
$$= -\phi(s)\phi^T(s)w_t + \gamma\phi(s)\sum_{s' \in \mathcal{S}} p(s'|s)\phi^T(s')w_t,$$

the second term in (8.24) becomes

$$\sum_{s \in \mathcal{S}} d_\pi(s)\mathbb{E}\Big[\phi(s_t)\big(\gamma\phi^T(s_{t+1}) - \phi^T(s_t)\big)w_t\big|s_t = s\Big]$$
$$= \sum_{s \in \mathcal{S}} d_\pi(s)\Big[-\phi(s)\phi^T(s)w_t + \gamma\phi(s)\sum_{s' \in \mathcal{S}} p(s'|s)\phi^T(s')w_t\Big]$$
$$= \sum_{s \in \mathcal{S}} d_\pi(s)\phi(s)\Big[-\phi(s) + \gamma\sum_{s' \in \mathcal{S}} p(s'|s)\phi(s')\Big]^T w_t$$
$$= \Phi^T D(-\Phi + \gamma P_\pi\Phi)w_t$$
$$= -\Phi^T D(I - \gamma P_\pi)\Phi w_t. \qquad (8.26)$$

Combining (8.25) and (8.26) gives

$$\mathbb{E}\Big[\big(r_{t+1} + \gamma\phi^T(s_{t+1})w_t - \phi^T(s_t)w_t\big)\phi(s_t)\Big] = \Phi^T D r_\pi - \Phi^T D(I - \gamma P_\pi)\Phi w_t$$
$$\doteq b - Aw_t, \qquad (8.27)$$

where $b \doteq \Phi^T D r_\pi$ and $A \doteq \Phi^T D(I - \gamma P_\pi)\Phi$.

Box 8.4: Proving that $A = \Phi^T D(I - \gamma P_\pi)\Phi$ is invertible and positive definite.

The matrix A is positive definite if $x^T A x > 0$ for any nonzero vector x with appropriate dimensions. If A is positive (or negative) definite, it is denoted as $A \succ 0$ (or $A \prec 0$). Here, $\succ$ and $\prec$ should be differentiated from $>$ and $<$, which indicate elementwise comparisons. Note that A may not be symmetric. Although positive definite matrices often refer to symmetric matrices, nonsymmetric ones can also be positive definite.

We next prove that $A \succ 0$ and hence A is invertible. The idea for proving $A \succ 0$ is to show that

$$D(I - \gamma P_\pi) \doteq M \succ 0. \tag{8.28}$$

It is clear that $M \succ 0$ implies $A = \Phi^T M \Phi \succ 0$ since Φ is a tall matrix with full column rank (suppose that the feature vectors are selected to be linearly independent). Note that

$$M = \frac{M + M^T}{2} + \frac{M - M^T}{2}.$$

Since $M - M^T$ is skew-symmetric and hence $x^T(M - M^T)x = 0$ for any x, we know that $M \succ 0$ if and only if $M + M^T \succ 0$. To show $M + M^T \succ 0$, we apply the fact that strictly diagonal dominant matrices are positive definite [4].

First, it holds that

$$(M + M^T)\mathbf{1}_n > 0, \tag{8.29}$$

where $\mathbf{1}_n = [1, \ldots, 1]^T \in \mathbb{R}^n$. The proof of (8.29) is given below. Since $P_\pi \mathbf{1}_n = \mathbf{1}_n$, we have $M\mathbf{1}_n = D(I - \gamma P_\pi)\mathbf{1}_n = D(\mathbf{1}_n - \gamma\mathbf{1}_n) = (1 - \gamma)d_\pi$. Moreover, $M^T\mathbf{1}_n = (I - \gamma P_\pi^T)D\mathbf{1}_n = (I - \gamma P_\pi^T)d_\pi = (1 - \gamma)d_\pi$, where the last equality is valid because $P_\pi^T d_\pi = d_\pi$. In summary, we have

$$(M + M^T)\mathbf{1}_n = 2(1 - \gamma)d_\pi.$$

Since all the entries of d_π are positive (see Box 8.1), we have $(M + M^T)\mathbf{1}_n > 0$.

Second, the elementwise form of (8.29) is

$$\sum_{j=1}^{n} [M + M^T]_{ij} > 0, \qquad i = 1, \ldots, n,$$

which can be further written as

$$[M + M^T]_{ii} + \sum_{j \neq i} [M + M^T]_{ij} > 0.$$

It can be verified according to the expression of M in (8.28) that the diagonal entries of M are positive and the off-diagonal entries of M are nonpositive. Therefore, the above inequality can be rewritten as

$$\left|[M + M^T]_{ii}\right| > \sum_{j \neq i} \left|[M + M^T]_{ij}\right|.$$

The above inequality indicates that the absolute value of the ith diagonal entry in $M + M^T$ is greater than the sum of the absolute values of the off-diagonal entries in the same row. Thus, $M + M^T$ is strictly diagonal dominant and the proof is complete.

TD learning minimizes the projected Bellman error

While we have shown that the TD-Linear algorithm converges to $w^* = A^{-1}b$, we next show that w^* is the optimal solution that minimizes the *projected Bellman error*. To do that, we review three objective functions.

⋄ The first objective function is

$$J_E(w) = \mathbb{E}[(v_\pi(S) - \hat{v}(S, w))^2],$$

which has been introduced in (8.3). By the definition of expectation, $J_E(w)$ can be reexpressed in a matrix-vector form as

$$J_E(w) = \|\hat{v}(w) - v_\pi\|_D^2,$$

where v_π is the true state value vector and $\hat{v}(w)$ is the approximated one. Here, $\|\cdot\|_D^2$ is a weighted norm: $\|x\|_D^2 = x^T D x = \|D^{1/2}x\|_2^2$, where D is given in (8.20).

This is the simplest objective function that we can imagine when talking about function approximation. However, it relies on the true state, which is unknown. To obtain an implementable algorithm, we must consider other objective functions such as the Bellman error and projected Bellman error [50–54].

⋄ The second objective function is the Bellman error. In particular, since v_π satisfies the Bellman equation $v_\pi = r_\pi + \gamma P_\pi v_\pi$, it is expected that the estimated value $\hat{v}(w)$ should also satisfy this equation to the greatest extent possible. Thus, the *Bellman error* is

$$J_{BE}(w) = \|\hat{v}(w) - (r_\pi + \gamma P_\pi \hat{v}(w))\|_D^2 \doteq \|\hat{v}(w) - T_\pi(\hat{v}(w))\|_D^2. \tag{8.30}$$

Here, $T_\pi(\cdot)$ is the Bellman operator. In particular, for any vector $x \in \mathbb{R}^n$, the Bellman operator is defined as

$$T_\pi(x) \doteq r_\pi + \gamma P_\pi x.$$

Minimizing the Bellman error is a standard least-squares problem. The details of the solution are omitted here.

◇ Third, it is notable that $J_{BE}(w)$ in (8.30) may not be minimized to *zero* due to the limited approximation ability of the approximator. By contrast, an objective function that can be minimized to zero is the *projected Bellman error*:

$$J_{PBE}(w) = \|\hat{v}(w) - MT_\pi(\hat{v}(w))\|_D^2,$$

where $M \in \mathbb{R}^{n \times n}$ is the orthogonal projection matrix that geometrically projects any vector onto the space of all approximations.

In fact, the TD learning algorithm in (8.13) aims to minimize the projected Bellman error J_{PBE} rather than J_E or J_{BE}. The reason is as follows. For the sake of simplicity, consider the linear case where $\hat{v}(w) = \Phi w$. Here, Φ is defined in (8.20). The range space of Φ is the set of all possible linear approximations. Then,

$$M = \Phi(\Phi^T D\Phi)^{-1}\Phi^T D \in \mathbb{R}^{n \times n}$$

is the projection matrix that geometrically projects any vector onto the range space Φ. Since $\hat{v}(w)$ is in the range space of Φ, we can always find a value of w that can minimize $J_{PBE}(w)$ to zero. It can be proven that the solution minimizing $J_{PBE}(w)$ is $w^* = A^{-1}b$. That is

$$w^* = A^{-1}b = \arg\min_w J_{PBE}(w),$$

The proof is given in Box 8.5.

Box 8.5: Showing that $w^* = A^{-1}b$ minimizes $J_{PBE}(w)$

We next show that $w^* = A^{-1}b$ is the optimal solution that minimizes $J_{PBE}(w)$. Since $J_{PBE}(w) = 0 \Leftrightarrow \hat{v}(w) - MT_\pi(\hat{v}(w)) = 0$, we only need to study the root of

$$\hat{v}(w) = MT_\pi(\hat{v}(w)).$$

In the linear case, substituting $\hat{v}(w) = \Phi w$ and the expression of M into the above equation gives

$$\Phi w = \Phi(\Phi^T D\Phi)^{-1}\Phi^T D(r_\pi + \gamma P_\pi \Phi w). \tag{8.31}$$

Since Φ has full column rank, we have $\Phi x = \Phi y \Leftrightarrow x = y$ for any x, y. Therefore, (8.31) implies

$$w = (\Phi^T D \Phi)^{-1} \Phi^T D (r_\pi + \gamma P_\pi \Phi w)$$
$$\Longleftrightarrow \Phi^T D (r_\pi + \gamma P_\pi \Phi w) = (\Phi^T D \Phi) w$$
$$\Longleftrightarrow \Phi^T D r_\pi + \gamma \Phi^T D P_\pi \Phi w = (\Phi^T D \Phi) w$$
$$\Longleftrightarrow \Phi^T D r_\pi = \Phi^T D (I - \gamma P_\pi) \Phi w$$
$$\Longleftrightarrow w = (\Phi^T D (I - \gamma P_\pi) \Phi)^{-1} \Phi^T D r_\pi = A^{-1} b,$$

where A, b are given in (8.21). Therefore, $w^* = A^{-1} b$ is the optimal solution that minimizes $J_{PBE}(w)$.

Since the TD algorithm aims to minimize J_{PBE} rather than J_E, it is natural to ask how close the estimated value $\hat{v}(w)$ is to the true state value v_π. In the linear case, the estimated value that minimizes the projected Bellman error is $\hat{v}(w^*) = \Phi w^*$. Its deviation from the true state value v_π satisfies

$$\|\hat{v}(w^*) - v_\pi\|_D = \|\Phi w^* - v_\pi\|_D \leq \frac{1}{1 - \gamma} \min_w \|\hat{v}(w) - v_\pi\|_D = \frac{1}{1 - \gamma} \min_w \sqrt{J_E(w)}.$$
(8.32)

The proof of this inequality is given in Box 8.6. Inequality (8.32) indicates that the discrepancy between Φw^* and v_π is bounded from above by the minimum value of $J_E(w)$. However, this bound is loose, especially when γ is close to one. It is thus mainly of theoretical value.

Box 8.6: Proof of the error bound in (8.32)

Note that

$$\|\Phi w^* - v_\pi\|_D = \|\Phi w^* - M v_\pi + M v_\pi - v_\pi\|_D$$
$$\leq \|\Phi w^* - M v_\pi\|_D + \|M v_\pi - v_\pi\|_D$$
$$= \|M T_\pi (\Phi w^*) - M T_\pi (v_\pi)\|_D + \|M v_\pi - v_\pi\|_D, \qquad (8.33)$$

where the last equality is due to $\Phi w^* = M T_\pi (\Phi w^*)$ and $v_\pi = T_\pi (v_\pi)$. Substituting

$$M T_\pi (\Phi w^*) - M T_\pi (v_\pi) = M (r_\pi + \gamma P_\pi \Phi w^*) - M (r_\pi + \gamma P_\pi v_\pi) = \gamma M P_\pi (\Phi w^* - v_\pi)$$

into (8.33) yields

$$
\begin{aligned}
\|\Phi w^* - v_\pi\|_D &\leq \|\gamma M P_\pi (\Phi w^* - v_\pi)\|_D + \|M v_\pi - v_\pi\|_D \\
&\leq \gamma \|M\|_D \|P_\pi (\Phi w^* - v_\pi)\|_D + \|M v_\pi - v_\pi\|_D \\
&= \gamma \|P_\pi (\Phi w^* - v_\pi)\|_D + \|M v_\pi - v_\pi\|_D \qquad (\text{because } \|M\|_D = 1) \\
&\leq \gamma \|\Phi w^* - v_\pi\|_D + \|M v_\pi - v_\pi\|_D. \qquad (\text{because } \|P_\pi x\|_D \leq \|x\|_D \text{ for all } x)
\end{aligned}
$$

The proof of $\|M\|_D = 1$ and $\|P_\pi x\|_D \leq \|x\|_D$ are postponed to the end of the box. Recognizing the above inequality gives

$$
\begin{aligned}
\|\Phi w^* - v_\pi\|_D &\leq \frac{1}{1-\gamma} \|M v_\pi - v_\pi\|_D \\
&= \frac{1}{1-\gamma} \min_w \|\hat{v}(w) - v_\pi\|_D,
\end{aligned}
$$

where the last equality is because $\|M v_\pi - v_\pi\|_D$ is the error between v_π and its orthogonal projection into the space of all possible approximations. Therefore, it is the minimum value of the error between v_π and any $\hat{v}(w)$.

We next prove some useful facts, which have already been used in the above proof.

◇ Properties of matrix weighted norms. By definition, $\|x\|_D = \sqrt{x^T D x} = \|D^{1/2} x\|_2$. The induced matrix norm is $\|A\|_D = \max_{x \neq 0} \|Ax\|_D / \|x\|_D = \|D^{1/2} A D^{-1/2}\|_2$. For matrices A, B with appropriate dimensions, we have $\|ABx\|_D \leq \|A\|_D \|B\|_D \|x\|_D$. To see that, $\|ABx\|_D = \|D^{1/2} ABx\|_2 = \|D^{1/2} A D^{-1/2} D^{1/2} B D^{-1/2} D^{1/2} x\|_2 \leq \|D^{1/2} A D^{-1/2}\|_2 \|D^{1/2} B D^{-1/2}\|_2 \|D^{1/2} x\|_2 = \|A\|_D \|B\|_D \|x\|_D$.

◇ Proof of $\|M\|_D = 1$. This is valid because $\|M\|_D = \|\Phi(\Phi^T D \Phi)^{-1} \Phi^T D\|_D = \|D^{1/2} \Phi(\Phi^T D \Phi)^{-1} \Phi^T D D^{-1/2}\|_2 = 1$, where the last equality is valid due to the fact that the matrix in the L_2-norm is an orthogonal projection matrix and the L_2-norm of any orthogonal projection matrix is equal to one.

◇ Proof of $\|P_\pi x\|_D \leq \|x\|_D$ for any $x \in \mathbb{R}^n$. First,

$$
\|P_\pi x\|_D^2 = x^T P_\pi^T D P_\pi x = \sum_{i,j} x_i [P_\pi^T D P_\pi]_{ij} x_j = \sum_{i,j} x_i \left(\sum_k [P_\pi^T]_{ik} [D]_{kk} [P_\pi]_{kj} \right) x_j.
$$

Reorganizing the above equation gives

$$
\begin{aligned}
\|P_\pi x\|_D^2 &= \sum_k [D]_{kk} \left(\sum_i [P_\pi]_{ki} x_i \right)^2 \\
&\leq \sum_k [D]_{kk} \left(\sum_i [P_\pi]_{ki} x_i^2 \right) \quad \text{(due to Jensen's inequality [55, 56])} \\
&= \sum_i \left(\sum_k [D]_{kk} [P_\pi]_{ki} \right) x_i^2 \\
&= \sum_i [D]_{ii} x_i^2 \quad \text{(due to } d_\pi^T P_\pi = d_\pi^T) \\
&= \|x\|_D^2.
\end{aligned}
$$

Least-squares TD

We next introduce an algorithm called *least-squares TD* (LSTD) [57]. Like the TD-Linear algorithm, LSTD aims to minimize the projected Bellman error. However, it has some advantages over the TD-Linear algorithm.

Recall that the optimal parameter for minimizing the projected Bellman error is $w^* = A^{-1}b$, where $A = \Phi^T D (I - \gamma P_\pi) \Phi$ and $b = \Phi^T D r_\pi$. In fact, it follows from (8.27) that A and b can also be written as

$$
\begin{aligned}
A &= \mathbb{E}\Big[\phi(s_t) \big(\phi(s_t) - \gamma \phi(s_{t+1}) \big)^T \Big], \\
b &= \mathbb{E}\Big[r_{t+1} \phi(s_t) \Big].
\end{aligned}
$$

The above two equations show that A and b are expectations of s_t, s_{t+1}, r_{t+1}. The *idea* of LSTD is simple: if we can use random samples to directly obtain the estimates of A and b, which are denoted as $\hat{A}$ and $\hat{b}$, then the optimal parameter can be directly estimated as $w^* \approx \hat{A}^{-1}\hat{b}$.

In particular, suppose that $(s_0, r_1, s_1, \ldots, s_t, r_{t+1}, s_{t+1}, \ldots)$ is a trajectory obtained by following a given policy π. Let $\hat{A}_t$ and $\hat{b}_t$ be the estimates of A and b at time t, respectively. They are calculated as the averages of the samples:

$$
\begin{aligned}
\hat{A}_t &= \sum_{k=0}^{t-1} \phi(s_k) \big(\phi(s_k) - \gamma \phi(s_{k+1}) \big)^T, \\
\hat{b}_t &= \sum_{k=0}^{t-1} r_{k+1} \phi(s_k).
\end{aligned}
\tag{8.34}
$$

Then, the estimated parameter is

$$
w_t = \hat{A}_t^{-1} \hat{b}_t.
$$

The reader may wonder if a coefficient of $1/t$ is missing on the right-hand side of (8.34). In fact, it is omitted for the sake of simplicity since the value of w_t remains the same when it is omitted. Since $\hat{A}_t$ may not be invertible especially when t is small, $\hat{A}_t$ is usually biased by a small constant matrix σI, where I is the identity matrix and σ is a small positive number.

The *advantage* of LSTD is that it uses experience samples more efficiently and converges faster than the TD method. That is because this algorithm is specifically designed based on the knowledge of the optimal solution's expression. The better we understand a problem, the better algorithms we can design.

The *disadvantages* of LSTD are as follows. First, it can only estimate state values. By contrast, the TD algorithm can be extended to estimate action values as shown in the next section. Moreover, while the TD algorithm allows nonlinear approximators, LSTD does not. That is because this algorithm is specifically designed based on the expression of w^*. Second, the computational cost of LSTD is higher than that of TD since LSTD updates an $m \times m$ matrix in each update step, whereas TD updates an m-dimensional vector. More importantly, in every step, LSTD needs to compute the inverse of $\hat{A}_t$, whose computational complexity is $O(m^3)$. The common method for resolving this problem is to directly update the inverse of $\hat{A}_t$ rather than updating $\hat{A}_t$. In particular, $\hat{A}_{t+1}$ can be calculated recursively as follows:

$$
\begin{aligned}
\hat{A}_{t+1} &= \sum_{k=0}^{t} \phi(s_k)\big(\phi(s_k) - \gamma\phi(s_{k+1})\big)^T \\
&= \sum_{k=0}^{t-1} \phi(s_k)\big(\phi(s_k) - \gamma\phi(s_{k+1})\big)^T + \phi(s_t)\big(\phi(s_t) - \gamma\phi(s_{t+1})\big)^T \\
&= \hat{A}_t + \phi(s_t)\big(\phi(s_t) - \gamma\phi(s_{t+1})\big)^T.
\end{aligned}
$$

The above expression decomposes $\hat{A}_{t+1}$ into the sum of two matrices. Its inverse can be calculated as [58]

$$
\begin{aligned}
\hat{A}_{t+1}^{-1} &= \left(\hat{A}_t + \phi(s_t)\big(\phi(s_t) - \gamma\phi(s_{t+1})\big)^T \right)^{-1} \\
&= \hat{A}_t^{-1} + \frac{\hat{A}_t^{-1}\phi(s_t)\big(\phi(s_t) - \gamma\phi(s_{t+1})\big)^T \hat{A}_t^{-1}}{1 + \big(\phi(s_t) - \gamma\phi(s_{t+1})\big)^T \hat{A}_t^{-1}\phi(s_t)}.
\end{aligned}
$$

Therefore, we can directly store and update $\hat{A}_t^{-1}$ to avoid the need to calculate the matrix inverse. This recursive algorithm does not require a step size. However, it requires setting the initial value of $\hat{A}_0^{-1}$. The initial value of such a recursive algorithm can be selected as $\hat{A}_0^{-1} = \sigma I$, where σ is a positive number. A good tutorial on the recursive least-squares approach can be found in [59].

8.3 TD learning of action values based on function approximation

While Section 8.2 introduced the problem of *state value* estimation, the present section introduces how to estimate *action values*. The tabular Sarsa and tabular Q-learning algorithms are extended to the case of value function approximation. Readers will see that the extension is straightforward.

8.3.1 Sarsa with function approximation

The Sarsa algorithm with function approximation can be readily obtained from (8.13) by replacing the state values with action values. In particular, suppose that $q_\pi(s, a)$ is approximated by $\hat{q}(s, a, w)$. Replacing $\hat{v}(s, w)$ in (8.13) by $\hat{q}(s, a, w)$ gives

$$w_{t+1} = w_t + \alpha_t \left[r_{t+1} + \gamma \hat{q}(s_{t+1}, a_{t+1}, w_t) - \hat{q}(s_t, a_t, w_t) \right] \nabla_w \hat{q}(s_t, a_t, w_t). \qquad (8.35)$$

The analysis of (8.35) is similar to that of (8.13) and is omitted here. When linear functions are used, we have

$$\hat{q}(s, a, w) = \phi^T(s, a)w,$$

where $\phi(s, a)$ is a feature vector. In this case, $\nabla_w \hat{q}(s, a, w) = \phi(s, a)$.

The value estimation step in (8.35) can be combined with a policy improvement step to learn optimal policies. The procedure is summarized in Algorithm 8.2. It should be noted that accurately estimating the action values of a given policy requires (8.35) to be run sufficiently many times. However, (8.35) is executed only once before switching to the policy improvement step. This is similar to the tabular Sarsa algorithm. Moreover, the implementation in Algorithm 8.2 aims to solve the task of finding a good path to the target state from a prespecified starting state. As a result, it cannot find the optimal policy for every state. However, if sufficient experience data are available, the implementation process can be easily adapted to find optimal policies for every state.

An illustrative example is shown in Figure 8.9. In this example, the task is to find a good policy that can lead the agent to the target when starting from the top-left state. Both the total reward and the length of each episode gradually converge to steady values. In this example, the linear feature vector is selected as the Fourier function of order 5. The expression of a Fourier feature vector is given in (8.18).

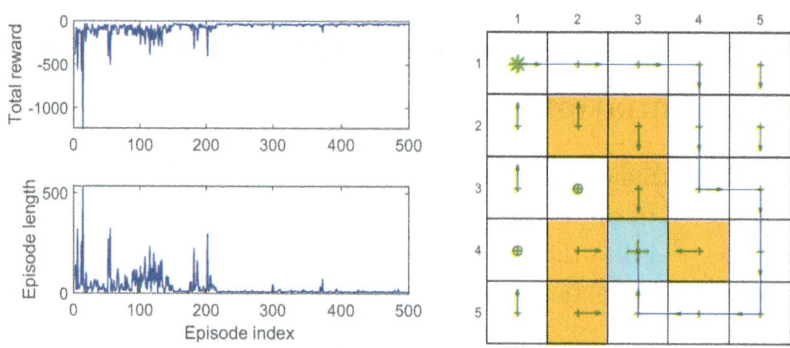

Figure 8.9: Sarsa with linear function approximation. Here, $\gamma = 0.9$, $\epsilon = 0.1$, $r_{\text{boundary}} = r_{\text{forbidden}} = -10$, $r_{\text{target}} = 1$, and $\alpha = 0.001$.

Algorithm 8.2: Sarsa with function approximation

Initialization: Initial parameter w_0. Initial policy π_0. $\alpha_t = \alpha > 0$ for all t. $\epsilon \in (0, 1)$.
Goal: Learn an optimal policy that can lead the agent to the target state from an initial state s_0.
For each episode, do
        Generate a_0 at s_0 following $\pi_0(s_0)$
        If s_t ($t = 0, 1, 2, \dots$) is not the target state, do
            Collect the experience sample $(r_{t+1}, s_{t+1}, a_{t+1})$ given (s_t, a_t): generate r_{t+1}, s_{t+1}
            by interacting with the environment; generate a_{t+1} following $\pi_t(s_{t+1})$.
            Update q-value:
$$w_{t+1} = w_t + \alpha_t \Big[r_{t+1} + \gamma \hat{q}(s_{t+1}, a_{t+1}, w_t) - \hat{q}(s_t, a_t, w_t) \Big] \nabla_w \hat{q}(s_t, a_t, w_t)$$
            Update policy:
$$\pi_{t+1}(a|s_t) = 1 - \frac{\epsilon}{|\mathcal{A}(s_t)|}(|\mathcal{A}(s_t)| - 1) \text{ if } a = \arg\max_{a \in \mathcal{A}(s_t)} \hat{q}(s_t, a, w_{t+1})$$
$$\pi_{t+1}(a|s_t) = \frac{\epsilon}{|\mathcal{A}(s_t)|} \text{ otherwise}$$
            $s_t \leftarrow s_{t+1}$, $a_t \leftarrow a_{t+1}$

8.3.2 Q-learning with function approximation

Tabular Q-learning can also be extended to the case of function approximation. The update rule is

$$w_{t+1} = w_t + \alpha_t \Big[r_{t+1} + \gamma \max_{a \in \mathcal{A}(s_{t+1})} \hat{q}(s_{t+1}, a, w_t) - \hat{q}(s_t, a_t, w_t) \Big] \nabla_w \hat{q}(s_t, a_t, w_t). \quad (8.36)$$

The above update rule is similar to (8.35) except that $\hat{q}(s_{t+1}, a_{t+1}, w_t)$ in (8.35) is replaced with $\max_{a \in \mathcal{A}(s_{t+1})} \hat{q}(s_{t+1}, a, w_t)$.

Similar to the tabular case, (8.36) can be implemented in either an on-policy or off-policy fashion. An on-policy version is given in Algorithm 8.3. An example for demonstrating the on-policy version is shown in Figure 8.10. In this example, the task is to find a good policy that can lead the agent to the target state from the top-left state.

Algorithm 8.3: Q-learning with function approximation (on-policy version)

Initialization: Initial parameter w_0. Initial policy π_0. $\alpha_t = \alpha > 0$ for all t. $\epsilon \in (0, 1)$.
Goal: Learn an optimal path that can lead the agent to the target state from an initial state s_0.

For each episode, do
    If s_t $(t = 0, 1, 2, \dots)$ is not the target state, do
        Collect the experience sample (a_t, r_{t+1}, s_{t+1}) given s_t: generate a_t following $\pi_t(s_t)$; generate r_{t+1}, s_{t+1} by interacting with the environment.
        Update q-value:
$$w_{t+1} = w_t + \alpha_t \left[r_{t+1} + \gamma \max_{a \in \mathcal{A}(s_{t+1})} \hat{q}(s_{t+1}, a, w_t) - \hat{q}(s_t, a_t, w_t) \right] \nabla_w \hat{q}(s_t, a_t, w_t)$$
        Update policy:
$$\pi_{t+1}(a|s_t) = 1 - \frac{\epsilon}{|\mathcal{A}(s_t)|}(|\mathcal{A}(s_t)| - 1) \text{ if } a = \arg\max_{a \in \mathcal{A}(s_t)} \hat{q}(s_t, a, w_{t+1})$$
$$\pi_{t+1}(a|s_t) = \frac{\epsilon}{|\mathcal{A}(s_t)|} \text{ otherwise}$$

As can be seen, Q-learning with linear function approximation can successfully learn an optimal policy. Here, linear Fourier basis functions of order five are used. The off-policy version will be demonstrated when we introduce deep Q-learning in Section 8.4.

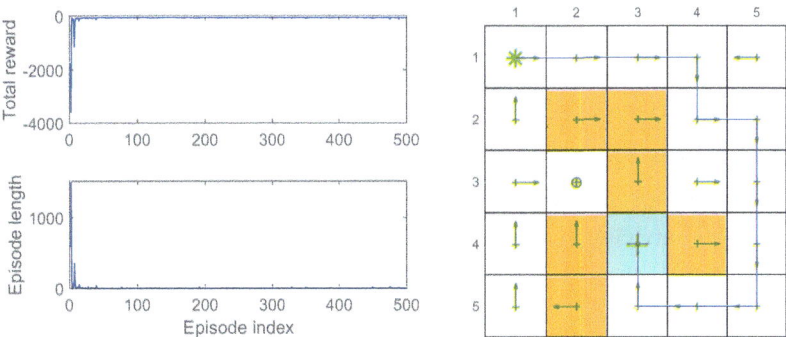

Figure 8.10: Q-learning with linear function approximation. Here, $\gamma = 0.9$, $\epsilon = 0.1$, $r_{\text{boundary}} = r_{\text{forbidden}} = -10$, $r_{\text{target}} = 1$, and $\alpha = 0.001$.

One may notice in Algorithm 8.2 and Algorithm 8.3 that, although the values are represented as functions, the policy $\pi(a|s)$ is still represented as a table. Thus, it still assumes finite numbers of states and actions. In Chapter 9, we will see that the policies can be represented as functions so that continuous state and action spaces can be handled.

8.4 Deep Q-learning

We can integrate deep neural networks into Q-learning to obtain an approach called *deep Q-learning* or *deep Q-network* (DQN) [22, 60, 61]. Deep Q-learning is one of the earliest and most successful deep reinforcement learning algorithms. Notably, the neural networks do not have to be deep. For simple tasks such as our grid world examples, shallow networks with one or two hidden layers may be sufficient.

Deep Q-learning can be viewed as an extension of the algorithm in (8.36). However, its mathematical formulation and implementation techniques are substantially different and deserve special attention.

8.4.1 Algorithm description

Mathematically, deep Q-learning aims to minimize the following objective function:

$$J = \mathbb{E}\left[\left(R + \gamma \max_{a \in \mathcal{A}(S')} \hat{q}(S', a, w) - \hat{q}(S, A, w)\right)^2\right], \tag{8.37}$$

where (S, A, R, S') are random variables that denote a state, an action, the immediate reward, and the next state, respectively. This objective function can be viewed as the squared Bellman optimality error. That is because

$$q(s, a) = \mathbb{E}\left[R_{t+1} + \gamma \max_{a \in \mathcal{A}(S_{t+1})} q(S_{t+1}, a)\Big| S_t = s, A_t = a\right], \quad \text{for all } s, a$$

is the Bellman optimality equation (the proof is given in Box 7.5). Therefore, $R + \gamma \max_{a \in \mathcal{A}(S')} \hat{q}(S', a, w) - \hat{q}(S, A, w)$ should equal zero in the expectation sense when $\hat{q}(S, A, w)$ can accurately approximate the optimal action values.

To minimize the objective function in (8.37), we can use the gradient descent algorithm. To that end, we need to calculate the gradient of J with respect to w. It is noted that the parameter w appears not only in $\hat{q}(S, A, w)$ but also in $y \doteq R + \gamma \max_{a \in \mathcal{A}(S')} \hat{q}(S', a, w)$. As a result, it is nontrivial to calculate the gradient. For the sake of simplicity, it is assumed that the value of w in y is fixed (for a short period of time) so that the calculation of the gradient becomes much easier. In particular, we introduce two networks: one is a *main network* representing $\hat{q}(s, a, w)$ and the other is a *target network* $\hat{q}(s, a, w_T)$. The objective function in this case becomes

$$J = \mathbb{E}\left[\left(R + \gamma \max_{a \in \mathcal{A}(S')} \hat{q}(S', a, w_T) - \hat{q}(S, A, w)\right)^2\right],$$

where w_T is the target network's parameter. When w_T is fixed, the gradient of J is

$$\nabla_w J = -\mathbb{E}\left[\left(R + \gamma \max_{a \in \mathcal{A}(S')} \hat{q}(S', a, w_T) - \hat{q}(S, A, w)\right) \nabla_w \hat{q}(S, A, w)\right], \tag{8.38}$$

where some constant coefficients are omitted without loss of generality.

To use the gradient in (8.38) to minimize the objective function, we need to pay attention to the following techniques.

◇ The first technique is to use two networks, a main network and a target network, as mentioned when we calculate the gradient in (8.38). The implementation details are explained below. Let w and w_T denote the parameters of the main and target networks, respectively. They are initially set to the same value.

In every iteration, we draw a mini-batch of samples $\{(s, a, r, s')\}$ from the replay buffer (the replay buffer will be explained soon). The inputs of the main network are s and a. The output $y = \hat{q}(s, a, w)$ is the estimated q-value. The target value of the output is $y_T \doteq r + \gamma \max_{a \in \mathcal{A}(s')} \hat{q}(s', a, w_T)$. The main network is updated to minimize the TD error (also called the loss function) $\sum(y - y_T)^2$ over the samples $\{(s, a, y_T)\}$.

Updating w in the main network does not explicitly use the gradient in (8.38). Instead, it relies on the existing software tools for training neural networks. As a result, we need a mini-batch of samples to train a network instead of using a single sample to update the main network based on (8.38). This is one notable difference between deep and nondeep reinforcement learning algorithms.

The main network is updated in every iteration. By contrast, the target network is set to be the same as the main network every certain number of iterations to satisfy the assumption that w_T is fixed when calculating the gradient in (8.38).

◇ The second technique is *experience replay* [22, 60, 62]. That is, after we have collected some experience samples, we do not use these samples in the order they were collected. Instead, we store them in a dataset called the *replay buffer*. In particular, let (s, a, r, s') be an experience sample and $\mathcal{B} \doteq \{(s, a, r, s')\}$ be the replay buffer. Every time we update the main network, we can draw a mini-batch of experience samples from the replay buffer. The draw of samples, or called *experience replay*, should follow a *uniform distribution*.

Why is experience replay necessary in deep Q-learning, and why must the replay follow a uniform distribution? The answer lies in the objective function in (8.37). In particular, to well define the objective function, we must specify the probability distributions for S, A, R, S'. The distributions of R and S' are determined by the system model once (S, A) is given. The simplest way to describe the distribution of the state-action pair (S, A) is to assume it to be *uniformly* distributed.

However, the state-action samples may *not* be uniformly distributed in practice since they are generated as a sample sequence according to the behavior policy. It is necessary to break the correlation between the samples in the sequence to satisfy the assumption of uniform distribution. To do this, we can use the experience replay technique by uniformly drawing samples from the replay buffer. This is the mathematical reason why experience replay is necessary and why experience replay must follow a uniform distribution. A benefit of random sampling is that each experience sample

---

Algorithm 8.3: Deep Q-learning (off-policy version)

Initialization: A main network and a target network with the same initial parameter.
Goal: Learn an optimal target network to approximate the *optimal* action values from the experience samples generated by a given behavior policy π_b.

Store the experience samples generated by π_b in a replay buffer $\mathcal{B} = \{(s, a, r, s')\}$
    For each iteration, do
        Uniformly draw a mini-batch of samples from $\mathcal{B}$
        For each sample (s, a, r, s'), calculate the target value as $y_T = r + \gamma \max_{a \in \mathcal{A}(s')} \hat{q}(s', a, w_T)$, where w_T is the parameter of the target network
        Update the main network to minimize $(y_T - \hat{q}(s, a, w))^2$ using the mini-batch of samples
    Set $w_T = w$ every C iterations

---

may be used multiple times, which can increase the data efficiency. This is especially important when we have a limited amount of data.

The implementation procedure of deep Q-learning is summarized in Algorithm 8.3. This implementation is off-policy. It can also be adapted to become on-policy if needed.

8.4.2 Illustrative examples

An example is given in Figure 8.11 to demonstrate Algorithm 8.3. This example aims to learn the optimal action values for *every* state-action pair. Once the optimal action values are obtained, the optimal greedy policy can be obtained immediately.

A single episode is generated by the behavior policy shown in Figure 8.11(a). This behavior policy is exploratory in the sense that it has the same probability of taking any action at any state. The episode has only 1,000 steps as shown in Figure 8.11(b). Although there are only 1,000 steps, almost all the state action pairs are visited in this episode due to the strong exploration ability of the behavior policy. The replay buffer is a set of 1,000 experience samples. The mini-batch size is 100, meaning that we uniformly draw 100 samples from the replay buffer every time we acquire samples.

The main and target networks have the same structure: a neural network with one hidden layer of 100 neurons (the numbers of layers and neurons can be tuned). The neural network has three inputs and one output. The first two inputs are the normalized row and column indexes of a state. The third input is the normalized action index. Here, "normalization" means converting a value to the interval of [0,1]. The output of the network is the estimated value. The reason why we design the inputs as the row and column of a state rather than a state index is that we know that a state corresponds to a two-dimensional location in the grid. The more information about the state we use when designing the network, the better the network can perform. Moreover, the neural

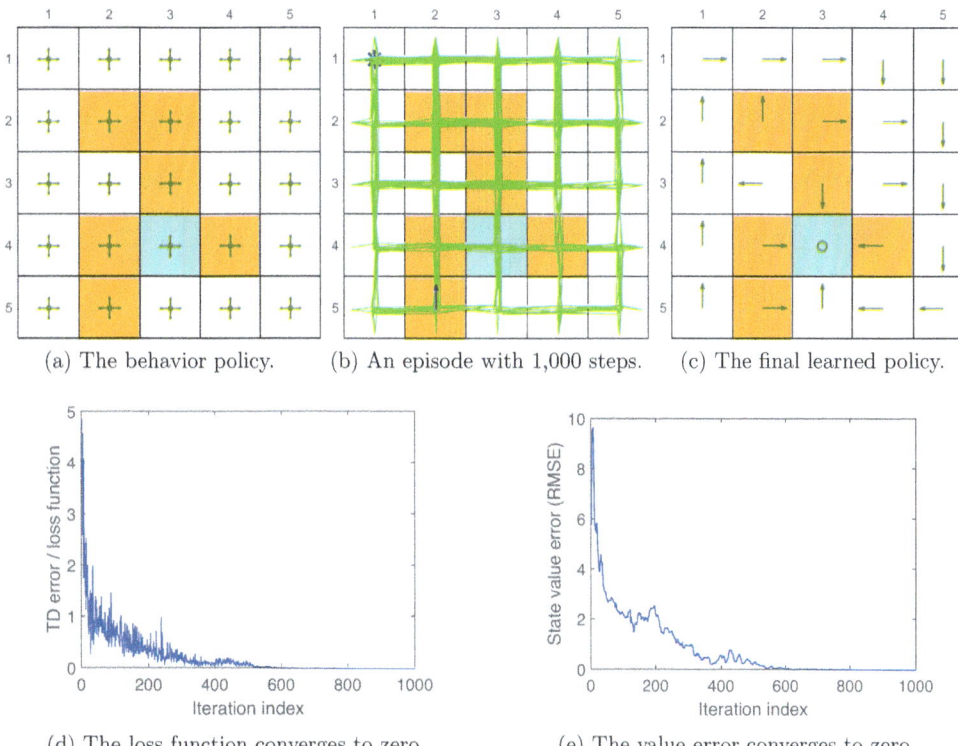

(a) The behavior policy. (b) An episode with 1,000 steps. (c) The final learned policy.

(d) The loss function converges to zero. (e) The value error converges to zero.

Figure 8.11: Optimal policy learning via deep Q-learning. Here, $\gamma = 0.9$, $r_{\text{boundary}} = r_{\text{forbidden}} = -10$, and $r_{\text{target}} = 1$. The batch size is 100.

network can also be designed in other ways. For example, it can have two inputs and five outputs, where the two inputs are the normalized row and column of a state and the outputs are the five estimated action values for the input state [22].

As shown in Figure 8.11(d), the loss function, defined as the average squared TD error of each mini-batch, converges to zero, meaning that the network can fit the training samples well. As shown in Figure 8.11(e), the state value estimation error also converges to zero, indicating that the estimates of the optimal action values become sufficiently accurate. Then, the corresponding greedy policy is optimal.

This example demonstrates the high efficiency of deep Q-learning. In particular, a short episode of 1,000 steps is sufficient for obtaining an optimal policy here. By contrast, an episode with 100,000 steps is required by tabular Q-learning, as shown in Figure 7.4. One reason for the high efficiency is that the function approximation method has a strong generalization ability. Another reason is that the experience samples can be repeatedly used.

We next deliberately challenge the deep Q-learning algorithm by considering a scenario with fewer experience samples. Figure 8.12 shows an example of an episode with merely 100 steps. In this example, although the network can still be well-trained in the sense

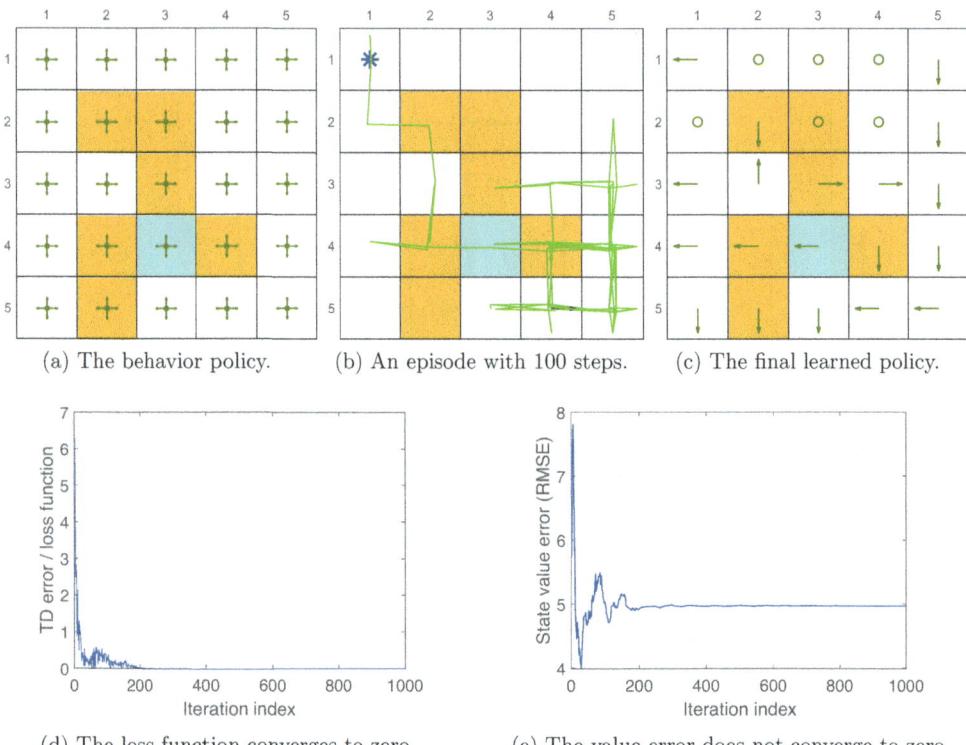

(a) The behavior policy.   (b) An episode with 100 steps.   (c) The final learned policy.

(d) The loss function converges to zero.   (e) The value error does not converge to zero.

Figure 8.12: Optimal policy learning via deep Q-learning. Here, $\gamma = 0.9$, $r_{\text{boundary}} = r_{\text{forbidden}} = -10$, and $r_{\text{target}} = 1$. The batch size is 50.

that the loss function converges to zero, the state estimation error cannot converge to zero. That means the network can properly fit the given experience samples, but the experience samples are too few to accurately estimate the optimal action values.

8.5 Summary

This chapter continued introducing TD learning algorithms. However, it switches from the tabular method to the function approximation method. The key to understanding the function approximation method is to know that it is an optimization problem. The simplest objective function is the squared error between the true state values and the estimated values. There are also other objective functions such as the Bellman error and the projected Bellman error. We have shown that the TD-Linear algorithm actually minimizes the projected Bellman error. Several optimization algorithms such as Sarsa and Q-learning with value approximation have been introduced.

One reason why the value function approximation method is important is that it allows artificial neural networks to be integrated with reinforcement learning. For example, deep Q-learning is one of the most successful deep reinforcement learning algorithms.

Although neural networks have been widely used as nonlinear function approximators, this chapter provides a comprehensive introduction to the linear function case. Fully understanding the linear case is important for better understanding the nonlinear case. Interested readers may refer to [63] for a thorough analysis of TD learning algorithms with function approximation. A more theoretical discussion on deep Q-learning can be found in [61].

An important concept named stationary distribution is introduced in this chapter. The stationary distribution plays an important role in defining an appropriate objective function in the value function approximation method. It also plays a key role in Chapter 9 when we use functions to approximate policies. An excellent introduction to this topic can be found in [49, Chapter IV]. The contents of this chapter heavily rely on matrix analysis. Some results are used without explanation. Excellent references regarding matrix analysis and linear algebra can be found in [4, 48].

8.6 Q&A

⋄ Q: What is the difference between the tabular and function approximation methods?

A: One important difference is how a value is updated and retrieved.

How to *retrieve* a value: When the values are represented by a table, if we would like to retrieve a value, we can directly read the corresponding entry in the table. However, when the values are represented by a function, we need to input the state index s into the function and calculate the function value. If the function is an artificial neural network, a forward prorogation process from the input to the output is needed.

How to *update* a value: When the values are represented by a table, if we would like to update one value, we can directly rewrite the corresponding entry in the table. However, when the values are represented by a function, we must update the function parameter to change the values indirectly.

⋄ Q: What are the advantages of the function approximation method over the tabular method?

A: Due to the way state values are retrieved, the function approximation method is more efficient in storage. In particular, while the tabular method needs to store $|\mathcal{S}|$ values, the function approximation method only needs to store a parameter vector whose dimension is usually much less than $|\mathcal{S}|$.

Due to the way in which state values are updated, the function approximation method has another merit: its generalization ability is stronger than that of the tabular method. The reason is as follows. With the tabular method, updating one state value would not change the other state values. However, with the function approximation method, updating the function parameter affects the values of many states.

Therefore, the experience sample for one state can generalize to help estimate the values of other states.

◇ Q: Can we unify the tabular and the function approximation methods?

A: Yes. The tabular method can be viewed as a special case of the function approximation method. The related details can be found in Box 8.2.

◇ Q: What is the stationary distribution and why is it important?

A: The stationary distribution describes the long-term behavior of a Markov decision process. More specifically, after the agent executes a given policy for a sufficiently long period, the probability of the agent visiting a state can be described by this stationary distribution. More information can be found in Box 8.1.

The reason why this concept emerges in this chapter is that it is necessary for defining a valid objective function. In particular, the objective function involves the probability distribution of the states, which is usually selected as the stationary distribution. The stationary distribution is important not only for the value approximation method but also for the policy gradient method, which will be introduced in Chapter 9.

◇ Q: What are the advantages and disadvantages of the linear function approximation method?

A: Linear function approximation is the simplest case whose theoretical properties can be thoroughly analyzed. However, the approximation ability of this method is limited. It is also nontrivial to select appropriate feature vectors for complex tasks. By contrast, artificial neural networks can be used to approximate values as black-box universal nonlinear approximators, which are more friendly to use. Nevertheless, it is still meaningful to study the linear case to better grasp the idea of the function approximation method. Moreover, the linear case is powerful in the sense that the tabular method can be viewed as a special linear case (Box 8.2).

◇ Q: Why does deep Q-learning require experience replay?

A: The reason lies in the objective function in (8.37). In particular, to well define the objective function, we must specify the probability distributions of S, A, R, S'. The distributions of R and S' are determined by the system model once (S, A) is given. The simplest way to describe the distribution of the state-action pair (S, A) is to assume it to be *uniformly* distributed. However, the state-action samples may *not* be uniformly distributed in practice since they are generated as a sequence by the behavior policy. It is necessary to *break the correlation* between the samples in the sequence to satisfy the assumption of uniform distribution. To do this, we can use the experience replay technique by uniformly drawing samples from the replay buffer. A benefit of experience replay is that each experience sample may be used multiple times, which can increase the data efficiency.

⋄ Q: Can tabular Q-learning use experience replay?

A: Although tabular Q-learning does not require experience replay, it can also use experience relay without encountering problems. That is because Q-learning has no requirements about how the samples are obtained due to its off-policy attribute. One benefit of using experience replay is that the samples can be used repeatedly and hence more efficiently.

⋄ Q: Why does deep Q-learning require two networks?

A: The fundamental reason is to simplify the calculation of the gradient of (8.37). Since w appears not only in $\hat{q}(S, A, w)$ but also in $R + \gamma \max_{a \in \mathcal{A}(S')} \hat{q}(S', a, w)$, it is nontrivial to calculate the gradient with respect to w. On the one hand, if we fix w in $R + \gamma \max_{a \in \mathcal{A}(S')} \hat{q}(S', a, w)$, the gradient can be easily calculated as shown in (8.38). This gradient suggests that two networks should be maintained. The main network's parameter is updated in every iteration. The target network's parameter is fixed within a certain period. On the other hand, the target network's parameter cannot be fixed forever. It should be updated every certain number of iterations.

⋄ Q: When an artificial neural network is used as a nonlinear function approximator, how should we update its parameter?

A: It must be noted that we should not directly update the parameter vector by using, for example, (8.36). Instead, we should follow the network training procedure to update the parameter. This procedure can be realized based on neural network training toolkits, which are currently mature and widely available.

Chapter 9

Policy Gradient Methods

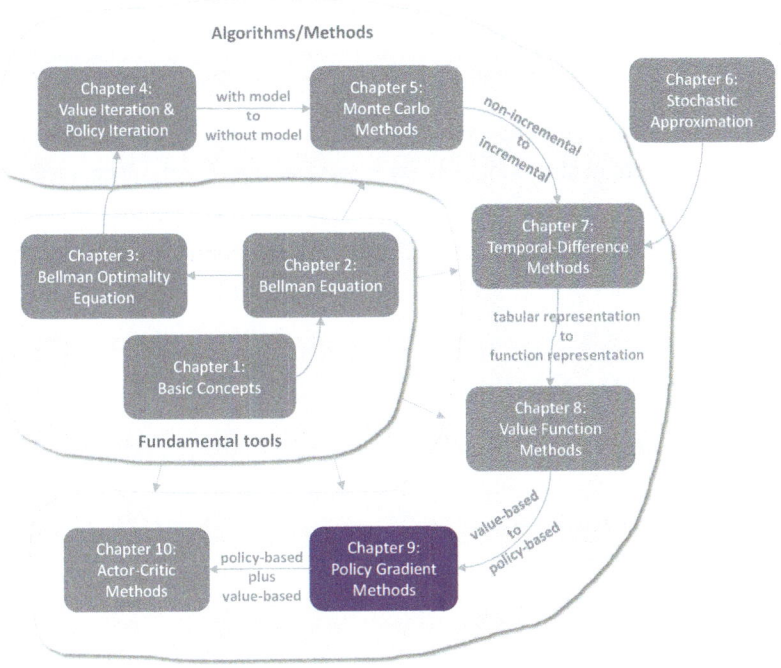

Figure 9.1: Where we are in this book.

The idea of function approximation can be applied not only to represent state/action values, as introduced in Chapter 8, but also to represent policies, as introduced in this chapter. So far in this book, policies have been represented by tables: the action probabilities of all states are stored in a table (e.g., Table 9.1). In this chapter, we show that policies can be represented by parameterized functions denoted as $\pi(a|s, \theta)$, where $\theta \in \mathbb{R}^m$ is a parameter vector. It can also be written in other forms such as $\pi_\theta(a|s)$, $\pi_\theta(a, s)$, or $\pi(a, s, \theta)$.

When policies are represented as functions, optimal policies can be obtained by optimizing certain scalar metrics. Such a method is called *policy gradient*. The policy

S. Zhao, *Mathematical Foundations of Reinforcement Learning*, https://doi.org/10.1007/978-981-97-3944-8_9

	a_1	a_2	a_3	a_4	a_5
s_1	$\pi(a_1\|s_1)$	$\pi(a_2\|s_1)$	$\pi(a_3\|s_1)$	$\pi(a_4\|s_1)$	$\pi(a_5\|s_1)$
$\vdots$	$\vdots$	$\vdots$	$\vdots$	$\vdots$	$\vdots$
s_9	$\pi(a_1\|s_9)$	$\pi(a_2\|s_9)$	$\pi(a_3\|s_9)$	$\pi(a_4\|s_9)$	$\pi(a_5\|s_9)$

Table 9.1: A tabular representation of a policy. There are nine states and five actions for each state.

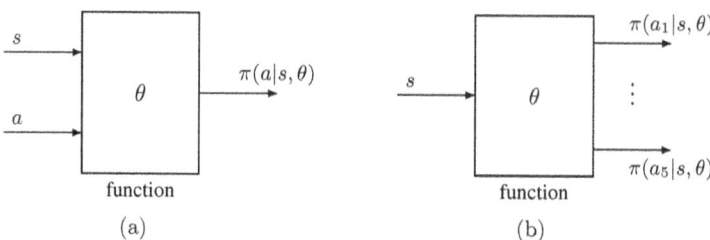

Figure 9.2: Function representations of policies. The functions may have different structures.

gradient method is a big step forward in this book because it is *policy-based*. By contrast, all the previous chapters in this book discuss *value-based* methods. The advantages of the policy gradient method are numerous. For example, it is more efficient for handling large state/action spaces. It has stronger generalization abilities and hence is more efficient in terms of sample usage.

9.1 Policy representation: From table to function

When the representation of a policy is switched from a table to a function, it is necessary to clarify the difference between the two representation methods.

⋄ First, how to define optimal policies? When represented as a table, a policy is defined as optimal if it can maximize *every state value*. When represented by a function, a policy is defined as optimal if it can maximize certain *scalar metrics*.

⋄ Second, how to update a policy? When represented by a table, a policy can be updated by directly changing the entries in the table. When represented by a parameterized function, a policy can no longer be updated in this way. Instead, it can only be updated by changing the parameter θ.

⋄ Third, how to retrieve the probability of an action? In the tabular case, the probability of an action can be directly obtained by looking up the corresponding entry in the table. In the case of function representation, we need to input (s, a) into the function to calculate its probability (see Figure 9.2(a)). Depending on the structure of the function, we can also input a state and then output the probabilities of all actions (see Figure 9.2(b)).

The basic idea of the policy gradient method is summarized below. Suppose that $J(\theta)$ is a scalar metric. Optimal policies can be obtained by optimizing this metric via the gradient-based algorithm:

$$\theta_{t+1} = \theta_t + \alpha \nabla_\theta J(\theta_t),$$

where $\nabla_\theta J$ is the gradient of J with respect to θ, t is the time step, and α is the optimization rate.

With this basic idea, we will answer the following three questions in the remainder of this chapter.

◇ What metrics should be used? (Section 9.2).

◇ How to calculate the gradients of the metrics? (Section 9.3)

◇ How to use experience samples to calculate the gradients? (Section 9.4)

9.2 Metrics for defining optimal policies

If a policy is represented by a function, there are two types of metrics for defining optimal policies. One is based on state values and the other is based on immediate rewards.

Metric 1: Average state value

The first metric is the *average state value* or simply called the *average value*. It is defined as

$$\bar{v}_\pi = \sum_{s \in \mathcal{S}} d(s) v_\pi(s),$$

where $d(s)$ is the weight of state s. It satisfies $d(s) \geq 0$ for any $s \in \mathcal{S}$ and $\sum_{s \in \mathcal{S}} d(s) = 1$. Therefore, we can interpret $d(s)$ as a probability distribution of s. Then, the metric can be written as

$$\bar{v}_\pi = \mathbb{E}_{S \sim d}[v_\pi(S)].$$

How to select the distribution d? This is an important question. There are two cases.

◇ The first and simplest case is that d is *independent* of the policy π. In this case, we specifically denote d as d_0 and $\bar{v}_\pi$ as $\bar{v}_\pi^0$ to indicate that the distribution is independent of the policy. One case is to treat all the states equally important and select $d_0(s) = 1/|\mathcal{S}|$. Another case is when we are only interested in a specific state s_0 (e.g., the agent always starts from s_0). In this case, we can design

$$d_0(s_0) = 1, \quad d_0(s \neq s_0) = 0.$$

◇ The second case is that d is *dependent* on the policy π. In this case, it is common to select d as d_π, which is the *stationary distribution* under π. One basic property of d_π is that it satisfies

$$d_\pi^T P_\pi = d_\pi^T,$$

where P_π is the state transition probability matrix. More information about the stationary distribution can be found in Box 8.1.

The interpretation of selecting d_π is as follows. The stationary distribution reflects the long-term behavior of a Markov decision process under a given policy. If one state is frequently visited in the long term, it is more important and deserves a higher weight; if a state is rarely visited, then its importance is low and deserves a lower weight.

As its name suggests, $\bar{v}_\pi$ is a weighted average of the state values. Different values of θ lead to different values of $\bar{v}_\pi$. Our ultimate goal is to find an optimal policy (or equivalently an optimal θ) to maximize $\bar{v}_\pi$.

We next introduce another two important equivalent expressions of $\bar{v}_\pi$.

◇ Suppose that an agent collects rewards $\{R_{t+1}\}_{t=0}^\infty$ by following a given policy $\pi(\theta)$. Readers may often see the following metric in the literature:

$$J(\theta) = \lim_{n \to \infty} \mathbb{E}\left[\sum_{t=0}^{n} \gamma^t R_{t+1}\right] = \mathbb{E}\left[\sum_{t=0}^{\infty} \gamma^t R_{t+1}\right]. \tag{9.1}$$

This metric may be nontrivial to interpret at first glance. In fact, it is equal to $\bar{v}_\pi$. To see that, we have

$$\mathbb{E}\left[\sum_{t=0}^{\infty} \gamma^t R_{t+1}\right] = \sum_{s \in \mathcal{S}} d(s) \mathbb{E}\left[\sum_{t=0}^{\infty} \gamma^t R_{t+1} | S_0 = s\right]$$
$$= \sum_{s \in \mathcal{S}} d(s) v_\pi(s)$$
$$= \bar{v}_\pi.$$

The first equality in the above equation is due to the law of total expectation. The second equality is by the definition of state values.

◇ The metric $\bar{v}_\pi$ can also be rewritten as the inner product of two vectors. In particular, let

$$v_\pi = [\ldots, v_\pi(s), \ldots]^T \in \mathbb{R}^{|\mathcal{S}|},$$
$$d = [\ldots, d(s), \ldots]^T \in \mathbb{R}^{|\mathcal{S}|}.$$

Then, we have

$$\bar{v}_\pi = d^T v_\pi.$$

This expression will be useful when we analyze its gradient.

Metric 2: Average reward

The second metric is the *average one-step reward* or simply called the *average reward* [2, 64, 65]. In particular, it is defined as

$$\bar{r}_\pi \doteq \sum_{s \in \mathcal{S}} d_\pi(s) r_\pi(s)$$

$$= \mathbb{E}_{S \sim d_\pi}[r_\pi(S)], \tag{9.2}$$

where d_π is the stationary distribution and

$$r_\pi(s) \doteq \sum_{a \in \mathcal{A}} \pi(a|s, \theta) r(s, a) = \mathbb{E}_{A \sim \pi(s, \theta)}[r(s, A)|s] \tag{9.3}$$

is the expectation of the immediate rewards. Here, $r(s, a) \doteq \mathbb{E}[R|s, a] = \sum_r r p(r|s, a)$.

We next present another two important equivalent expressions of $\bar{r}_\pi$.

◇ Suppose that the agent collects rewards $\{R_{t+1}\}_{t=0}^\infty$ by following a given policy $\pi(\theta)$. A common metric that readers may often see in the literature is

$$J(\theta) = \lim_{n \to \infty} \frac{1}{n} \mathbb{E}\left[\sum_{t=0}^{n-1} R_{t+1}\right]. \tag{9.4}$$

It may seem nontrivial to interpret this metric at first glance. In fact, it is equal to $\bar{r}_\pi$:

$$\lim_{n \to \infty} \frac{1}{n} \mathbb{E}\left[\sum_{t=0}^{n-1} R_{t+1}\right] = \sum_{s \in \mathcal{S}} d_\pi(s) r_\pi(s) = \bar{r}_\pi. \tag{9.5}$$

The proof of (9.5) is given in Box 9.1.

◇ The average reward $\bar{r}_\pi$ in (9.2) can also be written as the inner product of two vectors. In particular, let

$$r_\pi = [\ldots, r_\pi(s), \ldots]^T \in \mathbb{R}^{|\mathcal{S}|},$$
$$d_\pi = [\ldots, d_\pi(s), \ldots]^T \in \mathbb{R}^{|\mathcal{S}|},$$

where $r_\pi(s)$ is defined in (9.3). Then, it is clear that

$$\bar{r}_\pi = \sum_{s \in \mathcal{S}} d_\pi(s) r_\pi(s) = d_\pi^T r_\pi.$$

This expression will be useful when we derive its gradient.

Box 9.1: Proof of (9.5)

Step 1: We first prove that the following equation is valid for any starting state $s_0 \in \mathcal{S}$:

$$\bar{r}_\pi = \lim_{n \to \infty} \frac{1}{n} \mathbb{E} \left[\sum_{t=0}^{n-1} R_{t+1} | S_0 = s_0 \right]. \tag{9.6}$$

To do that, we notice

$$\lim_{n \to \infty} \frac{1}{n} \mathbb{E} \left[\sum_{t=0}^{n-1} R_{t+1} | S_0 = s_0 \right] = \lim_{n \to \infty} \frac{1}{n} \sum_{t=0}^{n-1} \mathbb{E} \left[R_{t+1} | S_0 = s_0 \right]$$

$$= \lim_{t \to \infty} \mathbb{E} \left[R_{t+1} | S_0 = s_0 \right], \tag{9.7}$$

where the last equality is due to the property of the Cesaro mean (also called the Cesaro summation). In particular, if $\{a_k\}_{k=1}^\infty$ is a convergent sequence such that $\lim_{k \to \infty} a_k$ exists, then $\{1/n \sum_{k=1}^n a_k\}_{n=1}^\infty$ is also a convergent sequence such that $\lim_{n \to \infty} 1/n \sum_{k=1}^n a_k = \lim_{k \to \infty} a_k$.

We next examine $\mathbb{E} \left[R_{t+1} | S_0 = s_0 \right]$ in (9.7) more closely. By the law of total expectation, we have

$$\mathbb{E} \left[R_{t+1} | S_0 = s_0 \right] = \sum_{s \in \mathcal{S}} \mathbb{E} \left[R_{t+1} | S_t = s, S_0 = s_0 \right] p^{(t)}(s|s_0)$$

$$= \sum_{s \in \mathcal{S}} \mathbb{E} \left[R_{t+1} | S_t = s \right] p^{(t)}(s|s_0)$$

$$= \sum_{s \in \mathcal{S}} r_\pi(s) p^{(t)}(s|s_0),$$

where $p^{(t)}(s|s_0)$ denotes the probability of transitioning from s_0 to s using exactly t steps. The second equality in the above equation is due to the Markov memoryless property: the reward obtained at the next time step depends only on the current state rather than the previous ones.

Note that

$$\lim_{t \to \infty} p^{(t)}(s|s_0) = d_\pi(s)$$

---

by the definition of the stationary distribution. As a result, the starting state s_0 does not matter. Then, we have

$$\lim_{t\to\infty} \mathbb{E}\left[R_{t+1}|S_0 = s_0\right] = \lim_{t\to\infty} \sum_{s\in\mathcal{S}} r_\pi(s)p^{(t)}(s|s_0) = \sum_{s\in\mathcal{S}} r_\pi(s)d_\pi(s) = \bar{r}_\pi.$$

Substituting the above equation into (9.7) gives (9.6).

Step 2: Consider an arbitrary state distribution d. By the law of total expectation, we have

$$\lim_{n\to\infty} \frac{1}{n}\mathbb{E}\left[\sum_{t=0}^{n-1} R_{t+1}\right] = \lim_{n\to\infty} \frac{1}{n}\sum_{s\in\mathcal{S}} d(s)\mathbb{E}\left[\sum_{t=0}^{n-1} R_{t+1}|S_0 = s\right]$$

$$= \sum_{s\in\mathcal{S}} d(s)\lim_{n\to\infty} \frac{1}{n}\mathbb{E}\left[\sum_{t=0}^{n-1} R_{t+1}|S_0 = s\right].$$

Since (9.6) is valid for any starting state, substituting (9.6) into the above equation yields

$$\lim_{n\to\infty} \frac{1}{n}\mathbb{E}\left[\sum_{t=0}^{n-1} R_{t+1}\right] = \sum_{s\in\mathcal{S}} d(s)\bar{r}_\pi = \bar{r}_\pi.$$

The proof is complete.

Some remarks

Metric	Expression 1	Expression 2	Expression 3
$\bar{v}_\pi$	$\sum_{s\in\mathcal{S}} d(s)v_\pi(s)$	$\mathbb{E}_{S\sim d}[v_\pi(S)]$	$\lim_{n\to\infty} \mathbb{E}\left[\sum_{t=0}^{n} \gamma^t R_{t+1}\right]$
$\bar{r}_\pi$	$\sum_{s\in\mathcal{S}} d_\pi(s)r_\pi(s)$	$\mathbb{E}_{S\sim d_\pi}[r_\pi(S)]$	$\lim_{n\to\infty} \frac{1}{n}\mathbb{E}\left[\sum_{t=0}^{n-1} R_{t+1}\right]$

Table 9.2: Summary of the different but equivalent expressions of $\bar{v}_\pi$ and $\bar{r}_\pi$.

Up to now, we have introduced two types of metrics: $\bar{v}_\pi$ and $\bar{r}_\pi$. Each metric has several different but equivalent expressions. They are summarized in Table 9.2. We sometimes use $\bar{v}_\pi$ to specifically refer to the case where the state distribution is the stationary distribution d_π and use $\bar{v}_\pi^0$ to refer to the case where d_0 is independent of π. Some remarks about the metrics are given below.

⋄ All these metrics are functions of π. Since π is parameterized by θ, these metrics are functions of θ. In other words, different values of θ can generate different metric values. Therefore, we can search for the optimal values of θ to maximize these metrics. This is the basic idea of policy gradient methods.

⋄ The two metrics $\bar{v}_\pi$ and $\bar{r}_\pi$ are equivalent in the discounted case where $\gamma < 1$. In particular, it can be shown that

$$\bar{r}_\pi = (1 - \gamma)\bar{v}_\pi.$$

The above equation indicates that these two metrics can be simultaneously maximized. The proof of this equation is given later in Lemma 9.1.

9.3 Gradients of the metrics

Given the metrics introduced in the last section, we can use gradient-based methods to maximize them. To do that, we need to first calculate the gradients of these metrics. The most important theoretical result in this chapter is the following theorem.

Theorem 9.1 (Policy gradient theorem). *The gradient of $J(\theta)$ is*

$$\nabla_\theta J(\theta) = \sum_{s \in \mathcal{S}} \eta(s) \sum_{a \in \mathcal{A}} \nabla_\theta \pi(a|s, \theta) q_\pi(s, a), \tag{9.8}$$

where η is a state distribution and $\nabla_\theta \pi$ is the gradient of π with respect to θ. Moreover, (9.8) has a compact form expressed in terms of expectation:

$$\nabla_\theta J(\theta) = \mathbb{E}_{S \sim \eta, A \sim \pi(S, \theta)} \Big[\nabla_\theta \ln \pi(A|S, \theta) q_\pi(S, A) \Big], \tag{9.9}$$

where $\ln$ is the natural logarithm.

Some important remarks about Theorem 9.1 are given below.

⋄ It should be noted that Theorem 9.1 is a summary of the results in Theorem 9.2, Theorem 9.3, and Theorem 9.5. These three theorems address different scenarios involving different metrics and discounted/undiscounted cases. The gradients in these scenarios all have similar expressions and hence are summarized in Theorem 9.1. The specific expressions of $J(\theta)$ and η are not given in Theorem 9.1 and can be found in Theorem 9.2, Theorem 9.3, and Theorem 9.5. In particular, $J(\theta)$ could be $\bar{v}_\pi^0$, $\bar{v}_\pi$, or $\bar{r}_\pi$. The equality in (9.8) may become a strict equality or an approximation. The distribution η also varies in different scenarios.

The derivation of the gradients is the most complicated part of the policy gradient method. For many readers, it is sufficient to be familiar with the result in Theorem 9.1 without knowing the proof. The derivation details presented in the rest of this section are mathematically intensive. Readers are suggested to study selectively based on their interests.

⬦ The expression in (9.9) is more favorable than (9.8) because it is expressed as an expectation. We will show in Section 9.4 that this true gradient can be approximated by a stochastic gradient.

Why can (9.8) be expressed as (9.9)? The proof is given below. By the definition of expectation, (9.8) can be rewritten as

$$\nabla_\theta J(\theta) = \sum_{s \in \mathcal{S}} \eta(s) \sum_{a \in \mathcal{A}} \nabla_\theta \pi(a|s, \theta) q_\pi(s, a)$$

$$= \mathbb{E}_{S \sim \eta} \left[\sum_{a \in \mathcal{A}} \nabla_\theta \pi(a|S, \theta) q_\pi(S, a) \right]. \tag{9.10}$$

Furthermore, the gradient of $\ln \pi(a|s, \theta)$ is

$$\nabla_\theta \ln \pi(a|s, \theta) = \frac{\nabla_\theta \pi(a|s, \theta)}{\pi(a|s, \theta)}.$$

It follows that

$$\nabla_\theta \pi(a|s, \theta) = \pi(a|s, \theta) \nabla_\theta \ln \pi(a|s, \theta). \tag{9.11}$$

Substituting (9.11) into (9.10) gives

$$\nabla_\theta J(\theta) = \mathbb{E} \left[\sum_{a \in \mathcal{A}} \pi(a|S, \theta) \nabla_\theta \ln \pi(a|S, \theta) q_\pi(S, a) \right]$$

$$= \mathbb{E}_{S \sim \eta, A \sim \pi(S, \theta)} \left[\nabla_\theta \ln \pi(A|S, \theta) q_\pi(S, A) \right].$$

⬦ It is notable that $\pi(a|s, \theta)$ must be *positive* for all (s, a) to ensure that $\ln \pi(a|s, \theta)$ is valid. This can be achieved by using *softmax functions*:

$$\pi(a|s, \theta) = \frac{e^{h(s,a,\theta)}}{\sum_{a' \in \mathcal{A}} e^{h(s,a',\theta)}}, \quad a \in \mathcal{A}, \tag{9.12}$$

where $h(s, a, \theta)$ is a function indicating the preference for selecting a at s. The policy in (9.12) satisfies $\pi(a|s, \theta) \in (0, 1)$ and $\sum_{a \in \mathcal{A}} \pi(a|s, \theta) = 1$ for any $s \in \mathcal{S}$. This policy can be realized by a neural network. The input of the network is s. The output layer is a softmax layer so that the network outputs $\pi(a|s, \theta)$ for all a and the sum of the outputs is equal to 1. See Figure 9.2(b) for an illustration.

Since $\pi(a|s, \theta) > 0$ for all a, the policy is *stochastic* and hence *exploratory*. The policy does not directly tell which action to take. Instead, the action should be generated according to the probability distribution of the policy.

9.3.1 Derivation of the gradients in the discounted case

We next derive the gradients of the metrics in the discounted case where $\gamma \in (0, 1)$. The state value and action value in the discounted case are defined as

$$v_\pi(s) = \mathbb{E}[R_{t+1} + \gamma R_{t+2} + \gamma^2 R_{t+3} + \dots | S_t = s],$$
$$q_\pi(s, a) = \mathbb{E}[R_{t+1} + \gamma R_{t+2} + \gamma^2 R_{t+3} + \dots | S_t = s, A_t = a].$$

It holds that $v_\pi(s) = \sum_{a \in \mathcal{A}} \pi(a|s, \theta) q_\pi(s, a)$ and the state value satisfies the Bellman equation.

First, we show that $\bar{v}_\pi(\theta)$ and $\bar{r}_\pi(\theta)$ are equivalent metrics.

Lemma 9.1 (Equivalence between $\bar{v}_\pi(\theta)$ and $\bar{r}_\pi(\theta)$). *In the discounted case where $\gamma \in (0, 1)$, it holds that*

$$\bar{r}_\pi = (1 - \gamma) \bar{v}_\pi. \tag{9.13}$$

Proof. Note that $\bar{v}_\pi(\theta) = d_\pi^T v_\pi$ and $\bar{r}_\pi(\theta) = d_\pi^T r_\pi$, where v_π and r_π satisfy the Bellman equation $v_\pi = r_\pi + \gamma P_\pi v_\pi$. Multiplying d_π^T on both sides of the Bellman equation yields

$$\bar{v}_\pi = \bar{r}_\pi + \gamma d_\pi^T P_\pi v_\pi = \bar{r}_\pi + \gamma d_\pi^T v_\pi = \bar{r}_\pi + \gamma \bar{v}_\pi,$$

which implies (9.13). $\qquad\square$

Second, the following lemma gives the gradient of $v_\pi(s)$ for any s.

Lemma 9.2 (Gradient of $v_\pi(s)$). *In the discounted case, it holds for any $s \in \mathcal{S}$ that*

$$\nabla_\theta v_\pi(s) = \sum_{s' \in \mathcal{S}} \Pr_\pi(s'|s) \sum_{a \in \mathcal{A}} \nabla_\theta \pi(a|s', \theta) q_\pi(s', a), \tag{9.14}$$

where

$$\Pr_\pi(s'|s) \doteq \sum_{k=0}^{\infty} \gamma^k [P_\pi^k]_{ss'} = \left[(I_n - \gamma P_\pi)^{-1} \right]_{ss'}$$

is the discounted total probability of transitioning from s to s' under policy π. Here, $[\cdot]_{ss'}$ denotes the entry in the sth row and s'th column, and $[P_\pi^k]_{ss'}$ is the probability of transitioning from s to s' using exactly k steps under π.

Box 9.2: Proof of Lemma 9.2

First, for any $s \in \mathcal{S}$, it holds that

$$\nabla_\theta v_\pi(s) = \nabla_\theta \left[\sum_{a \in \mathcal{A}} \pi(a|s, \theta) q_\pi(s, a) \right]$$
$$= \sum_{a \in \mathcal{A}} \left[\nabla_\theta \pi(a|s, \theta) q_\pi(s, a) + \pi(a|s, \theta) \nabla_\theta q_\pi(s, a) \right], \qquad (9.15)$$

where $q_\pi(s, a)$ is the action value given by

$$q_\pi(s, a) = r(s, a) + \gamma \sum_{s' \in \mathcal{S}} p(s'|s, a) v_\pi(s').$$

Since $r(s, a) = \sum_r r p(r|s, a)$ is independent of θ, we have

$$\nabla_\theta q_\pi(s, a) = 0 + \gamma \sum_{s' \in \mathcal{S}} p(s'|s, a) \nabla_\theta v_\pi(s').$$

Substituting this result into (9.15) yields

$$\nabla_\theta v_\pi(s) = \sum_{a \in \mathcal{A}} \left[\nabla_\theta \pi(a|s, \theta) q_\pi(s, a) + \pi(a|s, \theta) \gamma \sum_{s' \in \mathcal{S}} p(s'|s, a) \nabla_\theta v_\pi(s') \right]$$
$$= \sum_{a \in \mathcal{A}} \nabla_\theta \pi(a|s, \theta) q_\pi(s, a) + \gamma \sum_{a \in \mathcal{A}} \pi(a|s, \theta) \sum_{s' \in \mathcal{S}} p(s'|s, a) \nabla_\theta v_\pi(s'). \quad (9.16)$$

It is notable that $\nabla_\theta v_\pi$ appears on both sides of the above equation. One way to calculate it is to use the *unrolling technique* [64]. Here, we use another way based on the *matrix-vector form*, which we believe is more straightforward to understand. In particular, let

$$u(s) \doteq \sum_{a \in \mathcal{A}} \nabla_\theta \pi(a|s, \theta) q_\pi(s, a).$$

Since

$$\sum_{a \in \mathcal{A}} \pi(a|s, \theta) \sum_{s' \in \mathcal{S}} p(s'|s, a) \nabla_\theta v_\pi(s') = \sum_{s' \in \mathcal{S}} p(s'|s) \nabla_\theta v_\pi(s') = \sum_{s' \in \mathcal{S}} [P_\pi]_{ss'} \nabla_\theta v_\pi(s'),$$

equation (9.16) can be written in matrix-vector form as

$$\underbrace{\begin{bmatrix} \vdots \\ \nabla_\theta v_\pi(s) \\ \vdots \end{bmatrix}}_{\nabla_\theta v_\pi \in \mathbb{R}^{mn}} = \underbrace{\begin{bmatrix} \vdots \\ u(s) \\ \vdots \end{bmatrix}}_{u \in \mathbb{R}^{mn}} + \gamma (P_\pi \otimes I_m) \underbrace{\begin{bmatrix} \vdots \\ \nabla_\theta v_\pi(s') \\ \vdots \end{bmatrix}}_{\nabla_\theta v_\pi \in \mathbb{R}^{mn}},$$

which can be written concisely as

$$\nabla_\theta v_\pi = u + \gamma (P_\pi \otimes I_m) \nabla_\theta v_\pi.$$

Here, $n = |\mathcal{S}|$, and m is the dimension of the parameter vector θ. The reason that the Kronecker product $\otimes$ emerges in the equation is that $\nabla_\theta v_\pi(s)$ is a vector. The above equation is a linear equation of $\nabla_\theta v_\pi$, which can be solved as

$$
\begin{aligned}
\nabla_\theta v_\pi &= (I_{nm} - \gamma P_\pi \otimes I_m)^{-1} u \\
&= (I_n \otimes I_m - \gamma P_\pi \otimes I_m)^{-1} u \\
&= \left[(I_n - \gamma P_\pi)^{-1} \otimes I_m \right] u.
\end{aligned}
\tag{9.17}
$$

For any state s, it follows from (9.17) that

$$
\begin{aligned}
\nabla_\theta v_\pi(s) &= \sum_{s' \in \mathcal{S}} \left[(I_n - \gamma P_\pi)^{-1} \right]_{ss'} u(s') \\
&= \sum_{s' \in \mathcal{S}} \left[(I_n - \gamma P_\pi)^{-1} \right]_{ss'} \sum_{a \in \mathcal{A}} \nabla_\theta \pi(a|s', \theta) q_\pi(s', a).
\end{aligned}
\tag{9.18}
$$

The quantity $\left[(I_n - \gamma P_\pi)^{-1} \right]_{ss'}$ has a clear probabilistic interpretation. In particular, since $(I_n - \gamma P_\pi)^{-1} = I + \gamma P_\pi + \gamma^2 P_\pi^2 + \cdots$, we have

$$\left[(I_n - \gamma P_\pi)^{-1} \right]_{ss'} = [I]_{ss'} + \gamma [P_\pi]_{ss'} + \gamma^2 [P_\pi^2]_{ss'} + \cdots = \sum_{k=0}^{\infty} \gamma^k [P_\pi^k]_{ss'}.$$

Note that $[P_\pi^k]_{ss'}$ is the probability of transitioning from s to s' using exactly k steps (see Box 8.1). Therefore, $\left[(I_n - \gamma P_\pi)^{-1} \right]_{ss'}$ is the discounted total probability of transitioning from s to s' using any number of steps. By denoting $\left[(I_n - \gamma P_\pi)^{-1} \right]_{ss'} \doteq \mathrm{Pr}_\pi(s'|s)$, equation (9.18) becomes (9.14).

With the results in Lemma 9.2, we are ready to derive the gradient of $\bar{v}_\pi^0$.

Theorem 9.2 (Gradient of $\bar{v}_\pi^0$ in the discounted case). *In the discounted case where* $\gamma \in (0, 1)$, *the gradient of* $\bar{v}_\pi^0 = d_0^T v_\pi$ *is*

$$\nabla_\theta \bar{v}_\pi^0 = \mathbb{E} \left[\nabla_\theta \ln \pi(A|S, \theta) q_\pi(S, A) \right],$$

where $S \sim \rho_\pi$ *and* $A \sim \pi(S, \theta)$. *Here, the state distribution* ρ_π *is*

$$\rho_\pi(s) = \sum_{s' \in \mathcal{S}} d_0(s') \mathrm{Pr}_\pi(s|s'), \qquad s \in \mathcal{S}, \tag{9.19}$$

where $\mathrm{Pr}_\pi(s|s') = \sum_{k=0}^{\infty} \gamma^k [P_\pi^k]_{s's} = \left[(I - \gamma P_\pi)^{-1} \right]_{s's}$ *is the discounted total probability of* *transitioning from s' to s under policy π.*

Box 9.3: Proof of Theorem 9.2

Since $d_0(s)$ is independent of π, we have

$$\nabla_\theta \bar{v}_\pi^0 = \nabla_\theta \sum_{s \in \mathcal{S}} d_0(s) v_\pi(s) = \sum_{s \in \mathcal{S}} d_0(s) \nabla_\theta v_\pi(s).$$

Substituting the expression of $\nabla_\theta v_\pi(s)$ given in Lemma 9.2 into the above equation yields

$$\nabla_\theta \bar{v}_\pi^0 = \sum_{s \in \mathcal{S}} d_0(s) \nabla_\theta v_\pi(s) = \sum_{s \in \mathcal{S}} d_0(s) \sum_{s' \in \mathcal{S}} \mathrm{Pr}_\pi(s'|s) \sum_{a \in \mathcal{A}} \nabla_\theta \pi(a|s', \theta) q_\pi(s', a)$$

$$= \sum_{s' \in \mathcal{S}} \left(\sum_{s \in \mathcal{S}} d_0(s) \mathrm{Pr}_\pi(s'|s) \right) \sum_{a \in \mathcal{A}} \nabla_\theta \pi(a|s', \theta) q_\pi(s', a)$$

$$\doteq \sum_{s' \in \mathcal{S}} \rho_\pi(s') \sum_{a \in \mathcal{A}} \nabla_\theta \pi(a|s', \theta) q_\pi(s', a)$$

$$= \sum_{s \in \mathcal{S}} \rho_\pi(s) \sum_{a \in \mathcal{A}} \nabla_\theta \pi(a|s, \theta) q_\pi(s, a) \qquad \text{(change s' to s)}$$

$$= \sum_{s \in \mathcal{S}} \rho_\pi(s) \sum_{a \in \mathcal{A}} \pi(a|s, \theta) \nabla_\theta \ln \pi(a|s, \theta) q_\pi(s, a)$$

$$= \mathbb{E}\left[\nabla_\theta \ln \pi(A|S, \theta) q_\pi(S, A) \right],$$

where $S \sim \rho_\pi$ and $A \sim \pi(S, \theta)$. The proof is complete.

With Lemma 9.1 and Lemma 9.2, we can derive the gradients of $\bar{r}_\pi$ and $\bar{v}_\pi$.

Theorem 9.3 (Gradients of $\bar{r}_\pi$ and $\bar{v}_\pi$ in the discounted case). *In the discounted case where $\gamma \in (0, 1)$, the gradients of $\bar{r}_\pi$ and $\bar{v}_\pi$ are*

$$\nabla_\theta \bar{r}_\pi = (1 - \gamma) \nabla_\theta \bar{v}_\pi \approx \sum_{s \in \mathcal{S}} d_\pi(s) \sum_{a \in \mathcal{A}} \nabla_\theta \pi(a|s, \theta) q_\pi(s, a)$$

$$= \mathbb{E}\left[\nabla_\theta \ln \pi(A|S, \theta) q_\pi(S, A) \right],$$

where $S \sim d_\pi$ and $A \sim \pi(S, \theta)$. Here, the approximation is more accurate when γ is closer to 1.

Box 9.4: Proof of Theorem 9.3

It follows from the definition of $\bar{v}_\pi$ that

$$\nabla_\theta \bar{v}_\pi = \nabla_\theta \sum_{s \in \mathcal{S}} d_\pi(s) v_\pi(s)$$

$$= \sum_{s \in \mathcal{S}} \nabla_\theta d_\pi(s) v_\pi(s) + \sum_{s \in \mathcal{S}} d_\pi(s) \nabla_\theta v_\pi(s). \tag{9.20}$$

This equation contains two terms. On the one hand, substituting the expression of $\nabla_\theta v_\pi$ given in (9.17) into the second term gives

$$\sum_{s \in \mathcal{S}} d_\pi(s) \nabla_\theta v_\pi(s) = (d_\pi^T \otimes I_m) \nabla_\theta v_\pi$$

$$= (d_\pi^T \otimes I_m) \left[(I_n - \gamma P_\pi)^{-1} \otimes I_m \right] u$$

$$= \left[d_\pi^T (I_n - \gamma P_\pi)^{-1} \right] \otimes I_m u. \tag{9.21}$$

It is noted that

$$d_\pi^T (I_n - \gamma P_\pi)^{-1} = \frac{1}{1 - \gamma} d_\pi^T,$$

which can be easily verified by multiplying $(I_n - \gamma P_\pi)$ on both sides of the equation. Therefore, (9.21) becomes

$$\sum_{s \in \mathcal{S}} d_\pi(s) \nabla_\theta v_\pi(s) = \frac{1}{1 - \gamma} d_\pi^T \otimes I_m u$$

$$= \frac{1}{1 - \gamma} \sum_{s \in \mathcal{S}} d_\pi(s) \sum_{a \in \mathcal{A}} \nabla_\theta \pi(a|s, \theta) q_\pi(s, a).$$

On the other hand, the first term of (9.20) involves $\nabla_\theta d_\pi$. However, since the second term contains $\frac{1}{1-\gamma}$, the second term becomes dominant, and the first term becomes negligible when $\gamma \to 1$. Therefore,

$$\nabla_\theta \bar{v}_\pi \approx \frac{1}{1 - \gamma} \sum_{s \in \mathcal{S}} d_\pi(s) \sum_{a \in \mathcal{A}} \nabla_\theta \pi(a|s, \theta) q_\pi(s, a).$$

Furthermore, it follows from $\bar{r}_\pi = (1 - \gamma) \bar{v}_\pi$ that

$$\nabla_\theta \bar{r}_\pi = (1 - \gamma) \nabla_\theta \bar{v}_\pi \approx \sum_{s \in \mathcal{S}} d_\pi(s) \sum_{a \in \mathcal{A}} \nabla_\theta \pi(a|s, \theta) q_\pi(s, a)$$

$$= \sum_{s \in \mathcal{S}} d_\pi(s) \sum_{a \in \mathcal{A}} \pi(a|s, \theta) \nabla_\theta \ln \pi(a|s, \theta) q_\pi(s, a)$$

$$= \mathbb{E} \left[\nabla_\theta \ln \pi(A|S, \theta) q_\pi(S, A) \right].$$

The approximation in the above equation requires that the first term does not go to infinity when $\gamma \to 1$. More information can be found in [66, Section 4].

9.3.2 Derivation of the gradients in the undiscounted case

We next show how to calculate the gradients of the metrics in the undiscounted case where $\gamma = 1$. Readers may wonder why we suddenly start considering the undiscounted case while we have only considered the discounted case so far in this book. In fact, the definition of the average reward $\bar{r}_\pi$ is valid for both discounted and undiscounted cases. While the gradient of $\bar{r}_\pi$ in the discounted case is an approximation, we will see that its gradient in the undiscounted case is more elegant.

State values and the Poisson equation

In the undiscounted case, it is necessary to redefine state and action values. Since the undiscounted sum of the rewards, $\mathbb{E}[R_{t+1} + R_{t+2} + R_{t+3} + \dots | S_t = s]$, may diverge, the state and action values are defined in a special way [64]:

$$v_\pi(s) \doteq \mathbb{E}[(R_{t+1} - \bar{r}_\pi) + (R_{t+2} - \bar{r}_\pi) + (R_{t+3} - \bar{r}_\pi) + \dots | S_t = s],$$
$$q_\pi(s, a) \doteq \mathbb{E}[(R_{t+1} - \bar{r}_\pi) + (R_{t+2} - \bar{r}_\pi) + (R_{t+3} - \bar{r}_\pi) + \dots | S_t = s, A_t = a],$$

where $\bar{r}_\pi$ is the average reward, which is determined when π is given. There are different names for $v_\pi(s)$ in the literature such as the differential reward [65] or bias [2, Section 8.2.1]. It can be verified that the state value defined above satisfies the following Bellman-like equation:

$$v_\pi(s) = \sum_a \pi(a|s, \theta) \left[\sum_r p(r|s, a)(r - \bar{r}_\pi) + \sum_{s'} p(s'|s, a)v_\pi(s') \right]. \tag{9.22}$$

Since $v_\pi(s) = \sum_{a \in \mathcal{A}} \pi(a|s, \theta)q_\pi(s, a)$, it holds that $q_\pi(s, a) = \sum_r p(r|s, a)(r - \bar{r}_\pi) + \sum_{s'} p(s'|s, a)v_\pi(s')$. The matrix-vector form of (9.22) is

$$v_\pi = r_\pi - \bar{r}_\pi \mathbf{1}_n + P_\pi v_\pi, \tag{9.23}$$

where $\mathbf{1}_n = [1, \dots, 1]^T \in \mathbb{R}^n$. Equation (9.23) is similar to the Bellman equation and it has a specific name called the *Poisson equation* [65, 67].

How to solve v_π from the Poisson equation? The answer is given in the following theorem.

Theorem 9.4 (Solution of the Poisson equation). *Let*

$$v_\pi^* = (I_n - P_\pi + \mathbf{1}_n d_\pi^T)^{-1} r_\pi. \tag{9.24}$$

Then, v_π^* is a solution of the Poisson equation in (9.23). Moreover, any solution of the Poisson equation has the following form:

$$v_\pi = v_\pi^* + c\mathbf{1}_n,$$

where $c \in \mathbb{R}$.

This theorem indicates that the solution of the Poisson equation may not be unique.

Box 9.5: Proof of Theorem 9.4

We prove using three steps.

⬦ Step 1: Show that v_π^* in (9.24) is a solution of the Poisson equation.

For the sake of simplicity, let

$$A \doteq I_n - P_\pi + \mathbf{1}_n d_\pi^T.$$

Then, $v_\pi^* = A^{-1} r_\pi$. The fact that A is invertible will be proven in Step 3. Substituting $v_\pi^* = A^{-1} r_\pi$ into (9.23) gives

$$A^{-1} r_\pi = r_\pi - \mathbf{1}_n d_\pi^T r_\pi + P_\pi A^{-1} r_\pi.$$

This equation is valid as proven below. Recognizing this equation gives $(-A^{-1} + I_n - \mathbf{1}_n d_\pi^T + P_\pi A^{-1}) r_\pi = 0$, and consequently,

$$(-I_n + A - \mathbf{1}_n d_\pi^T A + P_\pi) A^{-1} r_\pi = 0.$$

The term in the brackets in the above equation is zero because $-I_n + A - \mathbf{1}_n d_\pi^T A + P_\pi = -I_n + (I_n - P_\pi + \mathbf{1}_n d_\pi^T) - \mathbf{1}_n d_\pi^T (I_n - P_\pi + \mathbf{1}_n d_\pi^T) + P_\pi = 0$. Therefore, v_π^* in (9.24) is a solution.

⬦ Step 2: General expression of the solutions.

Substituting $\bar{r}_\pi = d_\pi^T r_\pi$ into (9.23) gives

$$v_\pi = r_\pi - \mathbf{1}_n d_\pi^T r_\pi + P_\pi v_\pi \tag{9.25}$$

and consequently

$$(I_n - P_\pi) v_\pi = (I_n - \mathbf{1}_n d_\pi^T) r_\pi. \tag{9.26}$$

It is noted that $I_n - P_\pi$ is singular because $(I_n - P_\pi)\mathbf{1}_n = 0$ for any π. Therefore, the solution of (9.26) is not unique: if v_π^* is a solution, then $v_\pi^* + x$ is also a solution for any $x \in \text{Null}(I_n - P_\pi)$. When P_π is irreducible, $\text{Null}(I_n - P_\pi) = \text{span}\{\mathbf{1}_n\}$. Then, any solution of the Poisson equation has the expression $v_\pi^* + c\mathbf{1}_n$ where $c \in \mathbb{R}$.

◇ Step 3: Show that $A = I_n - P_\pi + \mathbf{1}_n d_\pi^T$ is invertible.

Since v_π^* involves A^{-1}, it is necessary to show that A is invertible. The analysis is summarized in the following lemma.

Lemma 9.3. *The matrix* $I_n - P_\pi + \mathbf{1}_n d_\pi^T$ *is invertible and its inverse is*

$$\left[I_n - (P_\pi - \mathbf{1}_n d_\pi^T) \right]^{-1} = \sum_{k=1}^{\infty} (P_\pi^k - \mathbf{1}_n d_\pi^T) + I_n.$$

Proof. First of all, we state some preliminary facts without proof. Let $\rho(M)$ be the spectral radius of a matrix M. Then, $I - M$ is invertible if $\rho(M) < 1$. Moreover, $\rho(M) < 1$ if and only if $\lim_{k \to \infty} M^k = 0$.

Based on the above facts, we next show that $\lim_{k \to \infty} (P_\pi - \mathbf{1}_n d_\pi^T)^k \to 0$, and then the invertibility of $I_n - (P_\pi - \mathbf{1}_n d_\pi^T)$ immediately follows. To do that, we notice that

$$(P_\pi - \mathbf{1}_n d_\pi^T)^k = P_\pi^k - \mathbf{1}_n d_\pi^T, \quad k \geq 1, \tag{9.27}$$

which can be proven by induction. For instance, when $k = 1$, the equation is valid. When $k = 2$, we have

$$\begin{aligned}
(P_\pi - \mathbf{1}_n d_\pi^T)^2 &= (P_\pi - \mathbf{1}_n d_\pi^T)(P_\pi - \mathbf{1}_n d_\pi^T) \\
&= P_\pi^2 - P_\pi \mathbf{1}_n d_\pi^T - \mathbf{1}_n d_\pi^T P_\pi + \mathbf{1}_n d_\pi^T \mathbf{1}_n d_\pi^T \\
&= P_\pi^2 - \mathbf{1}_n d_\pi^T,
\end{aligned}$$

where the last equality is due to $P_\pi \mathbf{1}_n = \mathbf{1}_n$, $d_\pi^T P_\pi = d_\pi^T$, and $d_\pi^T \mathbf{1}_n = 1$. The case of $k \geq 3$ can be proven similarly.

Since d_π is the stationary distribution of the state, it holds that $\lim_{k \to \infty} P_\pi^k = d_\pi^T \mathbf{1}_n$ (see Box 8.1). Therefore, (9.27) implies that

$$\lim_{k \to \infty} (P_\pi - \mathbf{1}_n d_\pi^T)^k = \lim_{k \to \infty} P_\pi^k - d_\pi^T \mathbf{1}_n = 0.$$

As a result, $\rho(P_\pi - \mathbf{1}_n d_\pi^T) < 1$ and hence $I_n - (P_\pi - \mathbf{1}_n d_\pi^T)$ is invertible. Furthermore, the inverse of this matrix is given by

$$(I_n - (P_\pi - \mathbf{1}_n d_\pi^T))^{-1} = \sum_{k=0}^{\infty}(P_\pi - \mathbf{1}_n d_\pi^T)^k$$

$$= I_n + \sum_{k=1}^{\infty}(P_\pi - \mathbf{1}_n d_\pi^T)^k$$

$$= I_n + \sum_{k=1}^{\infty}(P_\pi^k - \mathbf{1}_n d_\pi^T)$$

$$= \sum_{k=0}^{\infty}(P_\pi^k - \mathbf{1}_n d_\pi^T) + \mathbf{1}_n d_\pi^T.$$

The proof is complete. □

The proof of Lemma 9.3 is inspired by [66]. However, the result $(I_n - P_\pi + \mathbf{1}_n d_\pi^T)^{-1} = \sum_{k=0}^{\infty}(P_\pi^k - \mathbf{1}_n d_\pi^T)$ given in [66] (the statement above equation (16) in [66]) is inaccurate because $\sum_{k=0}^{\infty}(P_\pi^k - \mathbf{1}_n d_\pi^T)$ is singular since $\sum_{k=0}^{\infty}(P_\pi^k - \mathbf{1}_n d_\pi^T)\mathbf{1}_n = 0$. Lemma 9.3 corrects this inaccuracy.

Derivation of gradients

Although the value of v_π is not unique in the undiscounted case, as shown in Theorem 9.4, the value of $\bar{r}_\pi$ is unique. In particular, it follows from the Poisson equation that

$$\bar{r}_\pi \mathbf{1}_n = r_\pi + (P_\pi - I_n)v_\pi$$

$$= r_\pi + (P_\pi - I_n)(v_\pi^* + c\mathbf{1}_n)$$

$$= r_\pi + (P_\pi - I_n)v_\pi^*.$$

Notably, the undetermined value c is canceled and hence $\bar{r}_\pi$ is unique. Therefore, we can calculate the gradient of $\bar{r}_\pi$ in the undiscounted case. In addition, since v_π is not unique, $\bar{v}_\pi$ is not unique either. We do not study the gradient of $\bar{v}_\pi$ in the undiscounted case. For interested readers, it is worth mentioning that we can add more constraints to uniquely solve v_π from the Poisson equation. For example, by assuming that a recurrent state exists, the state value of this recurrent state can be determined [65, Section II], and hence c can be determined. There are also other ways to uniquely determine v_π. See, for example, equations (8.6.5)-(8.6.7) in [2].

The gradient of $\bar{r}_\pi$ in the undiscounted case is given below.

Theorem 9.5 (Gradient of $\bar{r}_\pi$ in the undiscounted case). *In the undiscounted case, the* *gradient of the average reward $\bar{r}_\pi$ is*

$$\nabla_\theta \bar{r}_\pi = \sum_{s \in S} d_\pi(s) \sum_{a \in A} \nabla_\theta \pi(a|s,\theta) q_\pi(s,a)$$

$$= \mathbb{E}\big[\nabla_\theta \ln \pi(A|S,\theta) q_\pi(S,A)\big], \tag{9.28}$$

where $S \sim d_\pi$ and $A \sim \pi(S,\theta)$.

Compared to the discounted case shown in Theorem 9.3, the gradient of $\bar{r}_\pi$ in the undiscounted case is more elegant in the sense that (9.28) is strictly valid and S obeys the stationary distribution.

Box 9.6: Proof of Theorem 9.5

First of all, it follows from $v_\pi(s) = \sum_{a \in A} \pi(a|s,\theta) q_\pi(s,a)$ that

$$\nabla_\theta v_\pi(s) = \nabla_\theta \left[\sum_{a \in A} \pi(a|s,\theta) q_\pi(s,a) \right]$$

$$= \sum_{a \in A} \big[\nabla_\theta \pi(a|s,\theta) q_\pi(s,a) + \pi(a|s,\theta) \nabla_\theta q_\pi(s,a) \big], \tag{9.29}$$

where $q_\pi(s,a)$ is the action value satisfying

$$q_\pi(s,a) = \sum_r p(r|s,a)(r - \bar{r}_\pi) + \sum_{s'} p(s'|s,a) v_\pi(s')$$

$$= r(s,a) - \bar{r}_\pi + \sum_{s'} p(s'|s,a) v_\pi(s').$$

Since $r(s,a) = \sum_r r p(r|s,a)$ is independent of θ, we have

$$\nabla_\theta q_\pi(s,a) = 0 - \nabla_\theta \bar{r}_\pi + \sum_{s' \in S} p(s'|s,a) \nabla_\theta v_\pi(s').$$

Substituting this result into (9.29) yields

$$\nabla_\theta v_\pi(s) = \sum_{a \in A} \left[\nabla_\theta \pi(a|s,\theta) q_\pi(s,a) + \pi(a|s,\theta) \left(-\nabla_\theta \bar{r}_\pi + \sum_{s' \in S} p(s'|s,a) \nabla_\theta v_\pi(s') \right) \right]$$

$$= \sum_{a \in A} \nabla_\theta \pi(a|s,\theta) q_\pi(s,a) - \nabla_\theta \bar{r}_\pi + \sum_{a \in A} \pi(a|s,\theta) \sum_{s' \in S} p(s'|s,a) \nabla_\theta v_\pi(s').$$

$$\tag{9.30}$$

Let

$$u(s) \doteq \sum_{a \in A} \nabla_\theta \pi(a|s,\theta) q_\pi(s,a).$$

Since $\sum_{a\in\mathcal{A}}\pi(a|s,\theta)\sum_{s'\in\mathcal{S}}p(s'|s,a)\nabla_\theta v_\pi(s') = \sum_{s'\in\mathcal{S}}p(s'|s)\nabla_\theta v_\pi(s')$, equation (9.30) can be written in matrix-vector form as

$$
\underbrace{\begin{bmatrix} \vdots \\ \nabla_\theta v_\pi(s) \\ \vdots \end{bmatrix}}_{\nabla_\theta v_\pi \in \mathbb{R}^{mn}} = \underbrace{\begin{bmatrix} \vdots \\ u(s) \\ \vdots \end{bmatrix}}_{u \in \mathbb{R}^{mn}} -\mathbf{1}_n \otimes \nabla_\theta \bar{r}_\pi + (P_\pi \otimes I_m) \underbrace{\begin{bmatrix} \vdots \\ \nabla_\theta v_\pi(s') \\ \vdots \end{bmatrix}}_{\nabla_\theta v_\pi \in \mathbb{R}^{mn}},
$$

where $n = |\mathcal{S}|$, m is the dimension of θ, and $\otimes$ is the Kronecker product. The above equation can be written concisely as

$$
\nabla_\theta v_\pi = u - \mathbf{1}_n \otimes \nabla_\theta \bar{r}_\pi + (P_\pi \otimes I_m)\nabla_\theta v_\pi,
$$

and hence

$$
\mathbf{1}_n \otimes \nabla_\theta \bar{r}_\pi = u + (P_\pi \otimes I_m)\nabla_\theta v_\pi - \nabla_\theta v_\pi.
$$

Multiplying $d_\pi^T \otimes I_m$ on both sides of the above equation gives

$$
\begin{aligned}
(d_\pi^T \mathbf{1}_n) \otimes \nabla_\theta \bar{r}_\pi &= d_\pi^T \otimes I_m u + (d_\pi^T P_\pi) \otimes I_m \nabla_\theta v_\pi - d_\pi^T \otimes I_m \nabla_\theta v_\pi \\
&= d_\pi^T \otimes I_m u,
\end{aligned}
$$

which implies

$$
\begin{aligned}
\nabla_\theta \bar{r}_\pi &= d_\pi^T \otimes I_m u \\
&= \sum_{s\in\mathcal{S}} d_\pi(s) u(s) \\
&= \sum_{s\in\mathcal{S}} d_\pi(s) \sum_{a\in\mathcal{A}} \nabla_\theta \pi(a|s,\theta) q_\pi(s,a).
\end{aligned}
$$

9.4 Monte Carlo policy gradient (REINFORCE)

With the gradient presented in Theorem 9.1, we next show how to use the gradient-based method to optimize the metrics to obtain optimal policies.

The gradient-ascent algorithm for maximizing $J(\theta)$ is

$$
\begin{aligned}
\theta_{t+1} &= \theta_t + \alpha\nabla_\theta J(\theta_t) \\
&= \theta_t + \alpha\mathbb{E}\Big[\nabla_\theta \ln \pi(A|S,\theta_t)q_\pi(S,A)\Big],
\end{aligned} \tag{9.31}
$$

where $\alpha > 0$ is a constant learning rate. Since the true gradient in (9.31) is unknown, we can replace the true gradient with a stochastic gradient to obtain the following algorithm:

$$\theta_{t+1} = \theta_t + \alpha \nabla_\theta \ln \pi(a_t|s_t, \theta_t) q_t(s_t, a_t), \tag{9.32}$$

where $q_t(s_t, a_t)$ is an approximation of $q_\pi(s_t, a_t)$. If $q_t(s_t, a_t)$ is obtained by Monte Carlo estimation, the algorithm is called *REINFORCE* [68] or *Monte Carlo policy gradient*, which is one of earliest and simplest policy gradient algorithms.

The algorithm in (9.32) is important since many other policy gradient algorithms can be obtained by extending it. We next examine the interpretation of (9.32) more closely. Since $\nabla_\theta \ln \pi(a_t|s_t, \theta_t) = \frac{\nabla_\theta \pi(a_t|s_t, \theta_t)}{\pi(a_t|s_t, \theta_t)}$, we can rewrite (9.32) as

$$\theta_{t+1} = \theta_t + \alpha \underbrace{\left(\frac{q_t(s_t, a_t)}{\pi(a_t|s_t, \theta_t)} \right)}_{\beta_t} \nabla_\theta \pi(a_t|s_t, \theta_t),$$

which can be further written concisely as

$$\theta_{t+1} = \theta_t + \alpha \beta_t \nabla_\theta \pi(a_t|s_t, \dot{\theta}_t). \tag{9.33}$$

Two important interpretations can be seen from this equation.

⋄ First, since (9.33) is a simple gradient-ascent algorithm, the following observations can be obtained.

- If $\beta_t \geq 0$, the probability of choosing (s_t, a_t) is enhanced. That is

$$\pi(a_t|s_t, \theta_{t+1}) \geq \pi(a_t|s_t, \theta_t).$$

  The greater β_t is, the stronger the enhancement is.
- If $\beta_t < 0$, the probability of choosing (s_t, a_t) decreases. That is

$$\pi(a_t|s_t, \theta_{t+1}) < \pi(a_t|s_t, \theta_t).$$

The above observations can be proven as follows. When $\theta_{t+1} - \theta_t$ is sufficiently small, it follows from the Taylor expansion that

$$\begin{aligned}
\pi(a_t|s_t, \theta_{t+1}) &\approx \pi(a_t|s_t, \theta_t) + (\nabla_\theta \pi(a_t|s_t, \theta_t))^T (\theta_{t+1} - \theta_t) \\
&= \pi(a_t|s_t, \theta_t) + \alpha \beta_t (\nabla_\theta \pi(a_t|s_t, \theta_t))^T (\nabla_\theta \pi(a_t|s_t, \theta_t)) \quad \text{(substituting (9.33))} \\
&= \pi(a_t|s_t, \theta_t) + \alpha \beta_t \| \nabla_\theta \pi(a_t|s_t, \theta_t) \|_2^2.
\end{aligned}$$

It is clear that $\pi(a_t|s_t, \theta_{t+1}) \geq \pi(a_t|s_t, \theta_t)$ when $\beta_t \geq 0$ and $\pi(a_t|s_t, \theta_{t+1}) < \pi(a_t|s_t, \theta_t)$ when $\beta_t < 0$.

⋄ Second, the algorithm can strike a balance between *exploration* and *exploitation* to a

---

Algorithm 9.1: Policy Gradient by Monte Carlo (REINFORCE)

Initialization: Initial parameter θ; $\gamma \in (0,1)$; $\alpha > 0$.
Goal: Learn an optimal policy for maximizing $J(\theta)$.

For each episode, do
    Generate an episode $\{s_0, a_0, r_1, \ldots, s_{T-1}, a_{T-1}, r_T\}$ following $\pi(\theta)$.
    For $t = 0, 1, \ldots, T-1$:
        Value update: $q_t(s_t, a_t) = \sum_{k=t+1}^{T} \gamma^{k-t-1} r_k$
        Policy update: $\theta \leftarrow \theta + \alpha \nabla_\theta \ln \pi(a_t|s_t, \theta) q_t(s_t, a_t)$

---

certain extent due to the expression of

$$\beta_t = \frac{q_t(s_t, a_t)}{\pi(a_t|s_t, \theta_t)}.$$

On the one hand, β_t is *proportional* to $q_t(s_t, a_t)$. As a result, if the action value of (s_t, a_t) is large, then $\pi(a_t|s_t, \theta_t)$ is enhanced so that the probability of selecting a_t increases. Therefore, the algorithm attempts to *exploit* actions with greater values. One the other hand, β_t is *inversely proportional* to $\pi(a_t|s_t, \theta_t)$ when $q_t(s_t, a_t) > 0$. As a result, if the probability of selecting a_t is small, then $\pi(a_t|s_t, \theta_t)$ is enhanced so that the probability of selecting a_t increases. Therefore, the algorithm attempts to *explore* actions with low probabilities.

Moreover, since (9.32) uses samples to approximate the true gradient in (9.31), it is important to understand how the samples should be obtained.

⋄ How to sample S? S in the true gradient $\mathbb{E}[\nabla_\theta \ln \pi(A|S, \theta_t) q_\pi(S, A)]$ should obey the distribution η which is either the stationary distribution d_π or the discounted total probability distribution ρ_π in (9.19). Either d_π or ρ_π represents the long-term behavior exhibited under π.

⋄ How to sample A? A in $\mathbb{E}[\nabla_\theta \ln \pi(A|S, \theta_t) q_\pi(S, A)]$ should obey the distribution of $\pi(A|S, \theta)$. The ideal way to sample A is to select a_t following $\pi(a|s_t, \theta_t)$. Therefore, the policy gradient algorithm is on-policy.

Unfortunately, the ideal ways for sampling S and A are not strictly followed in practice due to their low efficiency of sample usage. A more sample-efficient implementation of (9.32) is given in Algorithm 9.1. In this implementation, an episode is first generated by following $\pi(\theta)$. Then, θ is updated multiple times using every experience sample in the episode.

9.5 Summary

This chapter introduced the policy gradient method, which is the foundation of many modern reinforcement learning algorithms. Policy gradient methods are *policy-based*. It is a big step forward in this book because all the methods in the previous chapters are *value-based*. The basic idea of the policy gradient method is simple. That is to select an appropriate scalar metric and then optimize it via a gradient-ascent algorithm.

The most complicated part of the policy gradient method is the derivation of the gradients of the metrics. That is because we have to distinguish various scenarios with different metrics and discounted/undiscounted cases. Fortunately, the expressions of the gradients in different scenarios are similar. Hence, we summarized the expressions in Theorem 9.1, which is the most important theoretical result in this chapter. For many readers, it is sufficient to be aware of this theorem. Its proof is nontrivial, and it is not required for all readers to study.

The policy gradient algorithm in (9.32) must be properly understood since it is the foundation of many advanced policy gradient algorithms. In the next chapter, this algorithm will be extended to another important policy gradient method called actor-critic.

9.6 Q&A

◇ Q: What is the basic idea of the policy gradient method?

A: The basic idea is simple. That is to define an appropriate scalar metric, derive its gradient, and then use gradient-ascent methods to optimize the metric. The most important theoretical result regarding this method is the policy gradient given in Theorem 9.1.

◇ Q: What is the most complicated part of the policy gradient method?

A: The basic idea of the policy gradient method is simple. However, the derivation procedure of the gradients is quite complicated. That is because we have to distinguish numerous different scenarios. The mathematical derivation procedure in each scenario is nontrivial. It is sufficient for many readers to be familiar with the result in Theorem 9.1 without knowing the proof.

◇ Q: What metrics should be used in the policy gradient method?

A: We introduced three common metrics in this chapter: $\bar{v}_\pi$, $\bar{v}_\pi^0$, and $\bar{r}_\pi$. Since they all lead to similar policy gradients, they all can be adopted in the policy gradient method. More importantly, the expressions in (9.1) and (9.4) are often encountered in the literature.

◇ Q: Why is a natural logarithm function contained in the policy gradient?

A: A natural logarithm function is introduced to express the gradient as an expected value. In this way, we can approximate the true gradient with a stochastic one.

⋄ Q: Why do we need to study undiscounted cases when deriving the policy gradient?

A: The definition of the average reward $\bar{r}_\pi$ is valid for both discounted and undiscounted cases. While the gradient of $\bar{r}_\pi$ in the discounted case is an approximation, its gradient in the undiscounted case is more elegant.

⋄ Q: What does the policy gradient algorithm in (9.32) do mathematically?

A: To better understand this algorithm, readers are recommended to examine its concise expression in (9.33), which clearly shows that it is a gradient-ascent algorithm for updating the value of $\pi(a_t|s_t, \theta_t)$. That is, when a sample (s_t, a_t) is available, the policy can be updated so that $\pi(a_t|s_t, \theta_{t+1}) \geq \pi(a_t|s_t, \theta_t)$ or $\pi(a_t|s_t, \theta_{t+1}) < \pi(a_t|s_t, \theta_t)$ depending on the coefficients.

Chapter 10

Actor-Critic Methods

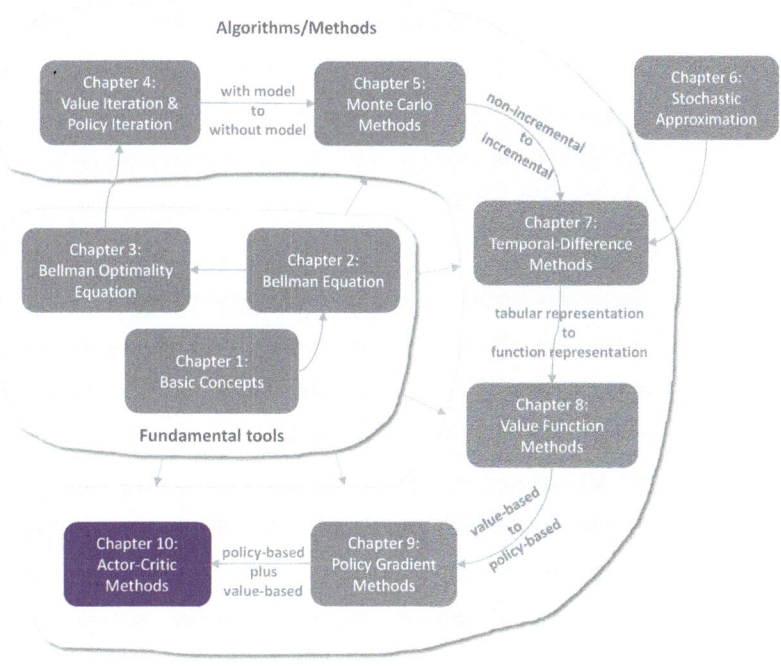

Figure 10.1: Where we are in this book.

This chapter introduces actor-critic methods. From one point of view, "actor-critic" refers to a structure that incorporates both policy-based and value-based methods. Here, an "actor" refers to a policy update step. The reason that it is called an actor is that the actions are taken by following the policy. Here, an "critic" refers to a value update step. It is called a critic because it criticizes the actor by evaluating its corresponding values. From another point of view, actor-critic methods are still policy gradient algorithms. They can be obtained by extending the policy gradient algorithm introduced in Chapter 9. It is important for the reader to well understand the contents of Chapters 8 and 9 before studying this chapter.

S. Zhao, *Mathematical Foundations of Reinforcement Learning*, https://doi.org/10.1007/978-981-97-3944-8_10

10.1 The simplest actor-critic algorithm (QAC)

This section introduces the simplest actor-critic algorithm. This algorithm can be easily obtained by extending the policy gradient algorithm in (9.32).

Recall that the idea of the policy gradient method is to search for an optimal policy by maximizing a scalar metric $J(\theta)$. The gradient-ascent algorithm for maximizing $J(\theta)$ is

$$
\begin{aligned}
\theta_{t+1} &= \theta_t + \alpha \nabla_\theta J(\theta_t) \\
&= \theta_t + \alpha \mathbb{E}_{S \sim \eta, A \sim \pi} \Big[\nabla_\theta \ln \pi(A|S, \theta_t) q_\pi(S, A) \Big],
\end{aligned}
\tag{10.1}
$$

where η is a distribution of the states (see Theorem 9.1 for more information). Since the true gradient is unknown, we can use a stochastic gradient to approximate it:

$$
\theta_{t+1} = \theta_t + \alpha \nabla_\theta \ln \pi(a_t|s_t, \theta_t) q_t(s_t, a_t).
\tag{10.2}
$$

This is the algorithm given in (9.32).

Equation (10.2) is important because it clearly shows how policy-based and value-based methods can be combined. On the one hand, it is a *policy-based* algorithm since it directly updates the policy parameter. On the other hand, this equation requires knowing $q_t(s_t, a_t)$, which is an estimate of the action value $q_\pi(s_t, a_t)$. As a result, another *value-based* algorithm is required to generate $q_t(s_t, a_t)$. So far, we have studied two ways to estimate action values in this book. The first is based on Monte Carlo learning and the second is temporal-difference (TD) learning.

⋄ If $q_t(s_t, a_t)$ is estimated by Monte Carlo learning, the corresponding algorithm is called *REINFORCE* or *Monte Carlo policy gradient*, which has already been introduced in Chapter 9.

⋄ If $q_t(s_t, a_t)$ is estimated by TD learning, the corresponding algorithms are usually called *actor-critic*. Therefore, actor-critic methods can be obtained by incorporating TD-based value estimation into policy gradient methods.

The procedure of the simplest actor-critic algorithm is summarized in Algorithm 10.1. The *critic* corresponds to the value update step via the Sarsa algorithm presented in (8.35). The action values are represented by a parameterized function $q(s, a, w)$. The *actor* corresponds to the policy update step in (10.2). This actor-citric algorithm is sometimes called *Q actor-critic* (QAC). Although it is simple, QAC reveals the core idea of actor-critic methods. It can be extended to generate many advanced ones as shown in the rest of this chapter.

Algorithm 10.1: The simplest actor-critic algorithm (QAC)

Initialization: A policy function $\pi(a|s, \theta_0)$ where θ_0 is the initial parameter. A value function $q(s, a, w_0)$ where w_0 is the initial parameter. $\alpha_w, \alpha_\theta > 0$.
Goal: Learn an optimal policy to maximize $J(\theta)$.

At time step t in each episode, do
    Generate a_t following $\pi(a|s_t, \theta_t)$, observe r_{t+1}, s_{t+1}, and then generate a_{t+1} following $\pi(a|s_{t+1}, \theta_t)$.
    Actor (policy update):
$$\theta_{t+1} = \theta_t + \alpha_\theta \nabla_\theta \ln \pi(a_t|s_t, \theta_t) q(s_t, a_t, w_t)$$
    Critic (value update):
$$w_{t+1} = w_t + \alpha_w \big[r_{t+1} + \gamma q(s_{t+1}, a_{t+1}, w_t) - q(s_t, a_t, w_t) \big] \nabla_w q(s_t, a_t, w_t)$$

10.2 Advantage actor-critic (A2C)

We now introduce the algorithm of *advantage actor-critic*. The core idea of this algorithm is to introduce a baseline to reduce estimation variance.

10.2.1 Baseline invariance

One interesting property of the policy gradient is that it is invariant to an additional *baseline*. That is

$$\mathbb{E}_{S \sim \eta, A \sim \pi} \Big[\nabla_\theta \ln \pi(A|S, \theta_t) q_\pi(S, A) \Big] = \mathbb{E}_{S \sim \eta, A \sim \pi} \Big[\nabla_\theta \ln \pi(A|S, \theta_t)(q_\pi(S, A) - b(S)) \Big],$$
$$(10.3)$$

where the additional baseline $b(S)$ is a scalar function of S. We next answer two questions about the baseline.

$\diamond$ First, why is (10.3) valid?

    Equation (10.3) holds if and only if

$$\mathbb{E}_{S \sim \eta, A \sim \pi} \Big[\nabla_\theta \ln \pi(A|S, \theta_t) b(S) \Big] = 0.$$

This equation is valid because

$$\mathbb{E}_{S\sim\eta,A\sim\pi}\left[\nabla_\theta \ln \pi(A|S,\theta_t)b(S)\right] = \sum_{s\in\mathcal{S}}\eta(s)\sum_{a\in\mathcal{A}}\pi(a|s,\theta_t)\nabla_\theta \ln \pi(a|s,\theta_t)b(s)$$

$$= \sum_{s\in\mathcal{S}}\eta(s)\sum_{a\in\mathcal{A}}\nabla_\theta\pi(a|s,\theta_t)b(s)$$

$$= \sum_{s\in\mathcal{S}}\eta(s)b(s)\sum_{a\in\mathcal{A}}\nabla_\theta\pi(a|s,\theta_t)$$

$$= \sum_{s\in\mathcal{S}}\eta(s)b(s)\nabla_\theta\sum_{a\in\mathcal{A}}\pi(a|s,\theta_t)$$

$$= \sum_{s\in\mathcal{S}}\eta(s)b(s)\nabla_\theta 1 = 0.$$

◇ Second, why is the baseline useful?

The baseline is useful because it can reduce the approximation variance when we use samples to approximate the true gradient. In particular, let

$$X(S,A) \doteq \nabla_\theta \ln \pi(A|S,\theta_t)[q_\pi(S,A) - b(S)]. \tag{10.4}$$

Then, the true gradient is $\mathbb{E}[X(S,A)]$. Since we need to use a stochastic sample x to approximate $\mathbb{E}[X]$, it would be favorable if the variance $\text{var}(X)$ is small. For example, if $\text{var}(X)$ is close to zero, then any sample x can accurately approximate $\mathbb{E}[X]$. On the contrary, if $\text{var}(X)$ is large, the value of a sample may be far from $\mathbb{E}[X]$.

Although $\mathbb{E}[X]$ is invariant to the baseline, the variance $\text{var}(X)$ is *not*. Our goal is to design a good baseline to minimize $\text{var}(X)$. In the algorithms of REINFORCE and QAC, we set $b = 0$, which is not guaranteed to be a good baseline.

In fact, the optimal baseline that minimizes $\text{var}(X)$ is

$$b^*(s) = \frac{\mathbb{E}_{A\sim\pi}\left[\|\nabla_\theta \ln \pi(A|s,\theta_t)\|^2 q_\pi(s,A)\right]}{\mathbb{E}_{A\sim\pi}\left[\|\nabla_\theta \ln \pi(A|s,\theta_t)\|^2\right]}, \quad s\in\mathcal{S}. \tag{10.5}$$

The proof is given in Box 10.1.

Although the baseline in (10.5) is optimal, it is too complex to be useful in practice. If the weight $\|\nabla_\theta \ln \pi(A|s,\theta_t)\|^2$ is removed from (10.5), we can obtain a suboptimal baseline that has a concise expression:

$$b^\dagger(s) = \mathbb{E}_{A\sim\pi}[q_\pi(s,A)] = v_\pi(s), \quad s\in\mathcal{S}.$$

Interestingly, this suboptimal baseline is the state value.

Box 10.1: Showing that $b^*(s)$ in (10.5) is the optimal baseline

Let $\bar{x} \doteq \mathbb{E}[X]$, which is invariant for any $b(s)$. If X is a vector, its variance is a matrix. It is common to select the trace of $\text{var}(X)$ as a scalar objective function for optimization:

$$
\begin{aligned}
\text{tr}[\text{var}(X)] &= \text{tr}\mathbb{E}[(X - \bar{x})(X - \bar{x})^T] \\
&= \text{tr}\mathbb{E}[XX^T - \bar{x}X^T - X\bar{x}^T + \bar{x}\bar{x}^T] \\
&= \mathbb{E}[X^TX - X^T\bar{x} - \bar{x}^TX + \bar{x}^T\bar{x}] \\
&= \mathbb{E}[X^TX] - \bar{x}^T\bar{x}.
\end{aligned}
\tag{10.6}
$$

When deriving the above equation, we use the trace property $\text{tr}(AB) = \text{tr}(BA)$ for any squared matrices A, B with appropriate dimensions. Since $\bar{x}$ is invariant, equation (10.6) suggests that we only need to minimize $\mathbb{E}[X^TX]$. With X defined in (10.4), we have

$$
\begin{aligned}
\mathbb{E}[X^TX] &= \mathbb{E}\left[(\nabla_\theta \ln \pi)^T (\nabla_\theta \ln \pi)(q_\pi(S, A) - b(S))^2\right] \\
&= \mathbb{E}\left[\|\nabla_\theta \ln \pi\|^2 (q_\pi(S, A) - b(S))^2\right],
\end{aligned}
$$

where $\pi(A|S, \theta)$ is written as π for short. Since $S \sim \eta$ and $A \sim \pi$, the above equation can be rewritten as

$$
\mathbb{E}[X^TX] = \sum_{s \in \mathcal{S}} \eta(s) \mathbb{E}_{A \sim \pi}\left[\|\nabla_\theta \ln \pi\|^2 (q_\pi(s, A) - b(s))^2\right].
$$

To ensure $\nabla_b \mathbb{E}[X^TX] = 0$, $b(s)$ for any $s \in \mathcal{S}$ should satisfy

$$
\mathbb{E}_{A \sim \pi}\left[\|\nabla_\theta \ln \pi\|^2 (b(s) - q_\pi(s, A))\right] = 0, \qquad s \in \mathcal{S}.
$$

The above equation can be easily solved to obtain the optimal baseline:

$$
b^*(s) = \frac{\mathbb{E}_{A \sim \pi}[\|\nabla_\theta \ln \pi\|^2 q_\pi(s, A)]}{\mathbb{E}_{A \sim \pi}[\|\nabla_\theta \ln \pi\|^2]}, \qquad s \in \mathcal{S}.
$$

More discussions on optimal baselines in policy gradient methods can be found in [69, 70].

10.2.2 Algorithm description

When $b(s) = v_\pi(s)$, the gradient-ascent algorithm in (10.1) becomes

$$\theta_{t+1} = \theta_t + \alpha \mathbb{E}\Big[\nabla_\theta \ln \pi(A|S, \theta_t)[q_\pi(S, A) - v_\pi(S)]\Big]$$
$$\doteq \theta_t + \alpha \mathbb{E}\Big[\nabla_\theta \ln \pi(A|S, \theta_t)\delta_\pi(S, A)\Big]. \tag{10.7}$$

Here,
$$\delta_\pi(S, A) \doteq q_\pi(S, A) - v_\pi(S)$$

is called the *advantage function*, which reflects the advantage of one action over the others. More specifically, note that $v_\pi(s) = \sum_{a \in \mathcal{A}} \pi(a|s)q_\pi(s, a)$ is the mean of the action values. If $\delta_\pi(s, a) > 0$, it means that the corresponding action has a greater value than the mean value.

The stochastic version of (10.7) is

$$\theta_{t+1} = \theta_t + \alpha \nabla_\theta \ln \pi(a_t|s_t, \theta_t)[q_t(s_t, a_t) - v_t(s_t)]$$
$$= \theta_t + \alpha \nabla_\theta \ln \pi(a_t|s_t, \theta_t)\delta_t(s_t, a_t), \tag{10.8}$$

where s_t, a_t are samples of S, A at time t. Here, $q_t(s_t, a_t)$ and $v_t(s_t)$ are approximations of $q_{\pi(\theta_t)}(s_t, a_t)$ and $v_{\pi(\theta_t)}(s_t)$, respectively. The algorithm in (10.8) updates the policy based on the *relative value* of q_t with respect to v_t rather than the *absolute value* of q_t. This is intuitively reasonable because, when we attempt to select an action at a state, we only care about which action has the greatest value *relative* to the others.

If $q_t(s_t, a_t)$ and $v_t(s_t)$ are estimated by Monte Carlo learning, the algorithm in (10.8) is called *REINFORCE with a baseline*. If $q_t(s_t, a_t)$ and $v_t(s_t)$ are estimated by TD learning, the algorithm is usually called *advantage actor-critic* (A2C). The implementation of A2C is summarized in Algorithm 10.2. It should be noted that the advantage function in this implementation is approximated by the TD error:

$$q_t(s_t, a_t) - v_t(s_t) \approx r_{t+1} + \gamma v_t(s_{t+1}) - v_t(s_t).$$

This approximation is reasonable because

$$q_\pi(s_t, a_t) - v_\pi(s_t) = \mathbb{E}\Big[R_{t+1} + \gamma v_\pi(S_{t+1}) - v_\pi(S_t)|S_t = s_t, A_t = a_t\Big],$$

which is valid due to the definition of $q_\pi(s_t, a_t)$. One merit of using the TD error is that we only need to use a single neural network to represent $v_\pi(s)$. Otherwise, if $\delta_t = q_t(s_t, a_t) - v_t(s_t)$, we need to maintain two networks to represent $v_\pi(s)$ and $q_\pi(s, a)$, respectively. When we use the TD error, the algorithm may also be called *TD actor-critic*. In addition, it is notable that the policy $\pi(\theta_t)$ is stochastic and hence exploratory. Therefore, it can be directly used to generate experience samples without relying on

---

Algorithm 10.2: Advantage actor-critic (A2C) or TD actor-critic

Initialization: A policy function $\pi(a|s, \theta_0)$ where θ_0 is the initial parameter. A value function $v(s, w_0)$ where w_0 is the initial parameter. $\alpha_w, \alpha_\theta > 0$.
Goal: Learn an optimal policy to maximize $J(\theta)$.

At time step t in each episode, do
    Generate a_t following $\pi(a|s_t, \theta_t)$ and then observe r_{t+1}, s_{t+1}.
    Advantage (TD error):
$$\delta_t = r_{t+1} + \gamma v(s_{t+1}, w_t) - v(s_t, w_t)$$
    Actor (policy update):
$$\theta_{t+1} = \theta_t + \alpha_\theta \delta_t \nabla_\theta \ln \pi(a_t|s_t, \theta_t)$$
    Critic (value update):
$$w_{t+1} = w_t + \alpha_w \delta_t \nabla_w v(s_t, w_t)$$

---

techniques such as ε-greedy. There are some variants of A2C such as *asynchronous advantage actor-critic* (A3C). Interested readers may check [71, 72].

10.3 Off-policy actor-critic

The policy gradient methods that we have studied so far, including REINFORCE, QAC, and A2C, are all *on-policy*. The reason for this can be seen from the expression of the true gradient:

$$\nabla_\theta J(\theta) = \mathbb{E}_{S \sim \eta, A \sim \pi} \Big[\nabla_\theta \ln \pi(A|S, \theta_t)(q_\pi(S, A) - v_\pi(S)) \Big].$$

To use samples to approximate this true gradient, we must generate the action samples by following $\pi(\theta)$. Hence, $\pi(\theta)$ is the behavior policy. Since $\pi(\theta)$ is also the target policy that we aim to improve, the policy gradient methods are on-policy.

In the case that we already have some samples generated by a given behavior policy, the policy gradient methods can still be applied to utilize these samples. To do that, we can employ a technique called *importance sampling*. It is worth mentioning that the importance sampling technique is not restricted to the field of reinforcement learning. It is a general technique for estimating expected values defined over one probability distribution using some samples drawn from another distribution.

10.3.1 Importance sampling

We next introduce the importance sampling technique. Consider a random variable $X \in \mathcal{X}$. Suppose that $p_0(X)$ is a probability distribution. Our goal is to estimate $\mathbb{E}_{X \sim p_0}[X]$. Suppose that we have some i.i.d. samples $\{x_i\}_{i=1}^n$.

◇ First, if the samples $\{x_i\}_{i=1}^n$ are generated by following p_0, then the average value $\bar{x} = \frac{1}{n}\sum_{i=1}^n x_i$ can be used to approximate $\mathbb{E}_{X\sim p_0}[X]$ because $\bar{x}$ is an unbiased estimate of $\mathbb{E}_{X\sim p_0}[X]$ and the estimation variance converges to zero as $n \to \infty$ (see the law of large numbers in Box 5.1 for more information).

◇ Second, consider a new scenario where the samples $\{x_i\}_{i=1}^n$ are *not* generated by p_0. Instead, they are generated by another distribution p_1. Can we still use these samples to approximate $\mathbb{E}_{X\sim p_0}[X]$? The answer is yes. However, we can no longer use $\bar{x} = \frac{1}{n}\sum_{i=1}^n x_i$ to approximate $\mathbb{E}_{X\sim p_0}[X]$ since $\bar{x} \approx \mathbb{E}_{X\sim p_1}[X]$ rather than $\mathbb{E}_{X\sim p_0}[X]$.

In the second scenario, $\mathbb{E}_{X\sim p_0}[X]$ can be approximated based on the *importance sampling* technique. In particular, $\mathbb{E}_{X\sim p_0}[X]$ satisfies

$$\mathbb{E}_{X\sim p_0}[X] = \sum_{x\in\mathcal{X}} p_0(x)x = \sum_{x\in\mathcal{X}} p_1(x) \underbrace{\frac{p_0(x)}{p_1(x)}x}_{f(x)} = \mathbb{E}_{X\sim p_1}[f(X)]. \tag{10.9}$$

Thus, estimating $\mathbb{E}_{X\sim p_0}[X]$ becomes the problem of estimating $\mathbb{E}_{X\sim p_1}[f(X)]$. Let

$$\bar{f} \doteq \frac{1}{n}\sum_{i=1}^n f(x_i).$$

Since $\bar{f}$ can effectively approximate $\mathbb{E}_{X\sim p_1}[f(X)]$, it then follows from (10.9) that

$$\mathbb{E}_{X\sim p_0}[X] = \mathbb{E}_{X\sim p_1}[f(X)] \approx \bar{f} = \frac{1}{n}\sum_{i=1}^n f(x_i) = \frac{1}{n}\sum_{i=1}^n \underbrace{\frac{p_0(x_i)}{p_1(x_i)}}_{\substack{\text{importance}\\\text{weight}}} x_i. \tag{10.10}$$

Equation (10.10) suggests that $\mathbb{E}_{X\sim p_0}[X]$ can be approximated by a weighted average of x_i. Here, $\frac{p_0(x_i)}{p_1(x_i)}$ is called the *importance weight*. When $p_1 = p_0$, the importance weight is 1 and $\bar{f}$ becomes $\bar{x}$. When $p_0(x_i) \geq p_1(x_i)$, x_i can be sampled more frequently by p_0 but less frequently by p_1. In this case, the importance weight, which is greater than one, emphasizes the importance of this sample.

Some readers may ask the following question: while $p_0(x)$ is required in (10.10), why do we not directly calculate $\mathbb{E}_{X\sim p_0}[X]$ using its definition $\mathbb{E}_{X\sim p_0}[X] = \sum_{x\in\mathcal{X}} p_0(x)x$? The answer is as follows. To use the definition, we need to know either the *analytical* expression of p_0 or the value of $p_0(x)$ for *every* $x \in \mathcal{X}$. However, it is difficult to obtain the *analytical* expression of p_0 when the distribution is represented by, for example, a neural network. It is also difficult to obtain the value of $p_0(x)$ for *every* $x \in \mathcal{X}$ when $\mathcal{X}$ is large. By contrast, (10.10) merely requires the values of $p_0(x_i)$ for some samples and is much easier to implement in practice.

An illustrative example

We next present an example to demonstrate the importance sampling technique. Consider $X \in \mathcal{X} \doteq \{+1, -1\}$. Suppose that p_0 is a probability distribution satisfying

$$p_0(X = +1) = 0.5, \quad p_0(X = -1) = 0.5.$$

The expectation of X over p_0 is

$$\mathbb{E}_{X \sim p_0}[X] = (+1) \cdot 0.5 + (-1) \cdot 0.5 = 0.$$

Suppose that p_1 is another distribution satisfying

$$p_1(X = +1) = 0.8, \quad p_1(X = -1) = 0.2.$$

The expectation of X over p_1 is

$$\mathbb{E}_{X \sim p_1}[X] = (+1) \cdot 0.8 + (-1) \cdot 0.2 = 0.6.$$

Suppose that we have some samples $\{x_i\}$ drawn over p_1. Our goal is to estimate $\mathbb{E}_{X \sim p_0}[X]$ using these samples. As shown in Figure 10.2, there are more samples of $+1$ than -1. That is because $p_1(X = +1) = 0.8 > p_1(X = -1) = 0.2$. If we directly calculate the average value $\sum_{i=1}^{n} x_i/n$ of the samples, this value converges to $\mathbb{E}_{X \sim p_1}[X] = 0.6$ (see the dotted line in Figure 10.2). By contrast, if we calculate the weighted average value as in (10.10), this value can successfully converge to $\mathbb{E}_{X \sim p_0}[X] = 0$ (see the solid line in Figure 10.2).

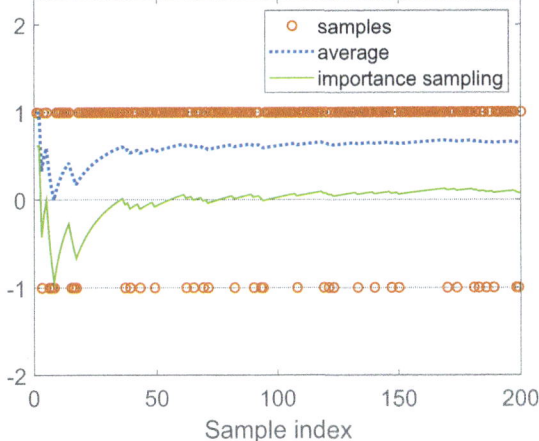

Figure 10.2: An example for demonstrating the importance sampling technique. Here, $X \in \{+1, -1\}$ and $p_0(X = +1) = p_0(X = -1) = 0.5$. The samples are generated according to p_1 where $p_1(X = +1) = 0.8$ and $p_1(X = -1) = 0.2$. The average of the samples converges to $\mathbb{E}_{X \sim p_1}[X] = 0.6$, but the weighted average calculated by the importance sampling technique in (10.10) converges to $\mathbb{E}_{X \sim p_0}[X] = 0$.

Finally, the distribution p_1, which is used to generate samples, must satisfy that $p_1(x) \neq 0$ when $p_0(x) \neq 0$. If $p_1(x) = 0$ while $p_0(x) \neq 0$, the estimation result may be problematic. For example, if

$$p_1(X = +1) = 1, \quad p_1(X = -1) = 0,$$

then the samples generated by p_1 are all positive: $\{x_i\} = \{+1, +1, \ldots, +1\}$. These samples cannot be used to correctly estimate $\mathbb{E}_{X \sim p_0}[X] = 0$ because

$$\frac{1}{n} \sum_{i=1}^{n} \frac{p_0(x_i)}{p_1(x_i)} x_i = \frac{1}{n} \sum_{i=1}^{n} \frac{p_0(+1)}{p_1(+1)} 1 = \frac{1}{n} \sum_{i=1}^{n} \frac{0.5}{1} 1 \equiv 0.5,$$

no matter how large n is.

10.3.2 The off-policy policy gradient theorem

With the importance sampling technique, we are ready to present the off-policy policy gradient theorem. Suppose that β is a behavior policy. Our goal is to use the samples generated by β to learn a target policy π that can maximize the following metric:

$$J(\theta) = \sum_{s \in \mathcal{S}} d_\beta(s) v_\pi(s) = \mathbb{E}_{S \sim d_\beta}[v_\pi(S)],$$

where d_β is the stationary distribution under policy β and v_π is the state value under policy π. The gradient of this metric is given in the following theorem.

Theorem 10.1 (Off-policy policy gradient theorem). *In the discounted case where $\gamma \in (0, 1)$, the gradient of $J(\theta)$ is*

$$\nabla_\theta J(\theta) = \mathbb{E}_{S \sim \rho, A \sim \beta} \left[\underbrace{\frac{\pi(A|S, \theta)}{\beta(A|S)}}_{\substack{importance \\ weight}} \nabla_\theta \ln \pi(A|S, \theta) q_\pi(S, A) \right], \tag{10.11}$$

where the state distribution ρ is

$$\rho(s) \doteq \sum_{s' \in \mathcal{S}} d_\beta(s') \mathrm{Pr}_\pi(s|s'), \qquad s \in \mathcal{S},$$

where $\mathrm{Pr}_\pi(s|s') = \sum_{k=0}^{\infty} \gamma^k [P_\pi^k]_{s's} = [(I - \gamma P_\pi)^{-1}]_{s's}$ is the discounted total probability of transitioning from s' to s under policy π.

The gradient in (10.11) is similar to that in the on-policy case in Theorem 9.1, but there are two differences. The first difference is the importance weight. The second difference is that $A \sim \beta$ instead of $A \sim \pi$. Therefore, we can use the action samples generated by following β to approximate the true gradient. The proof of the theorem is given in Box 10.2.

Box 10.2: Proof of Theorem 10.1

Since d_β is independent of θ, the gradient of $J(\theta)$ satisfies

$$\nabla_\theta J(\theta) = \nabla_\theta \sum_{s \in \mathcal{S}} d_\beta(s) v_\pi(s) = \sum_{s \in \mathcal{S}} d_\beta(s) \nabla_\theta v_\pi(s). \tag{10.12}$$

According to Lemma 9.2, the expression of $\nabla_\theta v_\pi(s)$ is

$$\nabla_\theta v_\pi(s) = \sum_{s' \in \mathcal{S}} \mathrm{Pr}_\pi(s'|s) \sum_{a \in \mathcal{A}} \nabla_\theta \pi(a|s', \theta) q_\pi(s', a), \tag{10.13}$$

where $\mathrm{Pr}_\pi(s'|s) \doteq \sum_{k=0}^\infty \gamma^k [P_\pi^k]_{ss'} = [(I_n - \gamma P_\pi)^{-1}]_{ss'}$. Substituting (10.13) into (10.12) yields

$$
\begin{aligned}
\nabla_\theta J(\theta) = \sum_{s \in \mathcal{S}} d_\beta(s) \nabla_\theta v_\pi(s) &= \sum_{s \in \mathcal{S}} d_\beta(s) \sum_{s' \in \mathcal{S}} \mathrm{Pr}_\pi(s'|s) \sum_{a \in \mathcal{A}} \nabla_\theta \pi(a|s', \theta) q_\pi(s', a) \\
&= \sum_{s' \in \mathcal{S}} \left(\sum_{s \in \mathcal{S}} d_\beta(s) \mathrm{Pr}_\pi(s'|s) \right) \sum_{a \in \mathcal{A}} \nabla_\theta \pi(a|s', \theta) q_\pi(s', a) \\
&\doteq \sum_{s' \in \mathcal{S}} \rho(s') \sum_{a \in \mathcal{A}} \nabla_\theta \pi(a|s', \theta) q_\pi(s', a) \\
&= \sum_{s \in \mathcal{S}} \rho(s) \sum_{a \in \mathcal{A}} \nabla_\theta \pi(a|s, \theta) q_\pi(s, a) \quad \text{(change } s' \text{ to } s\text{)} \\
&= \mathbb{E}_{S \sim \rho} \left[\sum_{a \in \mathcal{A}} \nabla_\theta \pi(a|S, \theta) q_\pi(S, a) \right].
\end{aligned}
$$

By using the importance sampling technique, the above equation can be further rewritten as

$$
\begin{aligned}
\mathbb{E}_{S \sim \rho} \left[\sum_{a \in \mathcal{A}} \nabla_\theta \pi(a|S, \theta) q_\pi(S, a) \right] &= \mathbb{E}_{S \sim \rho} \left[\sum_{a \in \mathcal{A}} \beta(a|S) \frac{\pi(a|S, \theta)}{\beta(a|S)} \frac{\nabla_\theta \pi(a|S, \theta)}{\pi(a|S, \theta)} q_\pi(S, a) \right] \\
&= \mathbb{E}_{S \sim \rho} \left[\sum_{a \in \mathcal{A}} \beta(a|S) \frac{\pi(a|S, \theta)}{\beta(a|S)} \nabla_\theta \ln \pi(a|S, \theta) q_\pi(S, a) \right] \\
&= \mathbb{E}_{S \sim \rho, A \sim \beta} \left[\frac{\pi(A|S, \theta)}{\beta(A|S)} \nabla_\theta \ln \pi(A|S, \theta) q_\pi(S, A) \right].
\end{aligned}
$$

The proof is complete. The above proof is similar to that of Theorem 9.1.

10.3.3 Algorithm description

Based on the off-policy policy gradient theorem, we are ready to present the off-policy actor-critic algorithm. Since the off-policy case is very similar to the on-policy case, we merely present some key steps.

First, the off-policy policy gradient is invariant to any additional baseline $b(s)$. In particular, we have

$$\nabla_\theta J(\theta) = \mathbb{E}_{S\sim\rho, A\sim\beta}\left[\frac{\pi(A|S,\theta)}{\beta(A|S)}\nabla_\theta \ln \pi(A|S,\theta)\big(q_\pi(S,A) - b(S)\big)\right],$$

because $\mathbb{E}\left[\frac{\pi(A|S,\theta)}{\beta(A|S)}\nabla_\theta \ln \pi(A|S,\theta)b(S)\right] = 0$. To reduce the estimation variance, we can select the baseline as $b(S) = v_\pi(S)$ and obtain

$$\nabla_\theta J(\theta) = \mathbb{E}\left[\frac{\pi(A|S,\theta)}{\beta(A|S)}\nabla_\theta \ln \pi(A|S,\theta)\big(q_\pi(S,A) - v_\pi(S)\big)\right].$$

The corresponding stochastic gradient-ascent algorithm is

$$\theta_{t+1} = \theta_t + \alpha_\theta \frac{\pi(a_t|s_t,\theta_t)}{\beta(a_t|s_t)}\nabla_\theta \ln \pi(a_t|s_t,\theta_t)\big(q_t(s_t,a_t) - v_t(s_t)\big),$$

where $\alpha_\theta > 0$. Similar to the on-policy case, the advantage function $q_t(s,a) - v_t(s)$ can be replaced by the TD error. That is

$$q_t(s_t,a_t) - v_t(s_t) \approx r_{t+1} + \gamma v_t(s_{t+1}) - v_t(s_t) \doteq \delta_t(s_t,a_t).$$

Then, the algorithm becomes

$$\theta_{t+1} = \theta_t + \alpha_\theta \frac{\pi(a_t|s_t,\theta)}{\beta(a_t|s_t)}\nabla_\theta \ln \pi(a_t|s_t,\theta)\delta_t(s_t,a_t).$$

The implementation of the off-policy actor-critic algorithm is summarized in Algorithm 10.3. As can be seen, the algorithm is the same as the advantage actor-critic algorithm except that an additional importance weight is included in both the critic and the actor. It must be noted that, in addition to the actor, the critic is also converted from on-policy to off-policy by the importance sampling technique. In fact, importance sampling is a general technique that can be applied to both policy-based and value-based algorithms. Finally, Algorithm 10.3 can be extended in various ways to incorporate more techniques such as eligibility traces [73].

Algorithm 10.3: Off-policy actor-critic based on importance sampling

Initialization: A given behavior policy $\beta(a|s)$. A target policy $\pi(a|s, \theta_0)$ where θ_0 is the initial parameter. A value function $v(s, w_0)$ where w_0 is the initial parameter. $\alpha_w, \alpha_\theta > 0$.
Goal: Learn an optimal policy to maximize $J(\theta)$.

At time step t in each episode, do
 Generate a_t following $\beta(s_t)$ and then observe r_{t+1}, s_{t+1}.
 Advantage (TD error):
  $\delta_t = r_{t+1} + \gamma v(s_{t+1}, w_t) - v(s_t, w_t)$
 Actor (policy update):
  $\theta_{t+1} = \theta_t + \alpha_\theta \frac{\pi(a_t|s_t,\theta_t)}{\beta(a_t|s_t)} \delta_t \nabla_\theta \ln \pi(a_t|s_t, \theta_t)$
 Critic (value update):
  $w_{t+1} = w_t + \alpha_w \frac{\pi(a_t|s_t,\theta_t)}{\beta(a_t|s_t)} \delta_t \nabla_w v(s_t, w_t)$

10.4 Deterministic actor-critic

Up to now, the policies used in the policy gradient methods are all *stochastic* since it is required that $\pi(a|s, \theta) > 0$ for every (s, a). This section shows that *deterministic* policies can also be used in policy gradient methods. Here, "deterministic" indicates that, for any state, a single action is given a probability of one and all the other actions are given probabilities of zero. It is important to study the deterministic case since it is naturally off-policy and can effectively handle continuous action spaces.

We have been using $\pi(a|s, \theta)$ to denote a general policy, which can be either stochastic or deterministic. In this section, we use

$$a = \mu(s, \theta)$$

to specifically denote a deterministic policy. Different from π which gives the probability of an action, μ directly gives the action since it is a mapping from $\mathcal{S}$ to $\mathcal{A}$. This deterministic policy can be represented by, for example, a neural network with s as its input, a as its output, and θ as its parameter. For the sake of simplicity, we often write $\mu(s, \theta)$ as $\mu(s)$ for short.

10.4.1 The deterministic policy gradient theorem

The policy gradient theorem introduced in the last chapter is only valid for stochastic policies. When we require the policy to be deterministic, a new policy gradient theorem must be derived.

Theorem 10.2 (Deterministic policy gradient theorem). *The gradient of $J(\theta)$ is*

$$\nabla_\theta J(\theta) = \sum_{s \in \mathcal{S}} \eta(s) \nabla_\theta \mu(s) \left(\nabla_a q_\mu(s, a)\right)|_{a=\mu(s)}$$
$$= \mathbb{E}_{S \sim \eta} \left[\nabla_\theta \mu(S) \left(\nabla_a q_\mu(S, a)\right)|_{a=\mu(S)}\right], \quad (10.14)$$

where η is a distribution of the states.

Theorem 10.2 is a summary of the results presented in Theorem 10.3 and Theorem 10.4 since the gradients in the two theorems have similar expressions. The specific expressions of $J(\theta)$ and η can be found in Theorems 10.3 and 10.4.

Unlike the stochastic case, the gradient in the deterministic case shown in (10.14) does not involve the action random variable A. As a result, when we use samples to approximate the true gradient, it is not required to sample actions. Therefore, the deterministic policy gradient method is *off-policy*. In addition, some readers may wonder why $\left(\nabla_a q_\mu(S, a)\right)|_{a=\mu(S)}$ cannot be written as $\nabla_a q_\mu(S, \mu(S))$, which seems more concise. That is simply because, if we do that, it is unclear how $q_\mu(S, \mu(S))$ is a function of a. A concise yet less confusing expression may be $\nabla_a q_\mu(S, a = \mu(S))$.

In the rest of this subsection, we present the derivation details of Theorem 10.2. In particular, we derive the gradients of two common metrics: the first is the average value and the second is the average reward. Since these two metrics have been discussed in detail in Section 9.2, we sometimes use their properties without proof. For most readers, it is sufficient to be familiar with Theorem 10.2 without knowing its derivation details. Interested readers can selectively examine the details in the remainder of this section.

Metric 1: Average value

We first derive the gradient of the average value:

$$J(\theta) = \mathbb{E}[v_\mu(s)] = \sum_{s \in \mathcal{S}} d_0(s) v_\mu(s), \quad (10.15)$$

where d_0 is the probability distribution of the states. Here, d_0 is selected to be *independent* of μ for simplicity. There are two special yet important cases of selecting d_0. The first case is that $d_0(s_0) = 1$ and $d_0(s \neq s_0) = 0$, where s_0 is a specific state of interest. In this case, the policy aims to maximize the discounted return that can be obtained when starting from s_0. The second case is that d_0 is the distribution of a given behavior policy that is different from the target policy.

To calculate the gradient of $J(\theta)$, we need to first calculate the gradient of $v_\mu(s)$ for any $s \in \mathcal{S}$. Consider the discounted case where $\gamma \in (0, 1)$.

Lemma 10.1 (Gradient of $v_\mu(s)$). *In the discounted case, it holds for any $s \in \mathcal{S}$ that*

$$\nabla_\theta v_\mu(s) = \sum_{s' \in \mathcal{S}} \Pr_\mu(s'|s) \nabla_\theta \mu(s') \big(\nabla_a q_\mu(s', a)\big)|_{a=\mu(s')}, \qquad (10.16)$$

where

$$\Pr_\mu(s'|s) \doteq \sum_{k=0}^{\infty} \gamma^k [P_\mu^k]_{ss'} = \big[(I - \gamma P_\mu)^{-1}\big]_{ss'}$$

is the discounted total probability of transitioning from s to s' under policy μ. Here, $[\cdot]_{ss'}$ denotes the entry in the sth row and s'th column of a matrix.

Box 10.3: Proof of Lemma 10.1

Since the policy is deterministic, we have

$$v_\mu(s) = q_\mu(s, \mu(s)).$$

Since both q_μ and μ are functions of θ, we have

$$\nabla_\theta v_\mu(s) = \nabla_\theta q_\mu(s, \mu(s)) = \big(\nabla_\theta q_\mu(s, a)\big)|_{a=\mu(s)} + \nabla_\theta \mu(s)\big(\nabla_a q_\mu(s, a)\big)|_{a=\mu(s)}. \quad (10.17)$$

By the definition of action values, for any given (s, a), we have

$$q_\mu(s, a) = r(s, a) + \gamma \sum_{s' \in \mathcal{S}} p(s'|s, a) v_\mu(s'),$$

where $r(s, a) = \sum_r r p(r|s, a)$. Since $r(s, a)$ is independent of μ, we have

$$\nabla_\theta q_\mu(s, a) = 0 + \gamma \sum_{s' \in \mathcal{S}} p(s'|s, a) \nabla_\theta v_\mu(s').$$

Substituting the above equation into (10.17) yields

$$\nabla_\theta v_\mu(s) = \gamma \sum_{s' \in \mathcal{S}} p(s'|s, \mu(s)) \nabla_\theta v_\mu(s') + \underbrace{\nabla_\theta \mu(s)\big(\nabla_a q_\mu(s, a)\big)|_{a=\mu(s)}}_{u(s)}, \quad s \in \mathcal{S}.$$

Since the above equation is valid for all $s \in \mathcal{S}$, we can combine these equations to obtain a matrix-vector form:

$$
\begin{bmatrix} \vdots \\ \nabla_\theta v_\mu(s) \\ \vdots \end{bmatrix} = \underbrace{\begin{bmatrix} \vdots \\ u(s) \\ \vdots \end{bmatrix}}_{} + \gamma(P_\mu \otimes I_m) \begin{bmatrix} \vdots \\ \nabla_\theta v_\mu(s') \\ \vdots \end{bmatrix},
$$

$$\underbrace{}_{\nabla_\theta v_\mu \in \mathbb{R}^{mn}} \qquad \underbrace{}_{u \in \mathbb{R}^{mn}} \qquad \underbrace{}_{\nabla_\theta v_\mu \in \mathbb{R}^{mn}}$$

where $n = |\mathcal{S}|$, m is the dimensionality of θ, P_μ is the state transition matrix with $[P_\mu]_{ss'} = p(s'|s, \mu(s))$, and $\otimes$ is the Kronecker product. The above matrix-vector form can be written concisely as

$$
\nabla_\theta v_\mu = u + \gamma(P_\mu \otimes I_m)\nabla_\theta v_\mu,
$$

which is a linear equation of $\nabla_\theta v_\mu$. Then, $\nabla_\theta v_\mu$ can be solved as

$$
\begin{aligned}
\nabla_\theta v_\mu &= (I_{mn} - \gamma P_\mu \otimes I_m)^{-1} u \\
&= (I_n \otimes I_m - \gamma P_\mu \otimes I_m)^{-1} u \\
&= \left[(I_n - \gamma P_\mu)^{-1} \otimes I_m \right] u.
\end{aligned} \tag{10.18}
$$

The elementwise form of (10.18) is

$$
\begin{aligned}
\nabla_\theta v_\mu(s) &= \sum_{s' \in \mathcal{S}} \left[(I - \gamma P_\mu)^{-1} \right]_{ss'} u(s') \\
&= \sum_{s' \in \mathcal{S}} \left[(I - \gamma P_\mu)^{-1} \right]_{ss'} \left[\nabla_\theta \mu(s') (\nabla_a q_\mu(s', a))|_{a=\mu(s')} \right].
\end{aligned} \tag{10.19}
$$

The quantity $\left[(I - \gamma P_\mu)^{-1} \right]_{ss'}$ has a clear probabilistic interpretation. Since $(I - \gamma P_\mu)^{-1} = I + \gamma P_\mu + \gamma^2 P_\mu^2 + \cdots$, we have

$$
\left[(I - \gamma P_\mu)^{-1} \right]_{ss'} = [I]_{ss'} + \gamma[P_\mu]_{ss'} + \gamma^2[P_\mu^2]_{ss'} + \cdots = \sum_{k=0}^{\infty} \gamma^k [P_\mu^k]_{ss'}.
$$

Note that $[P_\mu^k]_{ss'}$ is the probability of transitioning from s to s' using exactly k steps (see Box 8.1 for more information). Therefore, $\left[(I - \gamma P_\mu)^{-1} \right]_{ss'}$ is the discounted total probability of transitioning from s to s' using any number of steps. By denoting $\left[(I - \gamma P_\mu)^{-1} \right]_{ss'} \doteq \mathrm{Pr}_\mu(s'|s)$, equation (10.19) leads to (10.16).

With the preparation in Lemma 10.1, we are ready to derive the gradient of $J(\theta)$.

Theorem 10.3 (Deterministic policy gradient theorem in the discounted case)**.** *In the* *discounted case where $\gamma \in (0,1)$, the gradient of $J(\theta)$ in (10.15) is*

$$\nabla_\theta J(\theta) = \sum_{s\in\mathcal{S}} \rho_\mu(s)\nabla_\theta\mu(s)\big(\nabla_a q_\mu(s,a)\big)|_{a=\mu(s)}$$

$$= \mathbb{E}_{S\sim\rho_\mu}\left[\nabla_\theta\mu(S)\big(\nabla_a q_\mu(S,a)\big)|_{a=\mu(S)}\right],$$

where the state distribution ρ_μ is

$$\rho_\mu(s) = \sum_{s'\in\mathcal{S}} d_0(s')\mathrm{Pr}_\mu(s|s'), \qquad s \in \mathcal{S}.$$

Here, $\mathrm{Pr}_\mu(s|s') = \sum_{k=0}^\infty \gamma^k[P_\mu^k]_{s's} = [(I-\gamma P_\mu)^{-1}]_{s's}$ is the discounted total probability of transitioning from s' to s under policy μ.

Box 10.4: Proof of Theorem 10.3

Since d_0 is independent of μ, we have

$$\nabla_\theta J(\theta) = \sum_{s\in\mathcal{S}} d_0(s)\nabla_\theta v_\mu(s).$$

Substituting the expression of $\nabla_\theta v_\mu(s)$ given by Lemma 10.1 into the above equation yields

$$\nabla_\theta J(\theta) = \sum_{s\in\mathcal{S}} d_0(s)\nabla_\theta v_\mu(s)$$

$$= \sum_{s\in\mathcal{S}} d_0(s) \sum_{s'\in\mathcal{S}} \mathrm{Pr}_\mu(s'|s)\nabla_\theta\mu(s')\big(\nabla_a q_\mu(s',a)\big)|_{a=\mu(s')}$$

$$= \sum_{s'\in\mathcal{S}} \left(\sum_{s\in\mathcal{S}} d_0(s)\mathrm{Pr}_\mu(s'|s)\right)\nabla_\theta\mu(s')\big(\nabla_a q_\mu(s',a)\big)|_{a=\mu(s')}$$

$$\doteq \sum_{s'\in\mathcal{S}} \rho_\mu(s')\nabla_\theta\mu(s')\big(\nabla_a q_\mu(s',a)\big)|_{a=\mu(s')}$$

$$= \sum_{s\in\mathcal{S}} \rho_\mu(s)\nabla_\theta\mu(s)\big(\nabla_a q_\mu(s,a)\big)|_{a=\mu(s)} \qquad \text{(change } s' \text{ to } s)$$

$$= \mathbb{E}_{S\sim\rho_\mu}\left[\nabla_\theta\mu(S)\big(\nabla_a q_\mu(S,a)\big)|_{a=\mu(S)}\right].$$

The proof is complete. The above proof is consistent with the proof of Theorem 1 in [74]. Here, we consider the case in which the states and actions are finite. When they are continuous, the proof is similar, but the summations should be replaced by integrals [74].

Metric 2: Average reward

We next derive the gradient of the average reward:

$$J(\theta) = \bar{r}_\mu = \sum_{s \in \mathcal{S}} d_\mu(s) r_\mu(s)$$

$$= \mathbb{E}_{S \sim d_\mu}[r_\mu(S)], \tag{10.20}$$

where

$$r_\mu(s) = \mathbb{E}[R|s, a = \mu(s)] = \sum_r r p(r|s, a = \mu(s))$$

is the expectation of the immediate rewards. More information about this metric can be found in Section 9.2.

The gradient of $J(\theta)$ is given in the following theorem.

Theorem 10.4 (Deterministic policy gradient theorem in the undiscounted case). *In the undiscounted case, the gradient of $J(\theta)$ in (10.20) is*

$$\nabla_\theta J(\theta) = \sum_{s \in \mathcal{S}} d_\mu(s) \nabla_\theta \mu(s) \big(\nabla_a q_\mu(s, a) \big)\big|_{a=\mu(s)}$$

$$= \mathbb{E}_{S \sim d_\mu} \big[\nabla_\theta \mu(S) \big(\nabla_a q_\mu(S, a) \big)\big|_{a=\mu(S)} \big],$$

where d_μ is the stationary distribution of the states under policy μ.

Box 10.5: Proof of Theorem 10.4

Since the policy is deterministic, we have

$$v_\mu(s) = q_\mu(s, \mu(s)).$$

Since both q_μ and μ are functions of θ, we have

$$\nabla_\theta v_\mu(s) = \nabla_\theta q_\mu(s, \mu(s)) = \big(\nabla_\theta q_\mu(s, a) \big)\big|_{a=\mu(s)} + \nabla_\theta \mu(s) \big(\nabla_a q_\mu(s, a) \big)\big|_{a=\mu(s)}. \tag{10.21}$$

In the undiscounted case, it follows from the definition of action value (Section 9.3.2) that

$$q_\mu(s, a) = \mathbb{E}[R_{t+1} - \bar{r}_\mu + v_\mu(S_{t+1})|s, a]$$

$$= \sum_r p(r|s, a)(r - \bar{r}_\mu) + \sum_{s'} p(s'|s, a) v_\mu(s')$$

$$= r(s, a) - \bar{r}_\mu + \sum_{s'} p(s'|s, a) v_\mu(s').$$

Since $r(s,a) = \sum_r rp(r|s,a)$ is independent of θ, we have

$$\nabla_\theta q_\mu(s,a) = 0 - \nabla_\theta \bar{r}_\mu + \sum_{s'} p(s'|s,a)\nabla_\theta v_\mu(s').$$

Substituting the above equation into (10.21) gives

$$\nabla_\theta v_\mu(s) = -\nabla_\theta \bar{r}_\mu + \sum_{s'} p(s'|s,\mu(s))\nabla_\theta v_\mu(s') + \underbrace{\nabla_\theta \mu(s)(\nabla_a q_\mu(s,a))|_{a=\mu(s)}}_{u(s)}, \quad s \in \mathcal{S}.$$

While the above equation is valid for all $s \in \mathcal{S}$, we can combine these equations to obtain a matrix-vector form:

$$\underbrace{\begin{bmatrix} \vdots \\ \nabla_\theta v_\mu(s) \\ \vdots \end{bmatrix}}_{\nabla_\theta v_\mu \in \mathbb{R}^{mn}} = -\mathbf{1}_n \otimes \nabla_\theta \bar{r}_\mu + (P_\mu \otimes I_m) \underbrace{\begin{bmatrix} \vdots \\ \nabla_\theta v_\mu(s') \\ \vdots \end{bmatrix}}_{\nabla_\theta v_\mu \in \mathbb{R}^{mn}} + \underbrace{\begin{bmatrix} \vdots \\ u(s) \\ \vdots \end{bmatrix}}_{u \in \mathbb{R}^{mn}},$$

where $n = |\mathcal{S}|$, m is the dimension of θ, P_μ is the state transition matrix with $[P_\mu]_{ss'} = p(s'|s,\mu(s))$, and $\otimes$ is the Kronecker product. The above matrix-vector form can be written concisely as

$$\nabla_\theta v_\mu = u - \mathbf{1}_n \otimes \nabla_\theta \bar{r}_\mu + (P_\mu \otimes I_m)\nabla_\theta v_\mu,$$

and hence

$$\mathbf{1}_n \otimes \nabla_\theta \bar{r}_\mu = u + (P_\mu \otimes I_m)\nabla_\theta v_\mu - \nabla_\theta v_\mu. \tag{10.22}$$

Since d_μ is the stationary distribution, we have $d_\mu^T P_\mu = d_\mu^T$. Multiplying $d_\mu^T \otimes I_m$ on both sides of (10.22) gives

$$(d_\mu^T \mathbf{1}_n) \otimes \nabla_\theta \bar{r}_\mu = d_\mu^T \otimes I_m u + (d_\mu^T P_\mu) \otimes I_m \nabla_\theta v_\mu - d_\mu^T \otimes I_m \nabla_\theta v_\mu$$
$$= d_\mu^T \otimes I_m u + d_\mu^T \otimes I_m \nabla_\theta v_\mu - d_\mu^T \otimes I_m \nabla_\theta v_\mu$$
$$= d_\mu^T \otimes I_m u.$$

Since $d_\mu^T \mathbf{1}_n = 1$, the above equations become

$$\nabla_\theta \bar{r}_\mu = d_\mu^T \otimes I_m u$$

$$= \sum_{s \in \mathcal{S}} d_\mu(s) u(s)$$

$$= \sum_{s \in \mathcal{S}} d_\mu(s) \nabla_\theta \mu(s) \big(\nabla_a q_\mu(s, a) \big)|_{a = \mu(s)}$$

$$= \mathbb{E}_{S \sim d_\mu} \big[\nabla_\theta \mu(S) \big(\nabla_a q_\mu(S, a) \big)|_{a = \mu(S)} \big].$$

The proof is complete.

10.4.2 Algorithm description

Based on the gradient given in Theorem 10.2, we can apply the gradient-ascent algorithm to maximize $J(\theta)$:

$$\theta_{t+1} = \theta_t + \alpha_\theta \mathbb{E}_{S \sim \eta} \big[\nabla_\theta \mu(S) \big(\nabla_a q_\mu(S, a) \big)|_{a = \mu(S)} \big].$$

The corresponding stochastic gradient-ascent algorithm is

$$\theta_{t+1} = \theta_t + \alpha_\theta \nabla_\theta \mu(s_t) \big(\nabla_a q_\mu(s_t, a) \big)|_{a = \mu(s_t)}.$$

The implementation is summarized in Algorithm 10.4. It should be noted that this algorithm is *off-policy* since the behavior policy β may be different from μ. First, the actor is off-policy. We already explained the reason when presenting Theorem 10.2. Second, the critic is also off-policy. Special attention must be paid to why the critic is off-policy but does not require the importance sampling technique. In particular, the experience sample required by the critic is $(s_t, a_t, r_{t+1}, s_{t+1}, \tilde{a}_{t+1})$, where $\tilde{a}_{t+1} = \mu(s_{t+1})$. The generation of this experience sample involves two policies. The first is the policy for generating a_t at s_t, and the second is the policy for generating $\tilde{a}_{t+1}$ at s_{t+1}. The first policy that generates a_t is the behavior policy since a_t is used to interact with the environment. The second policy must be μ because it is the policy that the critic aims to evaluate. Hence, μ is the target policy. It should be noted that $\tilde{a}_{t+1}$ is not used to interact with the environment in the next time step. Hence, μ is not the behavior policy. Therefore, the critic is off-policy.

How to select the function $q(s, a, w)$? The original research work [74] that proposed the deterministic policy gradient method adopted linear functions: $q(s, a, w) = \phi^T(s, a) w$ where $\phi(s, a)$ is the feature vector. It is currently popular to represent $q(s, a, w)$ using neural networks, as suggested in the deep deterministic policy gradient (DDPG) method [75].

Algorithm 10.4: Deterministic policy gradient or deterministic actor-critic

Initialization: A given behavior policy $\beta(a|s)$. A deterministic target policy $\mu(s, \theta_0)$ where θ_0 is the initial parameter. A value function $q(s, a, w_0)$ where w_0 is the initial parameter. $\alpha_w, \alpha_\theta > 0$.

Goal: Learn an optimal policy to maximize $J(\theta)$.

At time step t in each episode, do

　　Generate a_t following β and then observe r_{t+1}, s_{t+1}.

　　TD error:
$$\delta_t = r_{t+1} + \gamma q(s_{t+1}, \mu(s_{t+1}, \theta_t), w_t) - q(s_t, a_t, w_t)$$

　　Actor (policy update):
$$\theta_{t+1} = \theta_t + \alpha_\theta \nabla_\theta \mu(s_t, \theta_t)\big(\nabla_a q(s_t, a, w_t)\big)|_{a=\mu(s_t)}$$

　　Critic (value update):
$$w_{t+1} = w_t + \alpha_w \delta_t \nabla_w q(s_t, a_t, w_t)$$

How to select the behavior policy β? It can be any exploratory policy. It can also be a stochastic policy obtained by adding noise to μ [75]. In this case, μ is also the behavior policy and hence this way is an on-policy implementation.

10.5　Summary

In this chapter, we introduced actor-critic methods. The contents are summarized as follows.

◇ Section 10.1 introduced the simplest actor-critic algorithm called QAC. This algorithm is similar to the policy gradient algorithm, REINFORCE, introduced in the last chapter. The only difference is that the q-value estimation in QAC relies on TD learning while REINFORCE relies on Monte Carlo estimation.

◇ Section 10.2 extended QAC to advantage actor-critic. It was shown that the policy gradient is invariant to any additional baseline. It was then shown that an optimal baseline could help reduce the estimation variance.

◇ Section 10.3 further extended the advantage actor-critic algorithm to the off-policy case. To do that, we introduced an important technique called importance sampling.

◇ Finally, while all the previously presented policy gradient algorithms rely on stochastic policies, we showed in Section 10.4 that the policy can be forced to be deterministic. The corresponding gradient was derived, and the deterministic policy gradient algorithm was introduced.

Policy gradient and actor-critic methods are widely used in modern reinforcement learning. There exist a large number of advanced algorithms in the literature such as SAC [76,77], TRPO [78], PPO [79], and TD3 [80]. In addition, the single-agent case can also be extended to the case of multi-agent reinforcement learning [81–85]. Experience samples can also be used to fit system models to achieve model-based reinforcement learning [15, 86, 87]. Distributional reinforcement learning provides a fundamentally different perspective from the conventional one [88, 89]. The relationships between reinforcement learning and control theory have been discussed in [90–95]. This book is not able to cover all these topics. Hopefully, the foundations laid by this book can help readers better study them in the future.

10.6 Q&A

⋄ Q: What is the relationship between actor-critic and policy gradient methods?

A: Actor-critic methods are actually policy gradient methods. Sometimes, we use them interchangeably. It is required to estimate action values in any policy gradient algorithm. When the action values are estimated using temporal-difference learning with value function approximation, such a policy gradient algorithm is called actor-critic. The name "actor-critic" highlights its algorithmic structure that combines the components of policy update and value update. This structure is also the fundamental structure used in all reinforcement learning algorithms.

⋄ Q: Why is it important to introduce additional baselines to actor-critic methods?

A: Since the policy gradient is invariant to any additional baseline, we can utilize the baseline to reduce estimation variance. The resulting algorithm is called advantage actor-critic.

⋄ Q: Can importance sampling be used in value-based algorithms other than policy-based ones?

A: The answer is yes. That is because importance sampling is a general technique for estimating the expectation of a random variable over *one distribution* using some samples drawn from *another distribution*. The reason why this technique is useful in reinforcement learning is that the many problems in reinforcement learning are to estimate expectations. For example, in value-based methods, the action or state values are defined as expectations. In the policy gradient method, the true gradient is also an expectation. As a result, importance sampling can be applied in both value-based and policy-based algorithms. In fact, it has been applied in the value-based component of Algorithm 10.3.

⋄ Q: Why is the deterministic policy gradient method off-policy?

A: The true gradient in the deterministic case does not involve the action random variable. As a result, when we use samples to approximate the true gradient, it is not required to sample actions and hence any policy can be used. Therefore, the deterministic policy gradient method is *off-policy*.

Appendix A

Preliminaries for Probability Theory

Reinforcement learning heavily relies on probability theory. We next summarize some concepts and results frequently used in this book.

⋄ *Random variable*: The term "variable" indicates that a random variable can take values from a set of numbers. The term "random" indicates that taking a value must follow a probability distribution.

A random variable is usually denoted by a capital letter. Its value is usually denoted by a lowercase letter. For example, X is a random variable, and x is a value that X can take.

This book mainly considers the case where a random variable can only take a finite number of values. A random variable can be a scalar or a vector.

Like normal variables, random variables have normal mathematical operations such as summation, product, and absolute value. For example, if X, Y are two random variables, we can calculate $X + Y$, $X + 1$, and XY.

⋄ A *stochastic sequence* is a sequence of random variables.

One scenario we often encounter is collecting a stochastic sampling sequence $\{x_i\}_{i=1}^{n}$ of a random variable X. For example, consider the task of tossing a die n times. Let x_i be a random variable representing the value obtained for the ith toss. Then, $\{x_1, x_2, \ldots, x_n\}$ is a stochastic process.

It may be confusing to beginners why x_i is a random variable instead of a deterministic value. In fact, if the sampling sequence is $\{1,6,3,5,...\}$, then this sequence is not a stochastic sequence because all the elements are already determined. However, if we use a variable x_i to represent the values that can possibly be sampled, it is a random variable since x_i can take any value in $\{1, \ldots, 6\}$. Although x_i is a lowercase letter, it still represents a random variable.

⋄ *Probability*: The notation $p(X = x)$ or $p_X(x)$ describes the probability of the random variable X taking the value x. When the context is clear, $p(X = x)$ is often written as $p(x)$ for short.

© The Author(s), under exclusive license to Springer Nature Singapore Pte Ltd. 2025   237
S. Zhao, *Mathematical Foundations of Reinforcement Learning*, https://doi.org/10.1007/978-981-97-3944-8

◇ *Joint probability*: The notation $p(X = x, Y = y)$ or $p(x, y)$ describes the probability of the random variable X taking the value x and Y taking the value y. One useful identity is as follows:

$$\sum_y p(x, y) = p(x).$$

◇ *Conditional probability*: The notation $p(X = x | A = a)$ describes the probability of the random variable X taking the value x given that the random variable A has already taken the value a. We often write $p(X = x | A = a)$ as $p(x|a)$ for short.

It holds that

$$p(x, a) = p(x|a)p(a)$$

and

$$p(x|a) = \frac{p(x, a)}{p(a)}.$$

Since $p(x) = \sum_a p(x, a)$, we have

$$p(x) = \sum_a p(x, a) = \sum_a p(x|a)p(a),$$

which is called the *law of total probability*.

◇ *Independence*: Two random variables are *independent* if the sampling value of one random variable does not affect the other. Mathematically, X and Y are independent if

$$p(x, y) = p(x)p(y).$$

Since $p(x, y) = p(x|y)p(y)$, the above equation implies

$$p(x|y) = p(x).$$

◇ *Conditional independence:* Let X, A, B be three random variables. X is said to be conditionally independent of A given B if

$$p(X = x | A = a, B = b) = p(X = x | B = b).$$

In the context of reinforcement learning, consider three consecutive states: s_t, s_{t+1}, s_{t+2}. Since they are obtained consecutively, s_{t+2} is dependent on s_{t+1} and also s_t. However, if s_{t+1} is already given, then s_{t+2} is conditionally independent of s_t. That is

$$p(s_{t+2} | s_{t+1}, s_t) = p(s_{t+2} | s_{t+1}).$$

This is also the memoryless property of Markov processes.

◇ *Law of total probability*: The law of total probability was already mentioned when we introduced the concept of conditional probability. Due to its importance, we list it again below:

$$p(x) = \sum_y p(x, y)$$

and

$$p(x|a) = \sum_y p(x, y|a).$$

◇ *Chain rule* of conditional probability and joint probability. By the definition of conditional probability, we have

$$p(a, b) = p(a|b)p(b).$$

This can be extended to

$$p(a, b, c) = p(a|b, c)p(b, c) = p(a|b, c)p(b|c)p(c),$$

and hence $p(a, b, c)/p(c) = p(a, b|c) = p(a|b, c)p(b|c)$. The fact that $p(a, b|c) = p(a|b, c)p(b|c)$ implies the following property:

$$p(x|a) = \sum_b p(x, b|a) = \sum_b p(x|b, a)p(b|a).$$

◇ *Expectation/expected value/mean*: Suppose that X is a random variable and the probability of taking the value x is $p(x)$. The expectation, expected value, or mean of X is defined as

$$\mathbb{E}[X] = \sum_x p(x)x.$$

The linearity property of expectation is

$$\mathbb{E}[X + Y] = \mathbb{E}[X] + \mathbb{E}[Y],$$
$$\mathbb{E}[aX] = a\mathbb{E}[X].$$

The second equation above can be trivially proven by definition. The first equation is proven below:

$$\mathbb{E}[X + Y] = \sum_x \sum_y (x + y)p(X = x, Y = y)$$
$$= \sum_x x \sum_y p(x, y) + \sum_y y \sum_x p(x, y)$$
$$= \sum_x xp(x) + \sum_y yp(y)$$
$$= \mathbb{E}[X] + \mathbb{E}[Y].$$

Due to the linearity of expectation, we have the following useful fact:

$$\mathbb{E}\left[\sum_i a_i X_i\right] = \sum_i a_i \mathbb{E}[X_i].$$

Similarly, it can be proven that

$$\mathbb{E}[AX] = A\mathbb{E}[X],$$

where $A \in \mathbb{R}^{n \times n}$ is a deterministic matrix and $X \in \mathbb{R}^n$ is a random vector.

◇ *Conditional expectation*: The definition of conditional expectation is

$$\mathbb{E}[X|A = a] = \sum_x x p(x|a).$$

Similar to the law of total probability, we have the *law of total expectation*:

$$\mathbb{E}[X] = \sum_a \mathbb{E}[X|A = a]p(a).$$

The proof is as follows. By the definition of expectation, it holds that

$$\sum_a \mathbb{E}[X|A = a]p(a) = \sum_a \left[\sum_x p(x|a)x\right] p(a)$$
$$= \sum_x \sum_a p(x|a)p(a)x$$
$$= \sum_x \left[\sum_a p(x|a)p(a)\right] x$$
$$= \sum_x p(x)x$$
$$= \mathbb{E}[X].$$

The law of total expectation is frequently used in reinforcement learning.

Similarly, conditional expectation satisfies

$$\mathbb{E}[X|A = a] = \sum_b \mathbb{E}[X|A = a, B = b]p(b|a).$$

This equation is useful in the derivation of the Bellman equation. A hint of its proof is the chain rule: $p(x|a, b)p(b|a) = p(x, b|a)$.

Finally, it is worth noting that $\mathbb{E}[X|A = a]$ is different from $\mathbb{E}[X|A]$. The former is a value, whereas the latter is a random variable. In fact, $\mathbb{E}[X|A]$ is a function of the random variable A. We need rigorous probability theory to define $\mathbb{E}[X|A]$.

◇ *Gradient of expectation*: Let $f(X, \beta)$ be a scalar function of a random variable X and a deterministic parameter vector β. Then,

$$\nabla_\beta \mathbb{E}[f(X, \beta)] = \mathbb{E}[\nabla_\beta f(X, \beta)].$$

Proof: Since $\mathbb{E}[f(X, \beta)] = \sum_x f(x, a)p(x)$, we have $\nabla_\beta \mathbb{E}[f(X, \beta)] = \nabla_\beta \sum_x f(x, a)p(x) = \sum_x \nabla_\beta f(x, a)p(x) = \mathbb{E}[\nabla_\beta f(X, \beta)]$.

◇ *Variance, covariance, covariance matrix*: For a single random variable X, its *variance* is defined as $\text{var}(X) = \mathbb{E}[(X - \bar{X})^2]$, where $\bar{X} = \mathbb{E}[X]$. For two random variables X, Y, their *covariance* is defined as $\text{cov}(X, Y) = \mathbb{E}[(X - \bar{X})(Y - \bar{Y})]$. For a random vector $X = [X_1, \ldots, X_n]^T$, the covariance matrix of X is defined as $\text{var}(X) \doteq \Sigma = \mathbb{E}[(X - \bar{X})(X - \bar{X})^T] \in \mathbb{R}^{n \times n}$. The *ij*th entry of Σ is $[\Sigma]_{ij} = \mathbb{E}[[X - \bar{X}]_i [X - \bar{X}]_j] = \mathbb{E}[(X_i - \bar{X}_i)(X_j - \bar{X}_j)] = \text{cov}(X_i, X_j)$. One trivial property is $\text{var}(a) = 0$ if a is deterministic. Moreover, it can be verified that $\text{var}(AX + a) = \text{var}(AX) = A\text{var}(X)A^T = A\Sigma A^T$.

Some useful facts are summarized below.

- Fact: $\mathbb{E}[(X - \bar{X})(Y - \bar{Y})] = \mathbb{E}[XY] - \bar{X}\bar{Y} = \mathbb{E}[XY] - \mathbb{E}[X]\mathbb{E}[Y]$.

  Proof: $\mathbb{E}[(X - \bar{X})(Y - \bar{Y})] = \mathbb{E}[XY - X\bar{Y} - \bar{X}Y + \bar{X}\bar{Y}] = \mathbb{E}[XY] - \mathbb{E}[X]\bar{Y} - \bar{X}\mathbb{E}[Y] + \bar{X}\bar{Y} = \mathbb{E}[XY] - \mathbb{E}[X]\mathbb{E}[Y] - \mathbb{E}[X]\mathbb{E}[Y] + \mathbb{E}[X]\mathbb{E}[Y] = \mathbb{E}[XY] - \mathbb{E}[X]\mathbb{E}[Y]$.

- Fact: $\mathbb{E}[XY] = \mathbb{E}[X]\mathbb{E}[Y]$ if X, Y are independent.

  Proof: $\mathbb{E}[XY] = \sum_x \sum_y p(x, y)xy = \sum_x \sum_y p(x)p(y)xy = \sum_x p(x)x \sum_y p(y)y = \mathbb{E}[X]\mathbb{E}[Y]$.

- Fact: $\text{cov}(X, Y) = 0$ if X, Y are independent.

  Proof: When X, Y are independent, $\text{cov}(X, Y) = \mathbb{E}[XY] - \mathbb{E}[X]\mathbb{E}[Y] = \mathbb{E}[X]\mathbb{E}[Y] - \mathbb{E}[X]\mathbb{E}[Y] = 0$.

Appendix B

Measure-Theoretic Probability Theory

We now briefly introduce measure-theoretic probability theory, which is also called rigorous probability theory. We only present basic notions and results. Comprehensive introductions can be found in [96–98]. Moreover, measure-theoretic probability theory requires some basic knowledge of measure theory, which is not covered here. Interested readers may refer to [99].

The reader may wonder if it is necessary to understand measure-theoretic probability theory before studying reinforcement learning. The answer is yes if the reader is interested in rigorously analyzing the convergence of stochastic sequences. For example, we often encounter the notion of *almost sure* convergence in Chapter 6 and Chapter 7. This notion is taken from measure-theoretic probability theory. If the reader is not interested in the convergence of stochastic sequences, it is okay to skip this part.

Probability triples

A *probability triple* is fundamental for establishing measure-theoretic probability theory. It is also called a probability space or probability measure space. A probability triple consists of three ingredients.

⋄ Ω: This is a set called the *sample space* (or outcome space). Any element (or point) in Ω, denoted as ω, is called an *outcome*. This set contains all the possible outcomes of a random sampling process.

Example: When playing a game of dice, we have six possible outcomes $\{1, 2, 3, 4, 5, 6\}$. Hence, $\Omega = \{1, 2, 3, 4, 5, 6\}$.

⋄ $\mathcal{F}$: This is a set called the *event space*. In particular, it is a σ-algebra (or σ-field) of Ω. The definition of a σ-algebra is given in Box B.1. An element in $\mathcal{F}$, denoted as A, is called an *event*. An *elementary event* refers to a single outcome in the sample space. An event may be an elementary event or a combination of multiple elementary events.

S. Zhao, *Mathematical Foundations of Reinforcement Learning*, https://doi.org/10.1007/978-981-97-3944-8

Example: Consider the game of dice. An example of an elementary event is "the number you get is i", where $i \in \{1, \ldots, 6\}$. An example of a nonelementary event is "the number you get is greater than 3". We care about such an event in practice because, for example, we can win the game if this event occurs. This event is mathematically expressed as $A = \{\omega \in \Omega : \omega > 3\}$. Since $\Omega = \{1, 2, 3, 4, 5, 6\}$ in this case, we have $A = \{4, 5, 6\}$.

◇ $\mathbb{P}$: This is a probability measure, which is a mapping from $\mathcal{F}$ to $[0, 1]$. Any $A \in \mathcal{F}$ is a set that contains some points in Ω. Then, $\mathbb{P}(A)$ is the measure of this set.

Example: If $A = \Omega$, which contains all ω values, then $\mathbb{P}(A) = 1$; if $A = \emptyset$, then $\mathbb{P}(A) = 0$. In the game of dice, consider the event "the number you get is greater than 3". In this case, $A = \{\omega \in \Omega : \omega > 3\}$, and $\Omega = \{1, 2, 3, 4, 5, 6\}$. Then, we have $A = \{4, 5, 6\}$ and hence $\mathbb{P}(A) = 1/2$. That is, the probability of us rolling a number greater than 3 is $1/2$.

Box B.1: Definition of a σ-algebra

An *algebra* of Ω is a set of some subsets of Ω that satisfy certain conditions. A σ-*algebra* is a specific and important type of algebra. In particular, denote $\mathcal{F}$ as a σ-algebra. Then, it must satisfy the following conditions.

◇ $\mathcal{F}$ contains $\emptyset$ and Ω;

◇ $\mathcal{F}$ is closed under complements;

◇ $\mathcal{F}$ is closed under countable unions and intersections.

The σ-algebras of a given Ω are not unique. $\mathcal{F}$ may contain all the subsets of Ω, and it may also merely contain some of them as long as it satisfies the above three conditions (see the examples below). Moreover, the three conditions are not independent. For example, if $\mathcal{F}$ contains Ω and is closed under complements, then it naturally contains $\emptyset$. More information can be found in [96–98].

◇ Example: When playing the dice game, we have $\Omega = \{1, 2, 3, 4, 5, 6\}$. Then, $\mathcal{F} = \{\Omega, \emptyset, \{1, 2, 3\}, \{4, 5, 6\}\}$ is a σ-algebra. The above three conditions can be easily verified. There are also other σ-algebras such as $\{\Omega, \emptyset, \{1, 2, 3, 4, 5\}, \{6\}\}$. Moreover, for any Ω with finite elements, the collection of all the subsets of Ω is a σ-algebra.

Random variables

Based on the notion of probability triples, we can formally define random variables. They are called variables, but they are actually functions that map from Ω to $\mathbb{R}$. In particular, a random variable assigns each outcome in Ω a numerical value, and hence it is a function: $X(\omega) : \Omega \to \mathbb{R}$.

Not all mappings from Ω to $\mathbb{R}$ are random variables. The formal definition of a random variable is as follows. A function $X : \Omega \to \mathbb{R}$ is a random variable if

$$A = \{\omega \in \Omega | X(\omega) \le x\} \in \mathcal{F}$$

for all $x \in \mathbb{R}$. This definition indicates that X is a random variable only if $X(\omega) \le x$ is an event in $\mathcal{F}$. More information can be found in [96, Section 3.1].

Expectation of random variables

The definition of the expectation of general random variables is sophisticated. Here, we only consider the special yet important case of simple random variables. In particular, a random variable is *simple* if $X(\omega)$ only takes a finite number of values. Let $\mathcal{X}$ be the set of all the possible values that X can take. A simple random variable is a function: $X(w) : \Omega \to \mathcal{X}$. It can be defined in a closed form as

$$X(\omega) \doteq \sum_{x \in \mathcal{X}} x \mathbb{1}_{A_x}(\omega),$$

where

$$A_x = \{\omega \in \Omega | X(\omega) = x\} \doteq X^{-1}(x)$$

and

$$\mathbb{1}_{A_x}(\omega) \doteq \begin{cases} 1, & \omega \in A_x, \\ 0, & \text{otherwise.} \end{cases} \tag{B.1}$$

Here, $\mathbb{1}_{A_x}(\omega)$ is an *indicator function* $\mathbb{1}_{A_x}(\omega) : \Omega \to \{0, 1\}$. If ω is mapped to x, the indicator function equals one; otherwise, it equals zero. It is possible that multiple ω's in Ω map to the same value in $\mathcal{X}$, but a single ω cannot be mapped to multiple values in $\mathcal{X}$.

With the above preparation, the expectation of a simple random variable is defined as

$$\mathbb{E}[X] \doteq \sum_{x \in \mathcal{X}} x \mathbb{P}(A_x), \tag{B.2}$$

where

$$A_x = \{\omega \in \Omega | X(\omega) = x\}.$$

The definition in (B.2) is similar to but more formal than the definition of expectation in the nonmeasure-theoretic case: $\mathbb{E}[X] = \sum_{x \in \mathcal{X}} x p(x)$.

As a demonstrative example, we next calculate the expectation of the indicator function in (B.1). It is notable that the indicator function is also a random variable that maps Ω to $\{0,1\}$ [96, Proposition 3.1.5]. As a result, we can calculate its expectation. In particular, consider the indicator function $\mathbb{1}_A$ where A denotes any event. We have

$$\mathbb{E}[\mathbb{1}_A] = \mathbb{P}(A).$$

To prove that, we have

$$\begin{aligned}
\mathbb{E}[\mathbb{1}_A] &= \sum_{z \in \{0,1\}} z\mathbb{P}(\mathbb{1}_A = z) \\
&= 0 \cdot \mathbb{P}(\mathbb{1}_A = 0) + 1 \cdot \mathbb{P}(\mathbb{1}_A = 1) \\
&= \mathbb{P}(\mathbb{1}_A = 1) \\
&= \mathbb{P}(A).
\end{aligned}$$

More properties of indicator functions can be found in [100, Chapter 24].

Conditional expectation as a random variable

While the expectation in (B.2) maps random variables to a specific value, we next introduce a conditional expectation that maps random variables to another random variable.

Suppose that X, Y, Z are all random variables. Consider three cases. First, a conditional expectation like $\mathbb{E}[X|Y = 2]$ or $\mathbb{E}[X|Y = 5]$ is specific *number*. Second, $\mathbb{E}[X|Y = y]$, where y is a variable, is a *function* of y. Third, $\mathbb{E}[X|Y]$, where Y is a random variable, is a function of Y and hence also a *random variable*. Since $\mathbb{E}[X|Y]$ is also a random variable, we can calculate, for example, its expectation.

We next examine the third case closely since it frequently emerges in the convergence analyses of stochastic sequences. The rigorous definition is not covered here and can be found in [96, Chapter 13]. We merely present some useful properties [101].

Lemma B.1 (Basic properties). *Let X, Y, Z be random variables. The following properties hold.*

(a) $\mathbb{E}[a|Y] = a$, *where a is a given number.*

(b) $\mathbb{E}[aX + bZ|Y] = a\mathbb{E}[X|Y] + b\mathbb{E}[Z|Y]$.

(c) $\mathbb{E}[X|Y] = \mathbb{E}[X]$ *if X, Y are independent.*

(d) $\mathbb{E}[Xf(Y)|Y] = f(Y)\mathbb{E}[X|Y]$.

(e) $\mathbb{E}[f(Y)|Y] = f(Y)$.

(f) $\mathbb{E}[X|Y, f(Y)] = \mathbb{E}[X|Y]$.

(g) *If $X \geq 0$, then $\mathbb{E}[X|Y] \geq 0$.*

(h) *If $X \geq Z$, then $\mathbb{E}[X|Y] \geq \mathbb{E}[Z|Y]$.*

Proof. We only prove some properties. The others can be proven similarly.

To prove $\mathbb{E}[a|Y] = a$ as in (a), we can show that $\mathbb{E}[a|Y = y] = a$ is valid for any y that Y can possibly take. This is clearly true, and the proof is complete.

To prove the property in (d), we can show that $\mathbb{E}[Xf(Y)|Y = y] = f(Y = y)\mathbb{E}[X|Y = y]$ for any y. This is valid because $\mathbb{E}[Xf(Y)|Y = y] = \sum_x xf(y)p(x|y) = f(y)\sum_x xp(x|y) = f(y)\mathbb{E}[X|Y = y]$. $\qquad\square$

Since $\mathbb{E}[X|Y]$ is a random variable, we can calculate its expectation. The related properties are presented below. These properties are useful for analyzing the convergence of stochastic sequences.

Lemma B.2. *Let X, Y, Z be random variables. The following properties hold.*

(a) $\mathbb{E}\big[\mathbb{E}[X|Y]\big] = \mathbb{E}[X]$.

(b) $\mathbb{E}\big[\mathbb{E}[X|Y, Z]\big] = \mathbb{E}[X]$.

(c) $\mathbb{E}\big[\mathbb{E}[X|Y]|Y\big] = \mathbb{E}[X|Y]$.

Proof. To prove the property in (a), we need to show that $\mathbb{E}\big[\mathbb{E}[X|Y = y]\big] = \mathbb{E}[X]$ for any y that Y can possibly take. To that end, considering that $\mathbb{E}[X|Y]$ is a function of Y, we denote it as $f(Y) = \mathbb{E}[X|Y]$. Then,

$$
\begin{aligned}
\mathbb{E}\big[\mathbb{E}[X|Y]\big] = \mathbb{E}\big[f(Y)\big] &= \sum_y f(Y = y)p(y) \\
&= \sum_y \mathbb{E}[X|Y = y]p(y) \\
&= \sum_y \left(\sum_x xp(x|y)\right) p(y) \\
&= \sum_x x \sum_y p(x|y)p(y) \\
&= \sum_x x \sum_y p(x, y) \\
&= \sum_x xp(x) \\
&= \mathbb{E}[X].
\end{aligned}
$$

The proof of the property in (b) is similar. In particular, we have

$$
\mathbb{E}\big[\mathbb{E}[X|Y, Z]\big] = \sum_{y,z} \mathbb{E}[X|y, z]p(y, z) = \sum_{y,z}\sum_x xp(x|y, z)p(y, z) = \sum_x xp(x) = \mathbb{E}[X].
$$

The proof of the property in (c) follows immediately from property (e) in Lemma B.1. That is because $\mathbb{E}[X|Y]$ is a function of Y. We denote this function as $f(Y)$. It then follows that $\mathbb{E}[\mathbb{E}[X|Y]|Y] = \mathbb{E}[f(Y)|Y] = f(Y) = \mathbb{E}[X|Y]$. $\qquad\square$

Definitions of stochastic convergence

One main reason why we care about measure-theoretic probability theory is that it can rigorously describe the convergence properties of stochastic sequences.

Consider the stochastic sequence $\{X_k\} \doteq \{X_1, X_2, \ldots, X_k, \ldots\}$. Each element in this sequence is a random variable defined on a triple $(\Omega, \mathcal{F}, \mathbb{P})$. When we say $\{X_k\}$ *converges* to a random variable X, we should be careful since there are different types of convergence as shown below.

◇ *Sure convergence:*

Definition: $\{X_k\}$ converges *surely* (or *everywhere* or *pointwise*) to X if

$$\lim_{k \to \infty} X_k(\omega) = X(\omega), \quad \text{for all } \omega \in \Omega.$$

It means that $\lim_{k \to \infty} X_k(\omega) = X(\omega)$ is valid for *all* points in Ω. This definition can be equivalently stated as

$$A = \Omega \quad \text{where} \quad A = \left\{ \omega \in \Omega : \lim_{k \to \infty} X_k(\omega) = X(\omega) \right\}.$$

◇ *Almost sure convergence:*

Definition: $\{X_k\}$ converges *almost surely* (or *almost everywhere* or *with probability 1* or *w.p.1*) to X if

$$\mathbb{P}(A) = 1 \quad \text{where} \quad A = \left\{ \omega \in \Omega : \lim_{k \to \infty} X_k(\omega) = X(\omega) \right\}. \tag{B.3}$$

It means that $\lim_{k \to \infty} X_k(\omega) = X(\omega)$ is valid for *almost all* points in Ω. The points, for which this limit is invalid, form a set of zero measure. For the sake of simplicity, (B.3) is often written as

$$\mathbb{P}\left(\lim_{k \to \infty} X_k = X \right) = 1.$$

Almost sure convergence can be denoted as $X_k \xrightarrow{a.s.} X$.

◇ *Convergence in probability:*

Definition: $\{X_k\}$ converges *in probability* to X if for any $\epsilon > 0$,

$$\lim_{k \to \infty} \mathbb{P}(A_k) = 0 \quad \text{where} \quad A_k = \{ \omega \in \Omega : |X_k(\omega) - X(\omega)| > \epsilon \}. \tag{B.4}$$

For simplicity, (B.4) can be written as

$$\lim_{k \to \infty} \mathbb{P}(|X_k - X| > \epsilon) = 0.$$

The difference between convergence in probability and (almost) sure convergence is as follows. Both sure convergence and almost sure convergence first evaluate the convergence of every point in Ω and then check the measure of these points that converge. By contrast, convergence in probability first checks the points that satisfy $|X_k - X| > \epsilon$ and then evaluates if the measure will converge to zero as $k \to \infty$.

⋄ *Convergence in mean:*

Definition: $\{X_k\}$ converges *in the r-th mean* (or *in the L^r norm*) to X if

$$\lim_{k \to \infty} \mathbb{E}[|X_k - X|^r] = 0.$$

The most frequently used cases are $r = 1$ and $r = 2$. It is worth mentioning that convergence in mean is not equivalent to $\lim_{k \to \infty} \mathbb{E}[X_k - X] = 0$ or $\lim_{k \to \infty} \mathbb{E}[X_k] = \mathbb{E}[X]$, which indicates that $\mathbb{E}[X_k]$ converges but the variance may not.

⋄ *Convergence in distribution:*

Definition: The *cumulative distribution function* of X_k is defined as $\mathbb{P}(X_k \leq a)$ where $a \in \mathbb{R}$. Then, $\{X_k\}$ converges to X *in distribution* if the cumulative distribution function converges:

$$\lim_{k \to \infty} \mathbb{P}(X_k \leq a) = \mathbb{P}(X \leq a), \quad \text{for all } a \in \mathbb{R}.$$

A compact expression is

$$\lim_{k \to \infty} \mathbb{P}(A_k) = \mathbb{P}(A),$$

where

$$A_k \doteq \{\omega \in \Omega : X_k(\omega) \leq a\}, \quad A \doteq \{\omega \in \Omega : X(\omega) \leq a\}.$$

The relationships between the above types of convergence are given below:

almost sure convergence $\Rightarrow$ convergence in probability $\Rightarrow$ convergence in distribution

convergence in mean $\Rightarrow$ convergence in probability $\Rightarrow$ convergence in distribution

Almost sure convergence and convergence in mean do not imply each other. More information can be found in [102].

Appendix C

Convergence of Sequences

We next introduce some results about the convergence of deterministic and stochastic sequences. These results are useful for analyzing the convergence of reinforcement learning algorithms such as those in Chapters 6 and 7.

We first consider deterministic sequences and then stochastic sequences.

C.1 Convergence of deterministic sequences

Convergence of monotonic sequences

Consider a sequence $\{x_k\} \doteq \{x_1, x_2, \ldots, x_k, \ldots\}$ where $x_k \in \mathbb{R}$. Suppose that this sequence is deterministic in the sense that x_k is not a random variable.

One of the most well-known convergence results is that a nonincreasing sequence with a lower bound converges. The following is a formal statement of this result.

Theorem C.1 (Convergence of monotonic sequences). *If the sequence $\{x_k\}$ is nonincreasing and bounded from below:*

◇ *Nonincreasing: $x_{k+1} \leq x_k$ for all k;*

◇ *Lower bound: $x_k \geq \alpha$ for all k;*

then x_k converges to a limit, which is the infimum of $\{x_k\}$, as $k \to \infty$.

Similarly, if $\{x_k\}$ is *nondecreasing* and bounded from above, then the sequence is convergent.

Convergence of nonmonotonic sequences

We next analyze the convergence of *nonmonotonic* sequences.

S. Zhao, *Mathematical Foundations of Reinforcement Learning*, https://doi.org/10.1007/978-981-97-3944-8

To analyze the convergence of nonmonotonic sequences, we introduce the following useful operator [103]. For any $z \in \mathbb{R}$, define

$$
z^+ \doteq \begin{cases} z, & \text{if } z \geq 0, \\ 0, & \text{if } z < 0, \end{cases}
$$

$$
z^- \doteq \begin{cases} z, & \text{if } z \leq 0, \\ 0, & \text{if } z > 0. \end{cases}
$$

It is obvious that $z^+ \geq 0$ and $z^- \leq 0$ for any z. Moreover, it holds that

$$
z = z^+ + z^-
$$

for all $z \in \mathbb{R}$.

To analyze the convergence of $\{x_k\}$, we rewrite x_k as

$$
\begin{aligned}
x_k &= x_k - x_{k-1} + x_{k-1} - x_{k-2} + \cdots - x_2 + x_2 - x_1 + x_1 \\
&= \sum_{i=1}^{k-1} (x_{i+1} - x_i) + x_1 \\
&\doteq S_k + x_1,
\end{aligned} \tag{C.1}
$$

where $S_k \doteq \sum_{i=1}^{k-1} (x_{i+1} - x_i)$. Note that S_k can be decomposed as

$$
S_k = \sum_{i=1}^{k-1} (x_{i+1} - x_i) = S_k^+ + S_k^-,
$$

where

$$
S_k^+ = \sum_{i=1}^{k-1} (x_{i+1} - x_i)^+ \geq 0, \qquad S_k^- = \sum_{i=1}^{k-1} (x_{i+1} - x_i)^- \leq 0.
$$

Some useful properties of S_k^+ and S_k^- are given below.

$\diamond$ $\{S_k^+ \geq 0\}$ is a nondecreasing sequence since $S_{k+1}^+ \geq S_k^+$ for all k.

$\diamond$ $\{S_k^- \leq 0\}$ is a nonincreasing sequence since $S_{k+1}^- \leq S_k^-$ for all k.

$\diamond$ If S_k^+ is bounded from above, then S_k^- is bounded from below. This is because $S_k^- \geq -S_k^+ - x_1$ due to the fact that $S_k^- + S_k^+ + x_1 = x_k \geq 0$.

With the above preparation, we can show the following result.

Theorem C.2 (Convergence of nonmonotonic sequences). *For any nonnegative sequence*

$\{x_k \geq 0\}$, *if*

$$\sum_{k=1}^{\infty}(x_{k+1} - x_k)^+ < \infty, \tag{C.2}$$

then $\{x_k\}$ *converges as* $k \to \infty$.

Proof. First, the condition $\sum_{k=1}^{\infty}(x_{k+1} - x_k)^+ < \infty$ indicates that $S_k^+ = \sum_{i=1}^{k-1}(x_{i+1} - x_i)^+$ is bounded from above for all k. Since $\{S_k^+\}$ is nondecreasing, the convergence of $\{S_k^+\}$ immediately follows from Theorem C.1. Suppose that S_k^+ converges to S_*^+.

Second, the boundedness of S_k^+ implies that S_k^- is bounded from below since $S_k^- \geq -S_k^+ - x_1$. Since $\{S_k^-\}$ is nonincreasing, the convergence of $\{S_k^-\}$ immediately follows from Theorem C.1. Suppose that S_k^- converges to S_*^-.

Finally, since $x_k = S_k^+ + S_k^- + x_1$, as shown in (C.1), the convergence of S_k^+ and S_k^- implies that $\{x_k\}$ converges to $S_*^+ + S_*^- + x_1$. $\qquad \square$

Theorem C.2 is more general than Theorem C.1 because it allows x_k to increase as long as the increase is damped as in (C.2). In the monotonic case, Theorem C.2 still applies. In particular, if $x_{k+1} \leq x_k$, then $\sum_{k=1}^{\infty}(x_{k+1} - x_k)^+ = 0$. In this case, (C.2) is still satisfied and the convergence follows.

We next consider a special yet importance case. Suppose that $\{x_k \geq 0\}$ is a nonnegative sequence satisfying

$$x_{k+1} \leq x_k + \eta_k.$$

When $\eta_k = 0$, we have $x_{k+1} \leq x_k$, meaning that the sequence is monotonic. When $\eta_k \geq 0$, the sequence is *not* monotonic because x_{k+1} may be greater than x_k. Nevertheless, we can still ensure the convergence of the sequence under some mild conditions. The following result is an immediate corollary of Theorem C.2.

Corollary C.1. *For any nonnegative sequence* $\{x_k \geq 0\}$, *if*

$$x_{k+1} \leq x_k + \eta_k$$

and $\{\eta_k \geq 0\}$ *satisfies*

$$\sum_{k=1}^{\infty}\eta_k < \infty,$$

then $\{x_k \geq 0\}$ *converges.*

Proof. Since $x_{k+1} \leq x_k + \eta_k$, we have $(x_{k+1} - x_k)^+ \leq \eta_k$ for all k. Then, we have

$$\sum_{k=1}^{\infty} (x_{k+1} - x_k)^+ \leq \sum_{k=1}^{\infty} \eta_k < \infty.$$

As a result, (C.2) is satisfied and the convergence follows from Theorem C.2. □

C.2   Convergence of stochastic sequences

We now consider stochastic sequences. While various definitions of stochastic sequences have been given in Appendix B, how to determine the convergence of a given stochastic sequence has not yet been discussed. We next present an important class of stochastic sequences called *martingales*. If a sequence can be classified as a martingale (or one of its variants), then the convergence of the sequence immediately follows.

Convergence of martingale sequences

◇ Definition: A stochastic sequence $\{X_k\}_{k=1}^{\infty}$ is called a *martingale* if $\mathbb{E}[\|X_k\|] < \infty$ and

$$\mathbb{E}[X_{k+1}|X_1, \ldots, X_k] = X_k \tag{C.3}$$

almost surely for all k.

Here, $\mathbb{E}[X_{k+1}|X_1, \ldots, X_k]$ is a random variable rather than a deterministic value. The term "almost surely" in the second condition is due to the definition of such expectations. In addition, $\mathbb{E}[X_{k+1}|X_1, \ldots, X_k]$ is often written as $\mathbb{E}[X_{k+1}|\mathcal{H}_k]$ for short where $\mathcal{H}_k = \{X_1, \ldots, X_k\}$ represents the "history" of the sequence. $\mathcal{H}_k$ has a specific name called a *filtration*. More information can be found in [96, Chapter 14] and [104].

◇ Example: An example that can demonstrate martingales is *random walk*, which is a stochastic process describing the position of a point that moves randomly. Specifically, let X_k denote the position of the point at time step k. Starting from X_k, the expectation of the next position X_{k+1} equals X_k if the mean of the one-step displacement is zero. In this case, we have $\mathbb{E}[X_{k+1}|X_1, \ldots, X_k] = X_k$ and hence $\{X_k\}$ is a martingale.

A basic property of martingales is that

$$\mathbb{E}[X_{k+1}] = \mathbb{E}[X_k]$$

for all k and hence

$$\mathbb{E}[X_k] = \mathbb{E}[X_{k-1}] = \cdots = \mathbb{E}[X_2] = \mathbb{E}[X_1].$$

This result can be obtained by calculating the expectation on both sides of (C.3) based on property (b) in Lemma B.2.

While the expectation of a martingale is constant, we next extend martingales to submartingales and supermartingales, whose expectations vary monotonically.

⋄ Definition: A stochastic sequence $\{X_k\}$ is called a *submartingale* if it satisfies $\mathbb{E}[|X_k|] < \infty$ and

$$\mathbb{E}[X_{k+1}|X_1, \ldots, X_k] \geq X_k \qquad (C.4)$$

for all k.

Taking the expectation on both sides of (C.4) yields $\mathbb{E}[X_{k+1}] \geq \mathbb{E}[X_k]$. In particular, the left-hand side leads to $\mathbb{E}[\mathbb{E}[X_{k+1}|X_1, \ldots, X_k]] = \mathbb{E}[X_{k+1}]$ due to property (b) in Lemma B.2. By induction, we have

$$\mathbb{E}[X_k] \geq \mathbb{E}[X_{k-1}] \geq \cdots \geq \mathbb{E}[X_2] \geq \mathbb{E}[X_1].$$

Therefore, the expectation of a submartingale is nondecreasing.

It may be worth mentioning that, for two random variables X and Y, $X \leq Y$ means $X(\omega) \leq Y(\omega)$ for all $\omega \in \Omega$. It does not mean the maximum of X is less than the minimum of Y.

⋄ Definition: A stochastic sequence $\{X_k\}$ is called a *supermartingale* if it satisfies $\mathbb{E}[|X_k|] < \infty$ and

$$\mathbb{E}[X_{k+1}|X_1, \ldots, X_k] \leq X_k \qquad (C.5)$$

for all k.

Taking expectation on both sides of (C.5) gives $\mathbb{E}[X_{k+1}] \leq \mathbb{E}[X_k]$. By induction, we have

$$\mathbb{E}[X_k] \leq \mathbb{E}[X_{k-1}] \leq \cdots \leq \mathbb{E}[X_2] \leq \mathbb{E}[X_1].$$

Therefore, the expectation of a supmartingale is nonincreasing.

The names "submartingale" and "supmartingale" are standard, but it may not be easy for beginners to distinguish them. Some tricks can be employed to do so. For example, since "supermartingale" has a letter "p" that points down, its expectation decreases; since submartingale has a letter "b" that points up, its expectation increases [104].

A supermartingale or submartingale is comparable to a deterministic monotonic sequence. While the convergence result for monotonic sequences has been given in Theorem C.1, we provide a similar convergence result for martingales as follows.

Theorem C.3 (Martingale convergence theorem). *If $\{X_k\}$ is a submartingale (or supermartingale), then there is a finite random variable X such that $X_k \to X$ almost surely.*

The proof is omitted. A comprehensive introduction to martingales can be found in [96, Chapter 14] and [104].

Convergence of quasimartingale sequences

We next introduce quasimartingales, which can be viewed as a generalization of martingales since their expectations are not monotonic. They are comparable to nonmonotonic deterministic sequences. The rigorous definition and convergence results of quasimartingales are nontrivial. We merely list some useful results.

The event A_k is defined as $A_k \doteq \{\omega \in \Omega : \mathbb{E}[X_{k+1} - X_k | \mathcal{H}_k] \geq 0\}$, where $\mathcal{H}_k = \{X_1, \ldots, X_k\}$. Intuitively, A_k indicates that X_{k+1} is greater than X_k in expectation. Let $\mathbb{1}_{A_k}$ be an indicator function:

$$\mathbb{1}_{A_k} = \begin{cases} 1, & \mathbb{E}[X_{k+1} - X_k | \mathcal{H}_k] \geq 0, \\ 0, & \mathbb{E}[X_{k+1} - X_k | \mathcal{H}_k] < 0. \end{cases}$$

The indicator function has a property that

$$1 = \mathbb{1}_A + \mathbb{1}_{A^c}$$

for any event A where A^c denotes the complementary event of A. As a result, it holds for any random variable that

$$X = \mathbb{1}_A X + \mathbb{1}_{A^c} X.$$

Although quasimartingales do not have monotonic expectations, their convergence is still ensured under some mild conditions as shown below.

Theorem C.4 (Quasimartingale convergence theorem). *For a nonnegative stochastic sequence $\{X_k \geq 0\}$, if*

$$\sum_{k=1}^{\infty} \mathbb{E}[(X_{k+1} - X_k)\mathbb{1}_{A_k}] < \infty,$$

then $\sum_{k=1}^{\infty} \mathbb{E}[(X_{k+1} - X_k)\mathbb{1}_{A_k^c}] > -\infty$ and there is a finite random variable such that $X_k \to X$ almost surely as $k \to \infty$.

Theorem C.4 can be viewed as an analogy of Theorem C.2, which is for nonmonotonic deterministic sequences. The proof of this theorem can be found in [105, Proposition 9.5]. Note that X_k here is required to be nonnegative. As a result, the boundedness of $\sum_{k=1}^{\infty} \mathbb{E}[(X_{k+1} - X_k)\mathbb{1}_{A_k}]$ implies the boundedness of $\sum_{k=1}^{\infty} \mathbb{E}[(X_{k+1} - X_k)\mathbb{1}_{A_k^c}]$.

Summary and comparison

We finally summarize and compare the results for deterministic and stochastic sequences.

◇ Deterministic sequences:

- Monotonic sequences: As shown in Theorem C.1, if a sequence is monotonic and bounded, then it converges.
- Nonmonotonic sequences: As shown in Theorem C.2, given a nonnegative sequence, even if it is nonmonotonic, it can still converge as long as its variation is damped in the sense that $\sum_{k=1}^{\infty}(x_{k+1} - x_k)^+ < \infty$.

◇ Stochastic sequences:

- Supermartingale/submartingale sequences: As shown in Theorem C.3, the expectation of a supermartingale or submartingale is monotonic. If a sequence is a supermartingale or submartingale, then the sequence converges almost surely.
- Quasimartingale sequences: As shown in Theorem C.4, even if a sequence's expectation is nonmonotonic, it can still converge as long as its variation is damped in the sense that $\sum_{k=1}^{\infty} \mathbb{E}[(X_{k+1} - X_k)\mathbf{1}_{\mathbb{E}[X_{k+1}-X_k|\mathcal{H}_k]>0}] < \infty$.

The above properties are summarized in Table C.1.

Variants of martingales	Monotonicity of $\mathbb{E}[X_k]$
Martingale	Constant: $\mathbb{E}[X_{k+1}] = \mathbb{E}[X_k]$
Submartingale	Increasing: $\mathbb{E}[X_{k+1}] \geq \mathbb{E}[X_k]$
Supermartingale	Decreasing: $\mathbb{E}[X_{k+1}] \leq \mathbb{E}[X_k]$
Quasimartingale	Non-monotonic

Table C.1: Summary of the monotonicity of different variants of martingales.

Appendix D

Preliminaries for Gradient Descent

We next present some preliminaries for the gradient descent method, which is one of the most frequently used optimization methods. The gradient descent method is also the foundation for the stochastic gradient descent method introduced in Chapter 6.

Convexity

◇ Definitions:

- Convex set: Suppose that $\mathcal{D}$ is a subset of $\mathbb{R}^n$. This set is *convex* if $z \doteq cx + (1 - c)y \in \mathcal{D}$ for any $x, y \in \mathcal{D}$ and any $c \in [0, 1]$.
- Convex function: Suppose $f : \mathcal{D} \to \mathbb{R}$ where $\mathcal{D}$ is convex. Then, the function $f(x)$ is *convex* if
$$f(cx + (1 - x)y) \leq cf(x) + (1 - c)f(y)$$
for any $x, y \in \mathcal{D}$ and $c \in [0, 1]$.

◇ Convex conditions:

- First-order condition: Consider a function $f : \mathcal{D} \to \mathbb{R}$ where $\mathcal{D}$ is convex. Then, f is convex if [106, 3.1.3]
$$f(y) - f(x) \geq \nabla f(x)^T (y - x), \quad \text{for all } x, y \in \mathcal{D}. \tag{D.1}$$
When x is a scalar, $\nabla f(x)$ represents the slope of the tangent line of $f(x)$ at x. The geometric interpretation of (D.1) is that the point $(y, f(y))$ is always located above the tangent line.
- Second-order condition: Consider a function $f : \mathcal{D} \to \mathbb{R}$ where $\mathcal{D}$ is convex. Then, f is convex if
$$\nabla^2 f(x) \succeq 0, \quad \text{for all } x \in \mathcal{D},$$
where $\nabla^2 f(x)$ is the Hessian matrix.

S. Zhao, *Mathematical Foundations of Reinforcement Learning*, https://doi.org/10.1007/978-981-97-3944-8

⋄ Degree of convexity:

Given a convex function, it is often of interest how strong its convexity is. The Hessian matrix is a useful tool for describing the degree of convexity. If $\nabla^2 f(x)$ is close to rank deficiency at a point, then the function is *flat* around that point and hence *weakly convex*. Otherwise, if the minimum singular value of $\nabla^2 f(x)$ is positive and large, the function is *curly* around that point and hence *strongly convex*. The degree of convexity influences the step size selection in gradient descent algorithms.

The lower and upper bounds of $\nabla^2 f(x)$ play an important role in characterizing the function convexity.

- Lower bound of $\nabla^2 f(x)$: A function is called *strongly convex* or *strictly convex* if $\nabla^2 f(x) \succeq \ell I_n$, where $\ell > 0$ for all x.
- Upper bound of $\nabla^2 f(x)$: If $\nabla^2 f(x)$ is bounded from above so that $\nabla^2 f(x) \preceq L I_n$, then the change in the first-order derivative $\nabla f(x)$ cannot be arbitrarily fast; equivalently, the function cannot be arbitrarily convex at a point.

  The upper bound can be implied by a Lipschitz condition of $\nabla f(x)$, as shown below.

  Lemma D.1. *Suppose that f is a convex function. If $\nabla f(x)$ is Lipschitz continuous with a constant L so that*

  $$\|\nabla f(x) - \nabla f(y)\| \le L\|x - y\|, \quad \text{for all } x, y,$$

  then $\nabla^2 f(x) \preceq L I_n$ for all x. Here, $\|\cdot\|$ denotes the Euclidean norm.

Gradient descent algorithms

Consider the following optimization problem:

$$\min_x f(x)$$

where $x \in \mathcal{D} \subseteq \mathbb{R}^n$ and $f : \mathcal{D} \to \mathbb{R}$. The gradient descent algorithm is

$$x_{k+1} = x_k - \alpha_k \nabla f(x_k), \quad k = 0, 1, 2, \ldots \tag{D.2}$$

where α_k is a positive coefficient that may be fixed or time-varying. Here, α_k is called the *step size* or *learning rate*. Some remarks about (D.2) are given below.

⋄ Direction of change: $\nabla f(x_k)$ is a vector that points in the direction along which $f(x_k)$ *increases* the fastest. Hence, the term $-\alpha_k \nabla f(x_k)$ changes x_k in the direction along which $f(x_k)$ *decreases* the fastest.

⋄ Magnitude of change: The magnitude of the change $-\alpha_k \nabla f(x_k)$ is jointly determined by the step size α_k and the magnitude of $\nabla f(x_k)$.

- Magnitude of $\nabla f(x_k)$:

  When x_k is close to the optimum x^* where $\nabla f(x^*) = 0$, the magnitude $\|\nabla f(x_k)\|$ is small. In this case, the update process of x_k is slow, which is reasonable because we do not want to update x too aggressively and miss the optimum.

  When x_k is far from the optimum, the magnitude of $\nabla f(x_k)$ may be large, and hence the update process of x_k is fast. This is also reasonable because we hope that the estimate can approach the optimum as quickly as possible.

- Step size α_k:

  If α_k is small, the magnitude of $-\alpha_k \nabla f(x_k)$ is small, and hence the convergence process is slow. If α_k is too large, the update process of x_k is aggressive, which leads to either fast convergence or divergence.

  How to select α_k? The selection of α_k should depend on the degree of convexity of $f(x_k)$. If the function is *curly* around the optimum (the degree of convexity is strong), then the step size α_k should be small to guarantee convergence. If the function is *flat* around the optimum (the degree of convexity is weak), then the step size could be large so that x_k can quickly approach the optimum. The above intuition will be verified in the following convergence analysis.

Convergence analysis

We next present a proof of the convergence of the gradient descent algorithm in (D.2). That is to show x_k converges to the optimum x^* where $\nabla f(x^*) = 0$. First of all, we make some assumptions.

$\diamond$ Assumption 1: $f(x)$ is strongly convex such that

$$\nabla^2 f(x) \succeq \ell I,$$

where $\ell > 0$.

$\diamond$ Assumption 2: $\nabla f(x)$ is Lipschitz continuous with a constant L. This assumption implies the following inequality according to Lemma D.1:

$$\nabla^2 f(x) \preceq L I_n.$$

The convergence proof is given below.

Proof. For any x_{k+1} and x_k, it follows from [106, Section 9.1.2] that

$$f(x_{k+1}) = f(x_k) + \nabla f(x_k)^T (x_{k+1} - x_k) + \frac{1}{2}(x_{k+1} - x_k)^T \nabla^2 f(z_k)(x_{k+1} - x_k), \quad \text{(D.3)}$$

where z_k is a convex combination of x_k and x_{k+1}. Since it is assumed that $\nabla^2 f(z_k) \preceq L I_n$, we have $\|\nabla^2 f(z_k)\| \leq L$. (D.3) implies

$$f(x_{k+1}) \leq f(x_k) + \nabla f(x_k)^T (x_{k+1} - x_k) + \frac{1}{2} \|\nabla^2 f(z_k)\| \|x_{k+1} - x_k\|^2$$
$$\leq f(x_k) + \nabla f(x_k)^T (x_{k+1} - x_k) + \frac{L}{2} \|x_{k+1} - x_k\|^2.$$

Substituting $x_{k+1} = x_k - \alpha_k \nabla f(x_k)$ into the above inequality yields

$$f(x_{k+1}) \leq f(x_k) + \nabla f(x_k)^T (-\alpha_k \nabla f(x_k)) + \frac{L}{2} \|\alpha_k \nabla f(x_k)\|^2$$
$$= f(x_k) - \alpha_k \|\nabla f(x_k)\|^2 + \frac{\alpha_k^2 L}{2} \|\nabla f(x_k)\|^2$$
$$= f(x_k) - \underbrace{\alpha_k \left(1 - \frac{\alpha_k L}{2}\right)}_{\eta_k} \|\nabla f(x_k)\|^2. \tag{D.4}$$

We next show that if we select

$$0 < \alpha_k < \frac{2}{L}, \tag{D.5}$$

then the sequence $\{f(x_k)\}_{k=1}^{\infty}$ converges to $f(x^*)$ where $\nabla f(x^*) = 0$. First, (D.5) implies that $\eta_k > 0$. Then, (D.4) implies that $f(x_{k+1}) \leq f(x_k)$. Therefore, $\{f(x_k)\}$ is a nonincreasing sequence. Second, since $f(x_k)$ is always bounded from below by $f(x^*)$, we know that $\{f(x_k)\}$ converges as $k \to \infty$ according to the monotone convergence theorem in Theorem C.1. Suppose that the limit of the sequence is f^*. Then, taking the limit on both sides of (D.4) gives

$$\lim_{k \to \infty} f(x_{k+1}) \leq \lim_{k \to \infty} f(x_k) - \lim_{k \to \infty} \eta_k \|\nabla f(x_k)\|^2$$
$$\Leftrightarrow f^* \leq f^* - \lim_{k \to \infty} \eta_k \|\nabla f(x_k)\|^2$$
$$\Leftrightarrow 0 \leq - \lim_{k \to \infty} \eta_k \|\nabla f(x_k)\|^2.$$

Since $\eta_k \|\nabla f(x_k)\|^2 \geq 0$, the above inequality implies that $\lim_{k \to \infty} \eta_k \|\nabla f(x_k)\|^2 = 0$. As a result, x converges to x^* where $\nabla f(x^*) = 0$. The proof is complete. The above proof is inspired by [107]. □

The inequality in (D.5) provides valuable insights into how α_k should be selected. If the function is flat (L is small), the step size can be large; otherwise, if the function is strongly convex (L is large), then the step size must be sufficiently small to ensure convergence. There are also many other ways to prove the convergence such as the contraction mapping theorem [108, Lemma 3]. A comprehensive introduction to convex optimization can be found in [106].

Bibliography

[1] M. Pinsky and S. Karlin, *An introduction to stochastic modeling (3rd Edition)*. Academic Press, 1998.

[2] M. L. Puterman, *Markov decision processes: Discrete stochastic dynamic programming*. John Wiley & Sons, 2014.

[3] R. S. Sutton and A. G. Barto, *Reinforcement learning: An introduction (2nd Edition)*. MIT Press, 2018.

[4] R. A. Horn and C. R. Johnson, *Matrix analysis*. Cambridge University Press, 2012.

[5] D. P. Bertsekas and J. N. Tsitsiklis, *Neuro-dynamic programming*. Athena Scientific, 1996.

[6] H. K. Khalil, *Nonlinear systems (3rd Edition)*. Patience Hall, 2002.

[7] G. Strang, *Calculus*. Wellesley-Cambridge Press, 1991.

[8] A. Besenyei, "A brief history of the mean value theorem," 2012. Lecture notes.

[9] A. Y. Ng, D. Harada, and S. Russell, "Policy invariance under reward transformations: Theory and application to reward shaping," in *International Conference on Machine Learning*, vol. 99, pp. 278–287, 1999.

[10] R. E. Bellman, *Dynamic programming*. Princeton University Press, 2010.

[11] R. E. Bellman and S. E. Dreyfus, *Applied dynamic programming*. Princeton University Press, 2015.

[12] J. Bibby, "Axiomatisations of the average and a further generalisation of monotonic sequences," *Glasgow Mathematical Journal*, vol. 15, no. 1, pp. 63–65, 1974.

[13] A. S. Polydoros and L. Nalpantidis, "Survey of model-based reinforcement learning: Applications on robotics," *Journal of Intelligent & Robotic Systems*, vol. 86, no. 2, pp. 153–173, 2017.

[14] T. M. Moerland, J. Broekens, A. Plaat, and C. M. Jonker, "Model-based reinforcement learning: A survey," *Foundations and Trends in Machine Learning*, vol. 16, no. 1, pp. 1–118, 2023.

[15] F.-M. Luo, T. Xu, H. Lai, X.-H. Chen, W. Zhang, and Y. Yu, "A survey on model-based reinforcement learning," *arXiv:2206.09328*, 2022.

[16] X. Wang, Z. Zhang, and W. Zhang, "Model-based multi-agent reinforcement learning: Recent progress and prospects," *arXiv:2203.10603*, 2022.

[17] M. Riedmiller, R. Hafner, T. Lampe, M. Neunert, J. Degrave, T. Wiele, V. Mnih, N. Heess, and J. T. Springenberg, "Learning by playing solving sparse reward tasks from scratch," in *International Conference on Machine Learning*, pp. 4344–4353, 2018.

[18] J. Ibarz, J. Tan, C. Finn, M. Kalakrishnan, P. Pastor, and S. Levine, "How to train your robot with deep reinforcement learning: Lessons we have learned," *The International Journal of Robotics Research*, vol. 40, no. 4-5, pp. 698–721, 2021.

[19] S. Narvekar, B. Peng, M. Leonetti, J. Sinapov, M. E. Taylor, and P. Stone, "Curriculum learning for reinforcement learning domains: A framework and survey," *The Journal of Machine Learning Research*, vol. 21, no. 1, pp. 7382–7431, 2020.

[20] C. Szepesvári, *Algorithms for reinforcement learning*. Springer, 2010.

[21] A. Maroti, "RBED: Reward based epsilon decay," *arXiv:1910.13701*, 2019.

[22] V. Mnih, K. Kavukcuoglu, D. Silver, A. A. Rusu, J. Veness, M. G. Bellemare, A. Graves, M. Riedmiller, A. K. Fidjeland, G. Ostrovski, S. Petersen, C. Beattie, A. Sadik, I. Antonoglou, H. King, D. Kumaran, D. Wierstra, S. Legg, and D. Hassabis, "Human-level control through deep reinforcement learning," *Nature*, vol. 518, no. 7540, pp. 529–533, 2015.

[23] W. Dabney, G. Ostrovski, and A. Barreto, "Temporally-extended epsilon-greedy exploration," *arXiv:2006.01782*, 2020.

[24] H.-F. Chen, *Stochastic approximation and its applications*, vol. 64. Springer Science & Business Media, 2006.

[25] H. Robbins and S. Monro, "A stochastic approximation method," *The Annals of Mathematical Statistics*, pp. 400–407, 1951.

[26] J. Venter, "An extension of the Robbins-Monro procedure," *The Annals of Mathematical Statistics*, vol. 38, no. 1, pp. 181–190, 1967.

[27] D. Ruppert, "Efficient estimations from a slowly convergent Robbins-Monro process," tech. rep., Cornell University Operations Research and Industrial Engineering, 1988.

[28] J. Lagarias, "Euler's constant: Euler's work and modern developments," *Bulletin of the American Mathematical Society*, vol. 50, no. 4, pp. 527–628, 2013.

[29] J. H. Conway and R. Guy, *The book of numbers*. Springer Science & Business Media, 1998.

[30] S. Ghosh, "The Basel problem," *arXiv:2010.03953*, 2020.

[31] A. Dvoretzky, "On stochastic approximation," in *The Third Berkeley Symposium on Mathematical Statistics and Probability*, 1956.

[32] T. Jaakkola, M. I. Jordan, and S. P. Singh, "On the convergence of stochastic iterative dynamic programming algorithms," *Neural Computation*, vol. 6, no. 6, pp. 1185–1201, 1994.

[33] T. Kailath, A. H. Sayed, and B. Hassibi, *Linear estimation*. Prentice Hall, 2000.

[34] C. K. Chui and G. Chen, *Kalman filtering*. Springer, 2017.

[35] G. A. Rummery and M. Niranjan, *On-line Q-learning using connectionist systems*. Technical Report, Cambridge University, 1994.

[36] H. Van Seijen, H. Van Hasselt, S. Whiteson, and M. Wiering, "A theoretical and empirical analysis of Expected Sarsa," in *IEEE Symposium on Adaptive Dynamic Programming and Reinforcement Learning*, pp. 177–184, 2009.

[37] M. Ganger, E. Duryea, and W. Hu, "Double Sarsa and double expected Sarsa with shallow and deep learning," *Journal of Data Analysis and Information Processing*, vol. 4, no. 4, pp. 159–176, 2016.

[38] C. J. C. H. Watkins, *Learning from delayed rewards*. PhD thesis, King's College, 1989.

[39] C. J. Watkins and P. Dayan, "Q-learning," *Machine learning*, vol. 8, no. 3-4, pp. 279–292, 1992.

[40] T. C. Hesterberg, *Advances in importance sampling*. PhD Thesis, Stanford University, 1988.

[41] H. Hasselt, "Double Q-learning," *Advances in Neural Information Processing Systems*, vol. 23, 2010.

[42] H. Van Hasselt, A. Guez, and D. Silver, "Deep reinforcement learning with double Q-learning," in *AAAI Conference on Artificial Intelligence*, vol. 30, 2016.

[43] C. Dann, G. Neumann, and J. Peters, "Policy evaluation with temporal differences: A survey and comparison," *Journal of Machine Learning Research*, vol. 15, pp. 809–883, 2014.

[44] J. Clifton and E. Laber, "Q-learning: Theory and applications," *Annual Review of Statistics and Its Application*, vol. 7, pp. 279–301, 2020.

[45] B. Jang, M. Kim, G. Harerimana, and J. W. Kim, "Q-learning algorithms: A comprehensive classification and applications," *IEEE Access*, vol. 7, pp. 133653–133667, 2019.

[46] R. S. Sutton, "Learning to predict by the methods of temporal differences," *Machine Learning*, vol. 3, no. 1, pp. 9–44, 1988.

[47] G. Strang, *Linear algebra and its applications (4th Edition)*. Belmont, CA: Thomson, Brooks/Cole, 2006.

[48] C. D. Meyer and I. Stewart, *Matrix analysis and applied linear algebra*. SIAM, 2023.

[49] M. Pinsky and S. Karlin, *An introduction to stochastic modeling*. Academic Press, 2010.

[50] M. G. Lagoudakis and R. Parr, "Least-squares policy iteration," *The Journal of Machine Learning Research*, vol. 4, pp. 1107–1149, 2003.

[51] R. Munos, "Error bounds for approximate policy iteration," in *International Conference on Machine Learning*, vol. 3, pp. 560–567, 2003.

[52] A. Geramifard, T. J. Walsh, S. Tellex, G. Chowdhary, N. Roy, and J. P. How, "A tutorial on linear function approximators for dynamic programming and reinforcement learning," *Foundations and Trends in Machine Learning*, vol. 6, no. 4, pp. 375–451, 2013.

[53] B. Scherrer, "Should one compute the temporal difference fix point or minimize the Bellman residual? the unified oblique projection view," in *International Conference on Machine Learning*, 2010.

[54] D. P. Bertsekas, *Dynamic programming and optimal control: Approximate dynamic programming (Volume II)*. Athena Scientific, 2011.

[55] S. Abramovich, G. Jameson, and G. Sinnamon, "Refining Jensen's inequality," *Bulletin mathématique de la Société des Sciences Mathématiques de Roumanie*, pp. 3–14, 2004.

[56] S. S. Dragomir, "Some reverses of the Jensen inequality with applications," *Bulletin of the Australian Mathematical Society*, vol. 87, no. 2, pp. 177–194, 2013.

[57] S. J. Bradtke and A. G. Barto, "Linear least-squares algorithms for temporal difference learning," *Machine Learning*, vol. 22, no. 1, pp. 33–57, 1996.

[58] K. S. Miller, "On the inverse of the sum of matrices," *Mathematics Magazine*, vol. 54, no. 2, pp. 67–72, 1981.

[59] S. A. U. Islam and D. S. Bernstein, "Recursive least squares for real-time implementation," *IEEE Control Systems Magazine*, vol. 39, no. 3, pp. 82–85, 2019.

[60] V. Mnih, K. Kavukcuoglu, D. Silver, A. Graves, I. Antonoglou, D. Wierstra, and M. Riedmiller, "Playing Atari with deep reinforcement learning," *arXiv preprint arXiv:1312.5602*, 2013.

[61] J. Fan, Z. Wang, Y. Xie, and Z. Yang, "A theoretical analysis of deep Q-learning," in *Learning for Dynamics and Control*, pp. 486–489, 2020.

[62] L.-J. Lin, *Reinforcement learning for robots using neural networks*. 1992. Technical report.

[63] J. N. Tsitsiklis and B. Van Roy, "An analysis of temporal-difference learning with function approximation," *IEEE Transactions on Automatic Control*, vol. 42, no. 5, pp. 674–690, 1997.

[64] R. S. Sutton, D. McAllester, S. Singh, and Y. Mansour, "Policy gradient methods for reinforcement learning with function approximation," *Advances in Neural Information Processing Systems*, vol. 12, 1999.

[65] P. Marbach and J. N. Tsitsiklis, "Simulation-based optimization of Markov reward processes," *IEEE Transactions on Automatic Control*, vol. 46, no. 2, pp. 191–209, 2001.

[66] J. Baxter and P. L. Bartlett, "Infinite-horizon policy-gradient estimation," *Journal of Artificial Intelligence Research*, vol. 15, pp. 319–350, 2001.

[67] X.-R. Cao, "A basic formula for online policy gradient algorithms," *IEEE Transactions on Automatic Control*, vol. 50, no. 5, pp. 696–699, 2005.

[68] R. J. Williams, "Simple statistical gradient-following algorithms for connectionist reinforcement learning," *Machine Learning*, vol. 8, no. 3, pp. 229–256, 1992.

[69] J. Peters and S. Schaal, "Reinforcement learning of motor skills with policy gradients," *Neural Networks*, vol. 21, no. 4, pp. 682–697, 2008.

[70] E. Greensmith, P. L. Bartlett, and J. Baxter, "Variance reduction techniques for gradient estimates in reinforcement learning," *Journal of Machine Learning Research*, vol. 5, no. 9, 2004.

[71] V. Mnih, A. P. Badia, M. Mirza, A. Graves, T. Lillicrap, T. Harley, D. Silver, and K. Kavukcuoglu, "Asynchronous methods for deep reinforcement learning," in *International Conference on Machine Learning*, pp. 1928–1937, 2016.

[72] M. Babaeizadeh, I. Frosio, S. Tyree, J. Clemons, and J. Kautz, "Reinforcement learning through asynchronous advantage actor-critic on a GPU," *arXiv:1611.06256*, 2016.

[73] T. Degris, M. White, and R. S. Sutton, "Off-policy actor-critic," *arXiv:1205.4839*, 2012.

[74] D. Silver, G. Lever, N. Heess, T. Degris, D. Wierstra, and M. Riedmiller, "Deterministic policy gradient algorithms," in *International Conference on Machine Learning*, pp. 387–395, 2014.

[75] T. P. Lillicrap, J. J. Hunt, A. Pritzel, N. Heess, T. Erez, Y. Tassa, D. Silver, and D. Wierstra, "Continuous control with deep reinforcement learning," *arXiv:1509.02971*, 2015.

[76] T. Haarnoja, A. Zhou, P. Abbeel, and S. Levine, "Soft actor-critic: Off-policy maximum entropy deep reinforcement learning with a stochastic actor," in *International Conference on Machine Learning*, pp. 1861–1870, 2018.

[77] T. Haarnoja, A. Zhou, K. Hartikainen, G. Tucker, S. Ha, J. Tan, V. Kumar, H. Zhu, A. Gupta, and P. Abbeel, "Soft actor-critic algorithms and applications," *arXiv:1812.05905*, 2018.

[78] J. Schulman, S. Levine, P. Abbeel, M. Jordan, and P. Moritz, "Trust region policy optimization," in *International Conference on Machine Learning*, pp. 1889–1897, 2015.

[79] J. Schulman, F. Wolski, P. Dhariwal, A. Radford, and O. Klimov, "Proximal policy optimization algorithms," *arXiv:1707.06347*, 2017.

[80] S. Fujimoto, H. Hoof, and D. Meger, "Addressing function approximation error in actor-critic methods," in *International Conference on Machine Learning*, pp. 1587–1596, 2018.

[81] J. Foerster, G. Farquhar, T. Afouras, N. Nardelli, and S. Whiteson, "Counterfactual multi-agent policy gradients," in *AAAI Conference on Artificial Intelligence*, vol. 32, 2018.

[82] R. Lowe, Y. I. Wu, A. Tamar, J. Harb, O. Pieter Abbeel, and I. Mordatch, "Multi-agent actor-critic for mixed cooperative-competitive environments," *Advances in Neural Information Processing Systems*, vol. 30, 2017.

[83] Y. Yang, R. Luo, M. Li, M. Zhou, W. Zhang, and J. Wang, "Mean field multi-agent reinforcement learning," in *International Conference on Machine Learning*, pp. 5571–5580, 2018.

[84] O. Vinyals, I. Babuschkin, W. M. Czarnecki, M. Mathieu, A. Dudzik, J. Chung, D. H. Choi, R. Powell, T. Ewalds, P. Georgiev, *et al.*, "Grandmaster level in StarCraft II using multi-agent reinforcement learning," *Nature*, vol. 575, no. 7782, pp. 350–354, 2019.

[85] Y. Yang and J. Wang, "An overview of multi-agent reinforcement learning from game theoretical perspective," *arXiv:2011.00583*, 2020.

[86] S. Levine and V. Koltun, "Guided policy search," in *International Conference on Machine Learning*, pp. 1–9, 2013.

[87] M. Janner, J. Fu, M. Zhang, and S. Levine, "When to trust your model: Model-based policy optimization," *Advances in Neural Information Processing Systems*, vol. 32, 2019.

[88] M. G. Bellemare, W. Dabney, and R. Munos, "A distributional perspective on reinforcement learning," in *International Conference on Machine Learning*, pp. 449–458, 2017.

[89] M. G. Bellemare, W. Dabney, and M. Rowland, *Distributional Reinforcement Learning*. MIT Press, 2023.

[90] H. Zhang, D. Liu, Y. Luo, and D. Wang, *Adaptive dynamic programming for control: algorithms and stability*. Springer Science & Business Media, 2012.

[91] F. L. Lewis, D. Vrabie, and K. G. Vamvoudakis, "Reinforcement learning and feedback control: Using natural decision methods to design optimal adaptive controllers," *IEEE Control Systems Magazine*, vol. 32, no. 6, pp. 76–105, 2012.

[92] F. L. Lewis and D. Liu, *Reinforcement learning and approximate dynamic programming for feedback control*. John Wiley & Sons, 2013.

[93] Z.-P. Jiang, T. Bian, and W. Gao, "Learning-based control: A tutorial and some recent results," *Foundations and Trends in Systems and Control*, vol. 8, no. 3, pp. 176–284, 2020.

[94] S. Meyn, *Control systems and reinforcement learning*. Cambridge University Press, 2022.

[95] S. E. Li, *Reinforcement learning for sequential decision and optimal control.* Springer, 2023.

[96] J. S. Rosenthal, *First look at rigorous probability theory (2nd Edition).* World Scientific Publishing Company, 2006.

[97] D. Pollard, *A user's guide to measure theoretic probability.* Cambridge University Press, 2002.

[98] P. J. Spreij, "Measure theoretic probability," *UvA Course Notes*, 2012.

[99] R. G. Bartle, *The elements of integration and Lebesgue measure.* John Wiley & Sons, 2014.

[100] M. Taboga, *Lectures on probability theory and mathematical statistics (2nd Edition).* CreateSpace Independent Publishing Platform, 2012.

[101] T. Kennedy, "Theory of probability," 2007. Lecture notes.

[102] A. W. Van der Vaart, *Asymptotic statistics.* Cambridge University Press, 2000.

[103] L. Bottou, "Online learning and stochastic approximations," *Online Learning in Neural Networks*, vol. 17, no. 9, p. 142, 1998.

[104] D. Williams, *Probability with martingales.* Cambridge University Press, 1991.

[105] M. Métivier, *Semimartingales: A course on stochastic processes.* Walter de Gruyter, 1982.

[106] S. Boyd, S. P. Boyd, and L. Vandenberghe, *Convex optimization.* Cambridge University Press, 2004.

[107] S. Bubeck *et al.*, "Convex optimization: Algorithms and complexity," *Foundations and Trends in Machine Learning*, vol. 8, no. 3-4, pp. 231–357, 2015.

[108] A. Jung, "A fixed-point of view on gradient methods for big data," *Frontiers in Applied Mathematics and Statistics*, vol. 3, p. 18, 2017.

Symbols

In this book, a matrix or a random variable is represented by capital letters. A vector, a scalar, or a sample is represented by a lowercase letter. The mathematical symbols that are frequently used in this book are listed below.

$=$	equality
$\approx$	approximation
$\doteq$	equality by definition
$\geq, >, \leq, <$	elementwise comparison
$\in$	is an element of
$\|\cdot\|_2$	Euclidean norm of a vector or the corresponding induced matrix norm
$\|\cdot\|_\infty$	maximum norm of a vector or the corresponding induced matrix norm
$\ln$	natural logarithm
$\mathbb{R}$	set of real numbers
$\mathbb{R}^n$	set of n-dimensional real vectors
$\mathbb{R}^{n \times m}$	set of all $n \times m$-dimensional real matrices
$A \succeq 0\ (A \succ 0)$	matrix A is positive semidefinite (definite)
$A \preceq 0\ (A \succ 0)$	matrix A is negative semidefinite (definite)
$|x|$	absolute value of real scalar x
$|\mathcal{S}|$	number of elements in set $\mathcal{S}$
$\nabla_x f(x)$	gradient of scalar function $f(x)$ with respect to vector x. It may be written as $\nabla f(x)$ for short.
$[A]_{ij}$	element in the ith row and jth column of matrix A
$[x]_i$	ith element of vector x
$X \sim p$	p is the probability distribution of random variable X.
$p(X = x), \Pr(X = x)$	probability of $X = x$. They are often written as $p(x)$ or $\Pr(x)$ for short.
$p(x|y)$	conditional probability
$\mathbb{E}_{X \sim p}[X]$	expectation or expected value of random variable X. It is often written as $\mathbb{E}[X]$ for short when the distribution of X is clear.

© The Author(s), under exclusive license to Springer Nature Singapore Pte Ltd. 2025
S. Zhao, *Mathematical Foundations of Reinforcement Learning*, https://doi.org/10.1007/978-981-97-3944-8

$\mathrm{var}(X)$	variance of random variable X
$\arg\max_x f(x)$	maximizer of function $f(x)$
$\mathbf{1}_n$	vector of all ones. It is often written as $\mathbf{1}$ for short when its dimension is clear.
I_n	$n \times n$-dimensional identity matrix. It is often written as I for short when its dimensions are clear.

Index

The manufacturer's authorised representative in the EU is Springer
Nature Customer Service Centre GmbH, Europaplatz 3, 69115 Heidelberg,
Germany. If you have any concerns regarding our products, please
contact ProductSafety@springernature.com

Printed and bound by CPI Group (UK) Ltd, Croydon, CR0 4YY
27/04/2026
02097566-0017